D1788801

WOODSMITH CUSTOM WOODWORKING

Heirloom Projects

SHOP SAFETY IS YOUR RESPONSIBILITY
Using hand or power tools improperly can result in
serious injury or death. Do not operate any tool until
you read the manual and understand how to operate the
tool safely. Always use all appropriate safety
equipment as well as the guards that come with your
tools and equipment and read the manuals that
accompany them. In some of the illustrations in this
book, the guards and safety equipment have been
removed only to provide a better view of the operation.
Do not attempt any procedure without using all
appropriate safety equipment or without ensuring that
all guards are in place. August Home Publishing
Company assumes no responsibility for injury, damage
or loss suffered as a result of your use of the material,
plans or illustrations contained in this book.

WOODSMITH CUSTOM WOODWORKING

Heirloom Projects

By the editors of *Woodsmith* magazine

CONTENTS

WOODSMITH
CUSTOM WOODWORKING

Heirloom Projects

CLOCKS   6

Tambour Clock

This clock features a quartz movement, a pendulum, and realistic-sounding chimes. The clockworks is mounted on a swing-out panel for easy access.

With its timeless design, hardwood cabinet, and brass clockworks, this clock is as beautiful as any piece of furniture. It's truly an heirloom project.

The double curves give this clock a classic, graceful shape. Creating the curves is easy with two techniques — kerf bending and template routing.

A weekend project with clean lines, this timepiece will still attract attention. We'll show how to make the eye-catching figure of curly maple "pop" out.

Jewelry Chest

Barrister's Bookcase

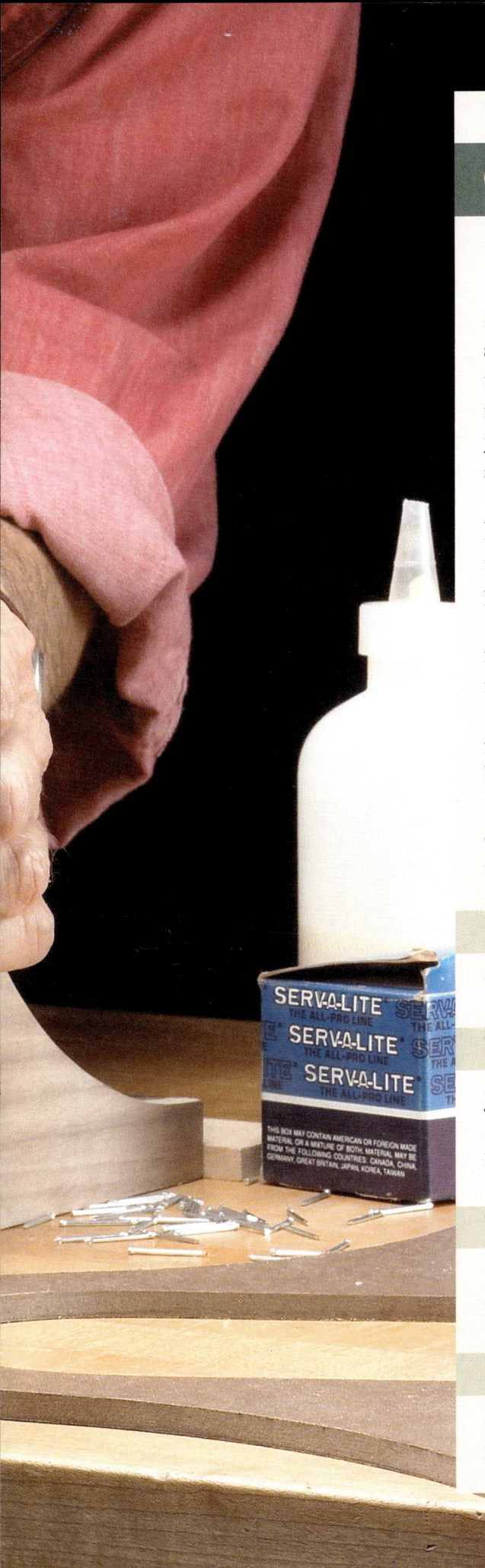

CLOCKS

Almost every woodworker worth his weight in sawdust has built (or wants to build) at least one clock. The four projects in this section run the gamut of styles: wall-hanging, floor-standing, and tabletop. Some use traditional mechanical clockworks, others use modern quartz movements. Some can be built with either. Whichever project you choose and however you decide to customize it, the appeal of a hand-crafted timepiece is universal.

Despite appearances (and sounds), the pendulum clock has a quartz movement. It brings to mind quiet hours at the library, where the only sounds are the slip of turning pages and the steady tick-tock of a swinging pendulum.

The tall case clock is as impressive as any piece of furniture. Construction is made easier thanks to a design that relies on a series of simple-to-build frames and cases.

The graceful, curving top of the tambour clock is formed from regular hardboard that's been shaped through kerf-bending and covered in handsome walnut veneer.

For a clock you can make in a weekend, the mantel clock will attract a lot of attention, thanks to the dramatic figured wood used to build it.

Pendulum Clock

This classically styled wall clock has a quartz movement with a realistic-sounding chime and a pendulum. For easy access, the clockworks is placed behind two doors — like a safe behind a picture.

There are a couple of things I had to consider when designing this Pendulum Clock. First, I needed to provide enough room inside the case for the clockworks I chose and the swinging pendulum. And second, I wanted enough space *behind* the works to reach inside to adjust the clock or replace a worn-out battery. And I wanted to do this without removing the clock from the wall and taking off the back.

TWO DOORS. The solution to both these problems was a door behind a door (see the photo on page 11). The outer door encloses the case. The inner door supports the clockworks. So it's easy to get at them if you need to adjust the time or change the battery.

CLOCKWORKS. Speaking of the clock parts, the movement I selected is battery-operated with a "bim-bam" chime. After selecting the movement and dial, the case was designed to fit around them.

But if you prefer another movement for a different sound, or maybe a different dial, hold off building the case until you have the parts in hand. That's because a different size dial and parts might require a different size case.

BUILT-UP MOLDINGS. Another interesting feature of this Pendulum Clock is the built-up moldings I used at the top and bottom. Instead of using store-bought molding, I've actually built up the molding using three pieces of solid wood with some routed profiles on each piece.

BARREL-TYPE MAGNET CATCH. The inner door is held shut with a surface mounted magnetic catch, but the outer door is held firmly shut with a barrel-type magnetic catch. The catch must be installed in the side of the clock case and a strike plate is attached to the inside of the door. Installing this catch can be difficult though. That's because it's important that the hole is drilled close to the same diameter and depth of the catch. A Technique article on page 13 will help you through the process.

EXPLODED VIEW

OVERALL DIMENSIONS:
13W x 5$\frac{13}{16}$D x 27$\frac{1}{2}$H

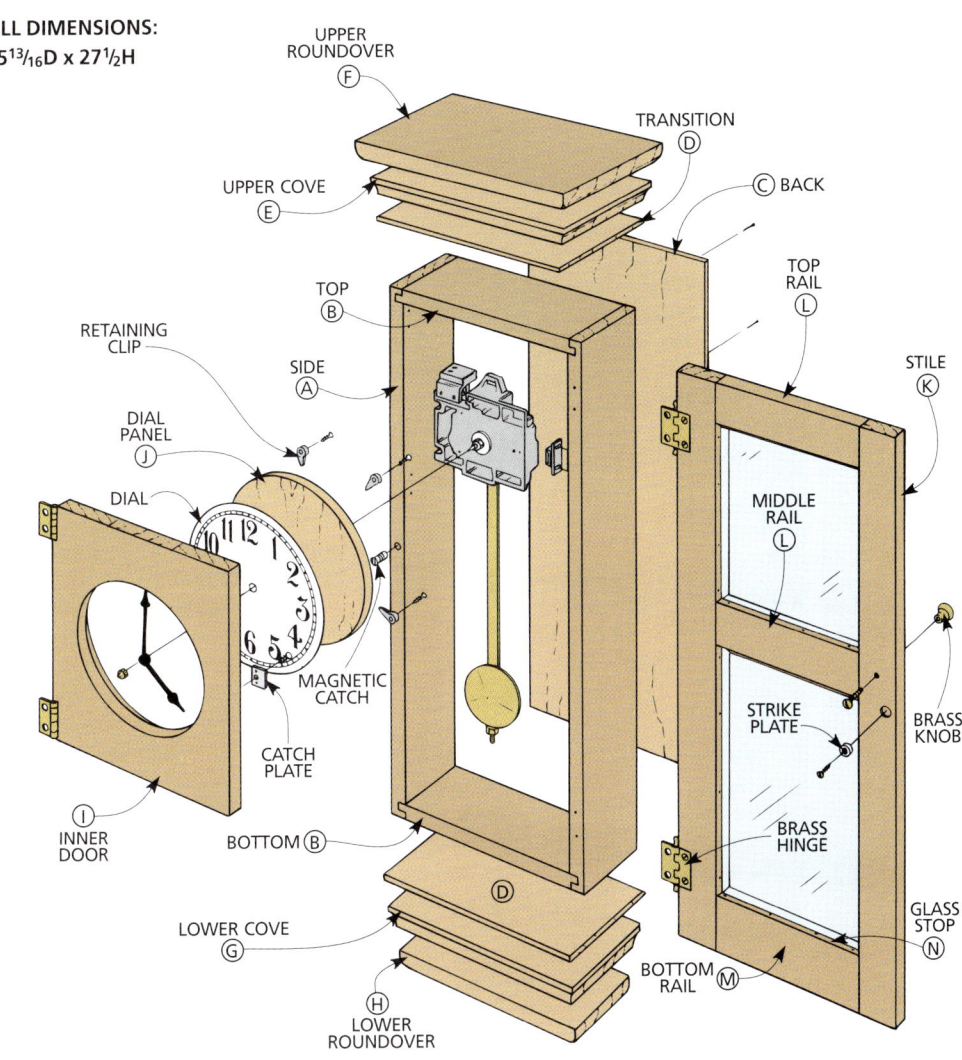

UPPER ROUNDOVER — F

TRANSITION — D

UPPER COVE — E

C BACK

TOP — B

TOP RAIL — L

RETAINING CLIP

SIDE — A

STILE — K

DIAL PANEL — J

DIAL

MIDDLE RAIL — L

MAGNETIC CATCH

STRIKE PLATE

BRASS KNOB

CATCH PLATE

BOTTOM — B

BRASS HINGE

INNER DOOR — I

GLASS STOP — N

LOWER COVE — G

BOTTOM RAIL — M

LOWER ROUNDOVER — H

CUTTING DIAGRAM

¾ x 6 - 84 (3.5 Bd. Ft.)

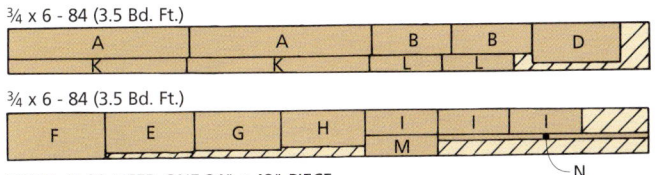

¾ x 6 - 84 (3.5 Bd. Ft.)

NOTE: ALSO NEED ONE 24" x 48" PIECE OF ¼" PLYWOOD FOR PARTS C AND J

MATERIALS LIST

WOOD

A	Sides (2)	¾ x 4 - 24
B	Top/Bottom (2)	¾ x 4 - 10¼
C	Back (1)	¼ ply - 10¼ x 23¼
D	Transitions (2)	¼ x 4$\frac{15}{16}$ - 11¼
E	Upper Cove (1)	¾ x 5$\frac{5}{16}$ - 12
F	Upper Roundover (1)	¾ x 5$\frac{13}{16}$ - 13
G	Lower Cove (1)	¾ x 4$\frac{13}{16}$ - 11
H	Lower Roundover (1)	¾ x 4$\frac{1}{16}$ - 9½
I	Inner Door (1)	¾ x 9⅜ - 9½
J	Dial Panel (1)	¼ ply - 8 dia.
K	Stiles (2)	¾ x 1¾ - 23⅞
L	Top/Middle Rails (2)	¾ x 1¾ - 9½
M	Bottom Rail (1)	¾ x 2½ - 9½
N	Glass Stops	¼ x ¼ - 70 rgh.

HARDWARE SUPPLIES

(2) 1½" x 1¼" brass butt hinges w/ screws
(2) 1½" x 1¼" brass ball-tipped hinges w/ screws
(1) Surface-mounted magnetic catch w/ catch plate
(2) Hanger plates w/ screws
(1) $\frac{5}{16}$"-dia. magnetic door catch w/ strike plate
(1) ¾" x ¾" brass door knob w/ screw
(1) ⅛" glass, 10" x 24" rough
(1) 8"-dia. clock face w/ hands
(4) Plastic retaining clips w/ screws
(1) Quartz clock movement w/ pendulum and mounting screws
(52) ½" wire brads

CASE

The case of the Pendulum Clock is just a tall box with a back. A tongue and dado joint holds the box together, while the back fits in a rabbet cut in the case sides, top, and bottom. Later, I added some molding to make the box look a little more like a clock case.

BOX PIECES. I started the case by cutting the case sides (A), top (B), and bottom (B) to finished length *(Fig. 1)*. Then I ripped all four pieces to the same finished width (4").

JOINTS. Before the case can be assembled, joints are cut on the ends of each piece *(Fig. 2)*. I decided to use rabbet and dado joints — they're simple to make, yet surprisingly strong.

Note: I'd like to clarify one thing. A rabbet and dado joint could also be called a tongue and dado joint. That's because the rabbets actually form short tongues on the ends of the top and bottom pieces. It's these tongues that support the weight of the clock case.

After cutting the joints on the case pieces, the box can be glued together, squared, and clamped.

BACK RABBET. Next, using the router table, cut a rabbet around the inside back edges of the case to accept the $1/4$" plywood back *(Figs. 3 and 3a)*. Then square up the corners with a chisel.

BACK PANEL. Finally, cut a plywood back (C) to fit in the rabbet *(Fig. 1)*. (Mine was $10^{1}/4$" x $23^{1}/4$".) But don't install the back just yet — it's easier to apply the finish to the case and install the clockworks if the back of the case is left open.

UPPER & LOWER MOLDING

Adding decorative molding turns the plain box into a finished case. Rather than use store-bought molding, I built up my molding from three pieces of matching hardwood with a couple of easy to rout profiles on the edges.

MOLDING. I started with two $1/4$"-thick pieces of transition (D) molding *(Fig. 4)*. First, cut these pieces to length so they extend $1/8$" beyond the sides of the case. Then rip them to width so they're flush at the back of the case and extend $15/16$" beyond the front of the case.

Note: The dimensions given for the molding assume the case is 4" deep.

Now, cut the upper cove (E) and the upper roundover (F) molding pieces for the upper assembly *(Fig. 4)*.

Then cut the the lower cove (G) and lower roundover (H) molding for the bottom of the case *(Fig. 4)*.

ROUT PROFILES. Next, decorative profiles can be routed on three edges of each molding piece *(Figs. 5, 6, and 7)*.

Note: The back edges of the molding pieces should not be routed.

ATTACH MOLDING. Lastly, the molding is attached to the case. A few brads with the heads nipped off keep the moldings from sliding around on the glue *(Fig. 4)*.

INNER DOOR

After the molding was attached to the case, I began work on the inner door.

INNER DOOR. First edge-glue three pieces of $3/4$" stock for the inner door panel (I) *(Fig. 8)*. (It's hard to find a single piece of stock that's wide enough.)

Then cut the panel to finished size.

Note: I allowed for a $1/16$" gap on the sides and top *(Fig. 11)*.

Next, use a jig saw to cut an opening in the middle of the panel to accommodate the clock dial *(Fig. 8)*.

RABBET AND CHAMFERS. After the opening for the dial has been cut, you can rout a rabbet around the back edge of the opening *(Figs. 9 and 9a)*. This will

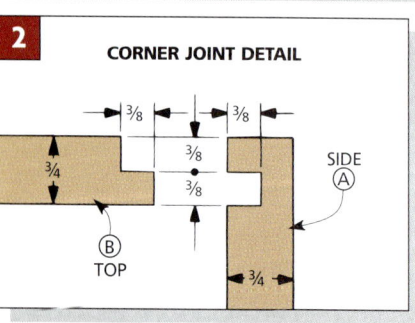

1

TOP
(B)

10¼

BACK
(C)

(A)
SIDE

23¼

(A)

24

10¼

(B)
BOTTOM

4

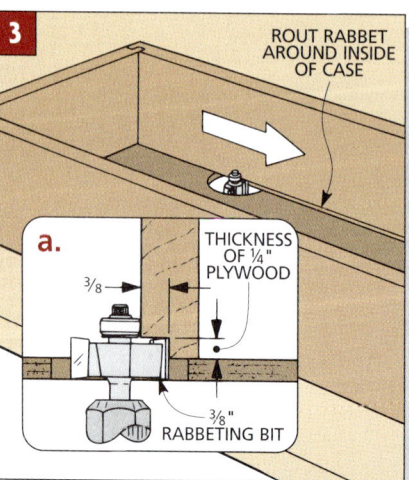

2 CORNER JOINT DETAIL

3/8 · 3/8

3/4 · 3/8 · 3/8 · 3/8

SIDE
(A)

(B)
TOP

3/4

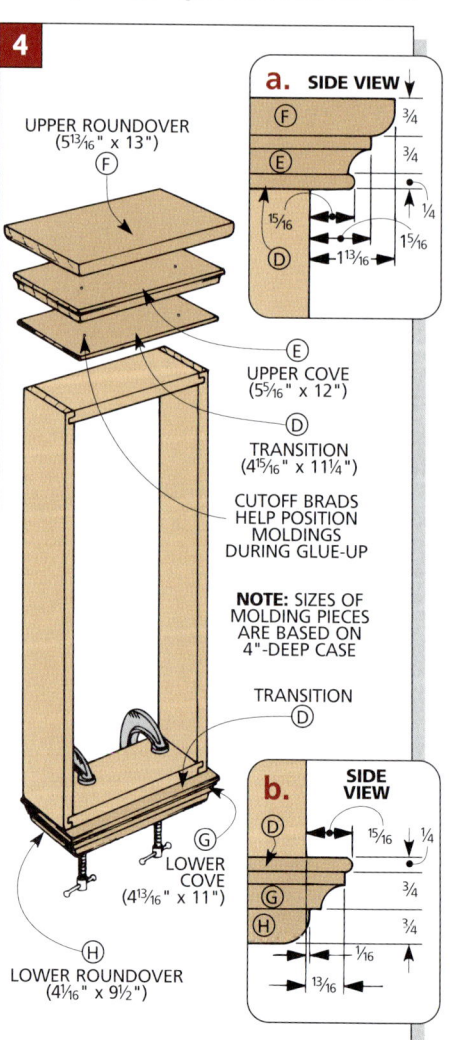

3

ROUT RABBET AROUND INSIDE OF CASE

a.

THICKNESS OF ¼" PLYWOOD

3/8

3/8"
RABBETING BIT

4

UPPER ROUNDOVER
($5^{13}/16$" x 13")
(F)

a. SIDE VIEW

(F) · 3/4
(E) · 3/4
· 1/4
$15/16$
$15/16$
(D)
$1^{13}/16$

(E)
UPPER COVE
($5^{5}/16$" x 12")

(D)
TRANSITION
($4^{15}/16$" x $11^{1}/4$")

CUTOFF BRADS HELP POSITION MOLDINGS DURING GLUE-UP

NOTE: SIZES OF MOLDING PIECES ARE BASED ON 4"-DEEP CASE

TRANSITION
(D)

(G)
LOWER COVE
($4^{13}/16$" x 11")

(H)
LOWER ROUNDOVER
($4^{1}/16$" x $9^{1}/2$")

b. SIDE VIEW

(D) · $15/16$ · 1/4
(G) · 3/4
(H) · 3/4
· 1/16
$13/16$

A door behind a door. That's the solution to the problem of how to make room for the clockworks and the pendulum, while providing a convenient way to get at the back of the clock to replace a battery.

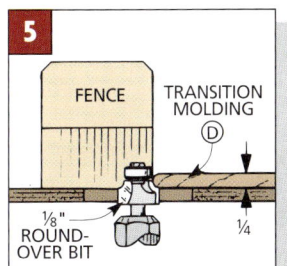

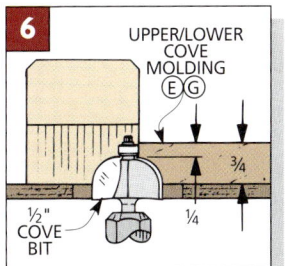

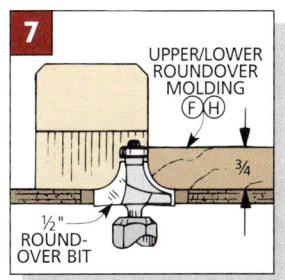

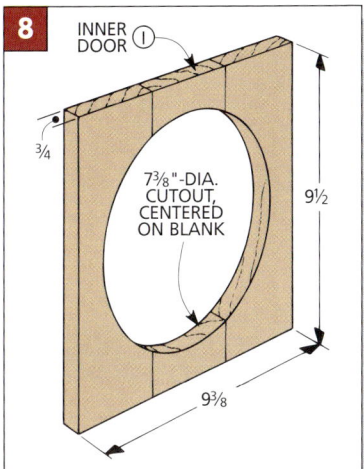

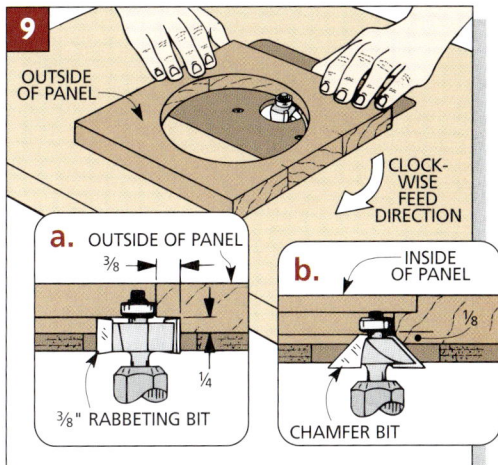

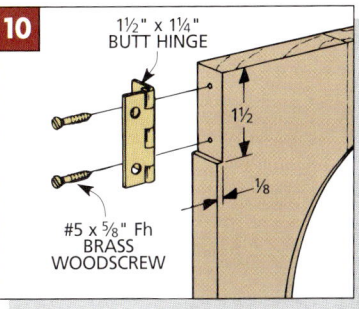

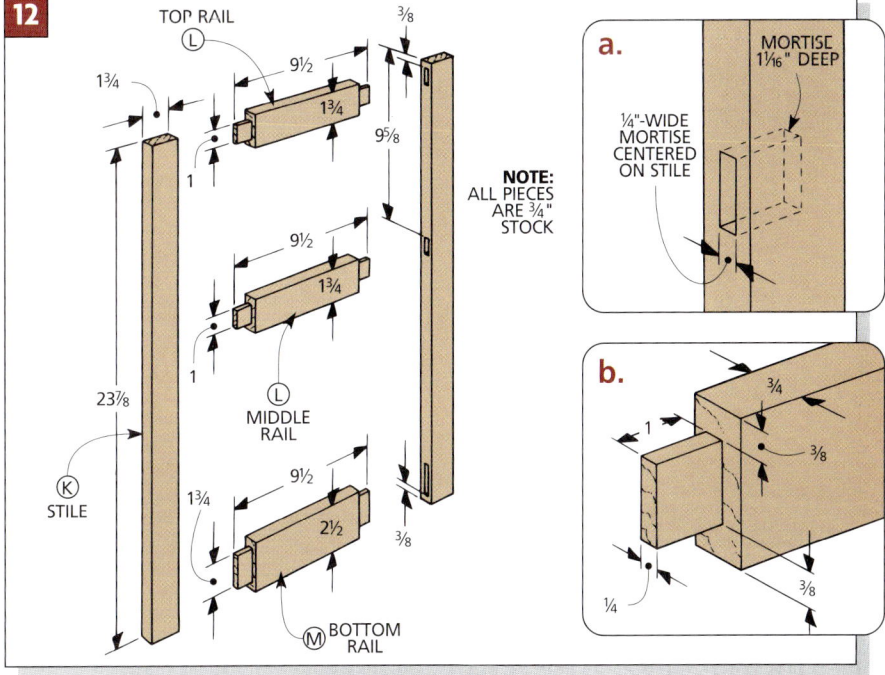

hold the clock dial and also a $\frac{1}{4}$" plywood dial panel that's made later.

Next, I routed a decorative chamfer on the front edge of the opening *(Fig. 9b)*.

HINGE MORTISES. To complete the inner door, cut a pair of shallow mortises on the left edge of the panel for the door hinges *(Fig. 10)*. Then attach the hinges to the door, but not yet to the case.

Finally, cut a dial panel (J) from $\frac{1}{4}$" plywood to fit the rabbet on the back of the door. Again, allow for a $\frac{1}{16}$" gap all around the panel for shrinking and swelling of the door.

OUTER DOOR

The outer door covers the front of the case, including the inner door. It's a mortise and tenon frame with two windows.

DOOR STILES. I started the outer door by ripping two vertical stiles (K) to width *(Fig. 12)*. Then I cut them to fit between the molding, allowing for a $\frac{1}{16}$" gap between the stiles and the transition moldings on the top and bottom.

DOOR RAILS. Next, rip three horizontal rails (L, M) to finished width *(Fig. 12)*.

Note: All three door rails are not the same width. If they were, the lower one would appear too narrow. So I cut it slightly wider than the others.

Now, cut the rails to fit between the stiles *(Fig. 12)*. Be sure to allow for the tenons on the ends of the rails *(Fig. 12b)*.

MORTISE AND TENON JOINTS. With the stiles and rails cut to size, centered mortises can be bored on the inside edges of the stiles *(Figs. 12 and 12a)*.

Then, cut the tenons on the ends of the rails to fit the mortises *(Fig. 12b)*. Now the door can be glued up and clamped.

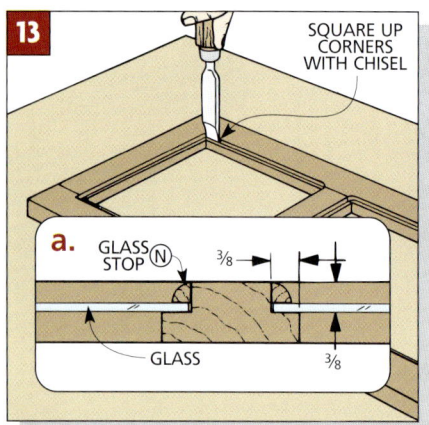

13 SQUARE UP CORNERS WITH CHISEL

a. GLASS STOP ⓝ 3/8

GLASS 3/8

14 NOTE: HINGE MORTISES ARE 1/8" DEEP

1 3/4

1 1/2

1/2

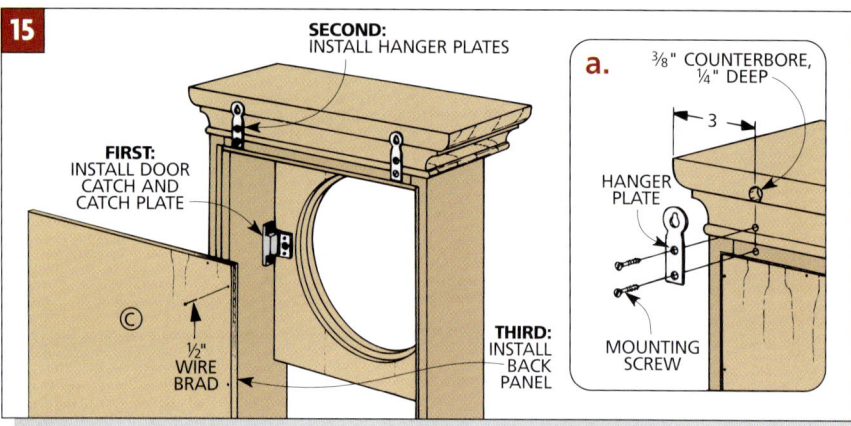

15 SECOND: INSTALL HANGER PLATES

FIRST: INSTALL DOOR CATCH AND CATCH PLATE

ⓒ

1/2" WIRE BRAD

THIRD: INSTALL BACK PANEL

a. 3/8" COUNTERBORE, 1/4" DEEP

3

HANGER PLATE

MOUNTING SCREW

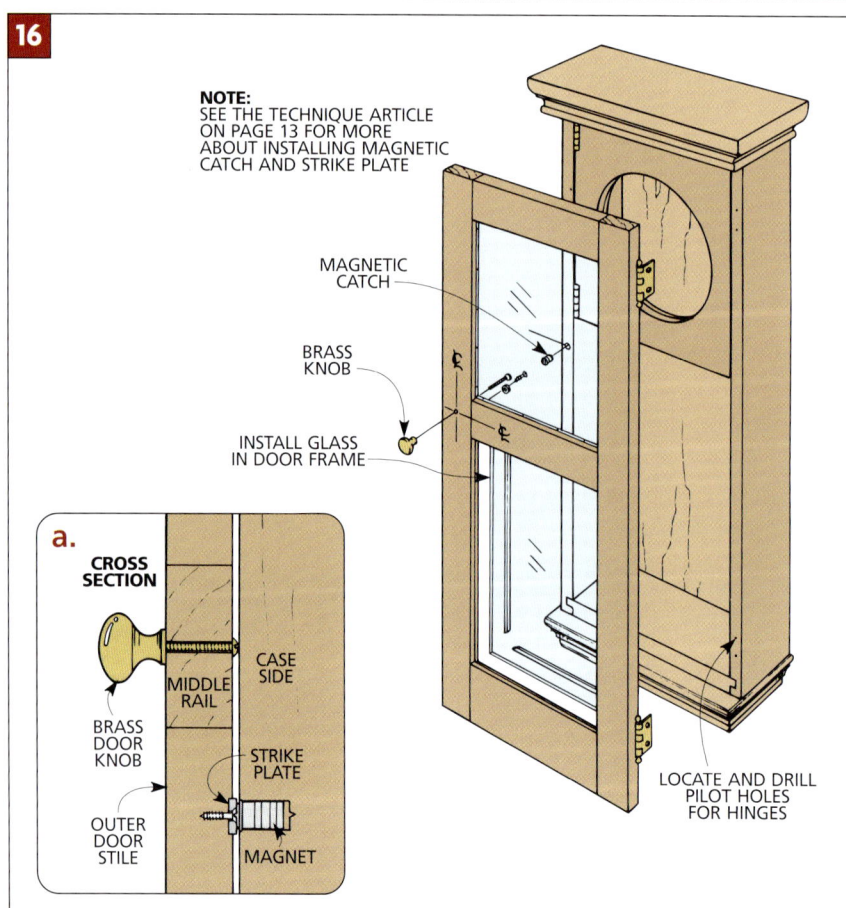

16 NOTE: SEE THE TECHNIQUE ARTICLE ON PAGE 13 FOR MORE ABOUT INSTALLING MAGNETIC CATCH AND STRIKE PLATE

MAGNETIC CATCH

BRASS KNOB

INSTALL GLASS IN DOOR FRAME

LOCATE AND DRILL PILOT HOLES FOR HINGES

a. CROSS SECTION

BRASS DOOR KNOB

OUTER DOOR STILE

MIDDLE RAIL

CASE SIDE

STRIKE PLATE

MAGNET

WINDOW RABBETS. The glass windows rest in rabbets routed in the back side of this door *(Fig. 13)*. And they're held in place with strips of quarter-round glass stops and short nails *(Fig. 13a)*.

After routing the rabbets, square up the corners with a chisel *(Fig. 13)*.

DOOR HINGES. Before installing the glass in the outer door, the hinges can be attached to the right edge of the door.

Note: It swings in the opposite direction of the inner door.

To install the hinges, I cut a mortise toward the top and bottom of the door *(Fig. 14)*. But don't hang the door until the clock case is complete.

FINISH & ASSEMBLY

It's easiest to apply finish to the clock before it's completely assembled. So first I rubbed on a coat of wood sealer. Then I applied two top coats of a satin finish.

When the top coat of finish has dried completely, the hardware can be installed.

INNER DOOR AND BACK. Start final assembly of the clock from the inside out, hanging the inner door first. The hinges are mortised into the left edge of this piece, so you can just screw the hinges to the case side *(Fig. 16)*. I aligned the door so there is a 1/16" gap between the top and the door.

DOOR CATCH. Then close the door and mark the position of a magnetic catch on the inside of the case *(Fig. 15)*. Now, screw the catch inside the case, then screw the catch plate to the door.

HANGER PLATES. Next, mark the positions for two hanger plates on the back of the case toward the top *(Fig. 15a)*.

Note: In order for the case to fit flush against the wall, I drilled a counterbore in the back edge of the upper moldings to accept the head of the screw in the wall. The diameter of the counterbore must be larger than the hole at the top of the hanger plate. This is so the notch in the plate can slip onto the shank of the screw you placed in the wall *(Fig. 15a)*.

The hanger plates can now be screwed to the back of the clock case.

BACK PANEL. At this point, the plywood back can be attached to the case *(Fig. 15)*. For this I used 1/2" brads, but no glue.

OUTER DOOR. Next, I turned to the outer door. First, mark the position of the holes for the hinges on the edge of the case side *(Fig. 16)*. Then drill pilot holes for the hinge screws and temporarily attach the door to the case.

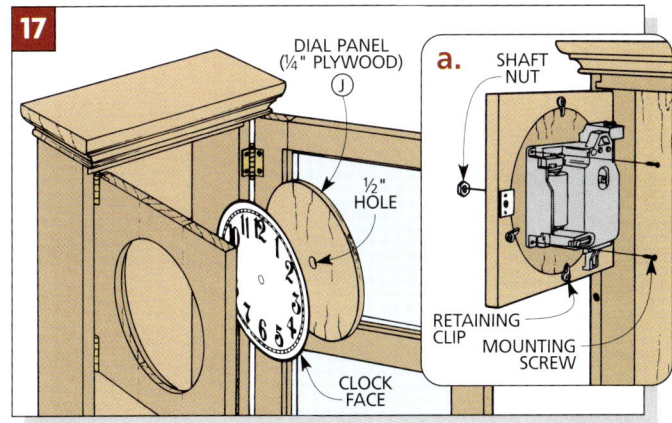

MAGNETIC CATCH. For the best appearance, I used a recessed magnetic catch for the outer door (*Fig. 16a*). To install the catch in the case, first drill a hole centered on the edge of the case side. Then press the magnet in the hole, and screw the strike plate to the door. (For more on this, see the Technique below.)

DOOR KNOB. Once the magnetic catch is installed, remove the door and install the door knob. It's centered on the outer stile and middle rail (*Figs. 16 and 16a*).

GLASS WINDOWS. Finally, I finished assembling the outer door by cutting the two glass windows to size and installing them each in their openings.

Note: I used quarter-round glass stop molding to hold the $\frac{1}{8}$"-thick (single strength) glass in place. The glass is cut $\frac{1}{8}$" smaller in both dimensions than the size of the window openings. (This is to allow for contraction of the door frame.) For more on making the glass stop, see the Shop Tip on page 36.

CLOCK DIAL & WORKS

Now that the clock case construction is complete, this is when the project becomes a working timepiece.

DIAL PANEL. Before installing the clock face (dial), a $\frac{1}{2}$"-dia. hole must be drilled

through the center of the dial panel for the shaft on the clock movement (*Fig. 17*).

CLOCK FACE. Then, place the clock face in the rabbet on the inside of the inner door. (Orient the markings on the dial so 12 o'clock is straight up.)

To complete the dial, I secured the clock face by installing the dial panel and four plastic retaining clips (*Fig. 17a*).

CLOCKWORKS. Now the clock movement can be screwed to the back of the dial panel (*Fig. 17a*). Just be sure you don't screw through the dial panel. Finally, to complete the Pendulum Clock, the hands are installed on the shaft, and the pendulum is hung below. ∎

TECHNIQUE . . . . *Installing a Magnetic Catch*

Installing barrel-type magnetic catches can be tricky. That's because some care must be taken when drilling the hole for the magnet.

If the the hole is too wide, the magnet comes out each time you open the door. And if the hole is too shallow, there's the possibility you won't be able to set the magnet properly, keeping the door from closing all the way.

HOLE DIAMETER. To install the magnet on the Pendulum Clock, I drilled a $\frac{5}{16}$"-dia.hole. (The hole should actually have been 8mm in diameter, but I didn't have a metric drill bit. My magnet was made to metric specifications.) The difference in diameter is minimal, though, so I was able to make do with what I had.

HOLE DEPTH. The depth of the hole is also critical. As I mentioned, if the hole is too shallow, the magnet will bottom out and the door won't close completely.

So I drilled the hole a little deeper than necessary ($\frac{5}{8}$" for the $\frac{9}{16}$"-long magnet).

Even though there's a small lip at the top of the catch that fits around the top of the hole, I used a small block when installing the magnet to keep from driving it in too far (detail 'a' in drawing).

STRIKE PLATE. The easy part is attaching the strike plate. First, stick the metal strike plate to the magnet. Then shut the door against the plate and press firmly. This creates a small "dimple" on the door stile (detail 'b'). This indicates

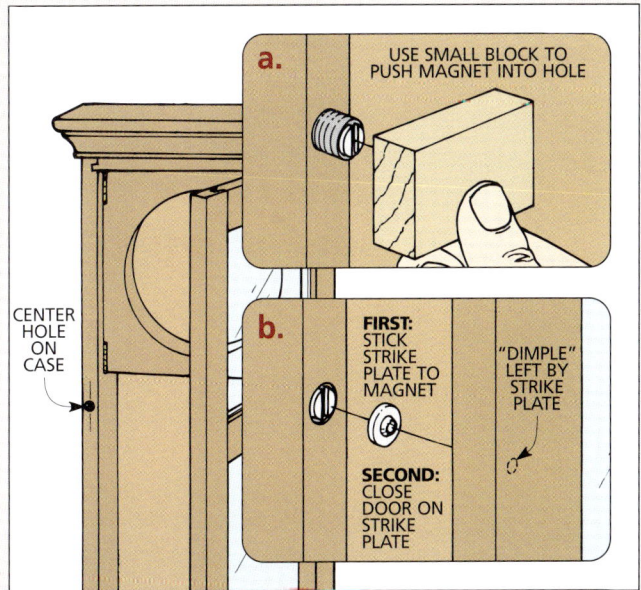

where to drill the pilot hole and a shallow couterbore (so that the plate is flush with the door frame). Now when the plate is attached, it will align perfectly with the magnet.

Tall Case Clock

This Tall Case Clock is built out of cherry and designed to be a welcome addition to any home. With its timeless design, beautiful hardwood case, and brass clockworks, this is truly an heirloom project.

Most of the clocks I've built have been designed to hang on a wall or sit on a table. But this Tall Case Clock is more than just another timepiece — it features brass clockworks and is an impressive piece of "furniture." Yet, it's surprisingly easy to build.

FRAMES. I was able to keep the construction relatively simple by designing the case with similar components.

Much of the construction involves making frames with molded edges. In fact, there are six frames that separate the three main sections of the case, as well as the crown molding at the top of the clock and the kickboard base.

EQUIPMENT. At first glance, you might expect that you would need a lot of tools to build a clock like this — especially with all the molding. But I cut all the molding with a router — then stacked it to look more detailed.

WOOD. Its warm luster and rich tones make cherry a perfect choice for this clock. And except for the case backs and dust panels, it's all cut from ³⁄₄"-thick stock.

CLOCK KIT. As for the clockworks, I purchased a kit with a high-quality brass clockworks (see Sources, page 126).

As you can see, the Tall Case Clock has a glass door so you can proudly display the brass pendulum and weights (main photo). However, if you'd prefer to use a less expensive quartz timepiece in your clock, you could substitute a solid-cherry hardwood panel for the glass (inset photo).

FINISH. When finishing projects with a lot of molding, I like to use a wipe-on tung oil finish.

BRACKET. One last little detail. Because this is a tall piece of furniture, it has a high center of gravity and could be easy to tip over. So to keep it stable, I used an L-bracket to anchor the top of the clock to the wall.

EXPLODED VIEW

OVERALL DIMENSIONS:
17½W x 10½D x 74H

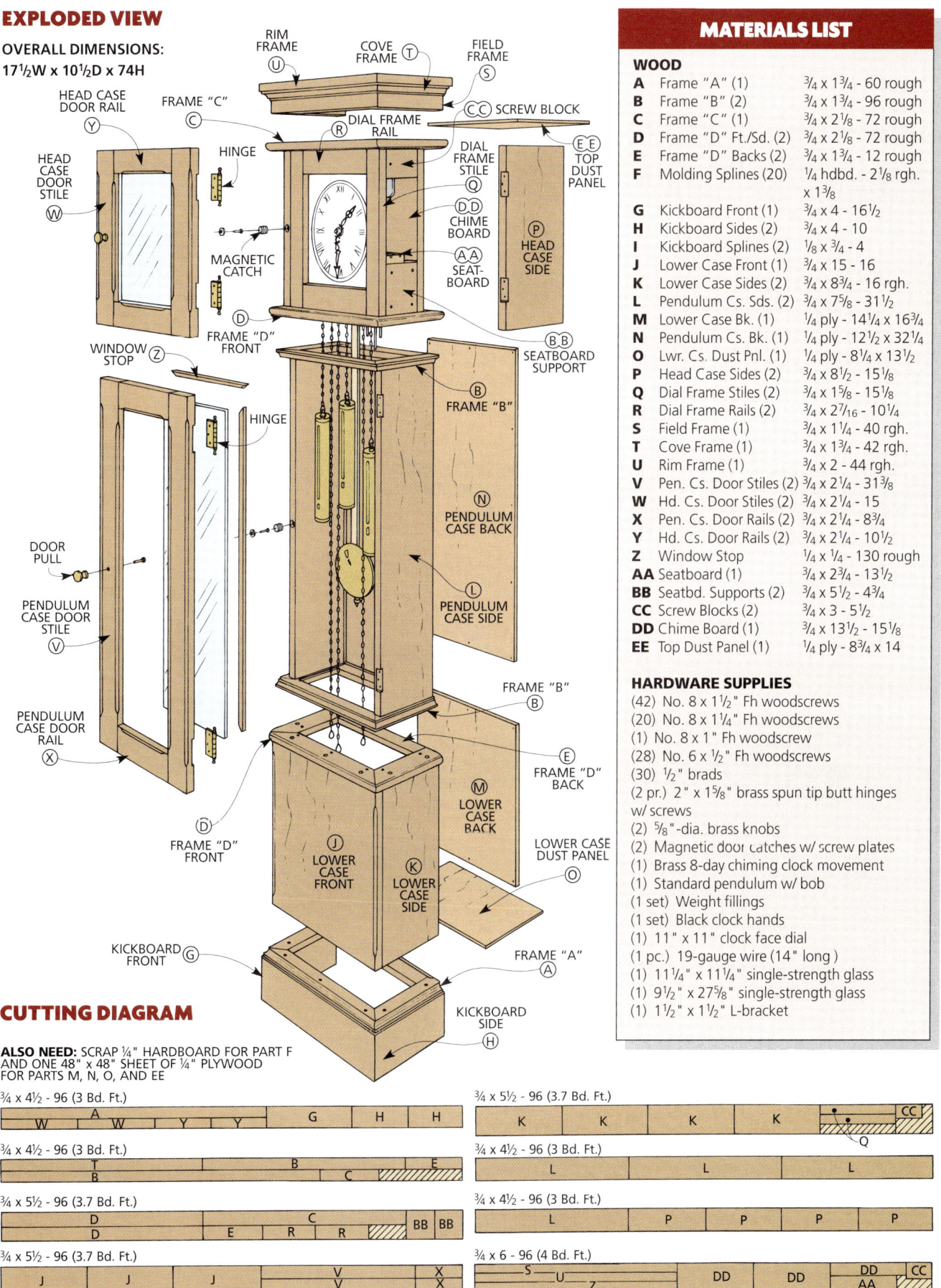

MATERIALS LIST

WOOD

A	Frame "A" (1)	¾ x 1¾ - 60 rough
B	Frame "B" (2)	¾ x 1¾ - 96 rough
C	Frame "C" (1)	¾ x 2⅛ - 72 rough
D	Frame "D" Ft./Sd. (2)	¾ x 2⅛ - 72 rough
E	Frame "D" Backs (2)	¾ x 1¾ - 12 rough
F	Molding Splines (20)	¼ hdbd. - 2⅛ rgh. x 1⅜
G	Kickboard Front (1)	¾ x 4 - 16½
H	Kickboard Sides (2)	¾ x 4 - 10
I	Kickboard Splines (2)	⅛ x ¾ - 4
J	Lower Case Front (1)	¾ x 15 - 16
K	Lower Case Sides (2)	¾ x 8¾ - 16 rgh.
L	Pendulum Cs. Sds. (2)	¾ x 7⅝ - 31½
M	Lower Case Bk. (1)	¼ ply - 14¼ x 16¾
N	Pendulum Cs. Bk. (1)	¼ ply - 12½ x 32¼
O	Lwr. Cs. Dust Pnl. (1)	¼ ply - 8¼ x 13½
P	Head Case Sides (2)	¾ x 8½ - 15⅛
Q	Dial Frame Stiles (2)	¾ x 1⅝ - 15⅛
R	Dial Frame Rails (2)	¾ x 2⁷⁄₁₆ - 10¼
S	Field Frame (1)	¾ x 1¼ - 40 rgh.
T	Cove Frame (1)	¾ x 1¾ - 42 rgh.
U	Rim Frame (1)	¾ x 2 - 44 rgh.
V	Pen. Cs. Door Stiles (2)	¾ x 2¼ - 31⅜
W	Hd. Cs. Door Stiles (2)	¾ x 2¼ - 15
X	Pen. Cs. Door Rails (2)	¾ x 2¼ - 8¾
Y	Hd. Cs. Door Rails (2)	¾ x 2¼ - 10½
Z	Window Stop	¼ x ¼ - 130 rough
AA	Seatboard (1)	¾ x 2¾ - 13½
BB	Seatbd. Supports (2)	¾ x 5½ - 4¾
CC	Screw Blocks (2)	¾ x 3 - 5½
DD	Chime Board (1)	¾ x 13½ - 15⅛
EE	Top Dust Panel (1)	¼ ply - 8¾ x 14

HARDWARE SUPPLIES

(42) No. 8 x 1½" Fh woodscrews
(20) No. 8 x 1¼" Fh woodscrews
(1) No. 8 x 1" Fh woodscrew
(28) No. 6 x ½" Fh woodscrews
(30) ½" brads
(2 pr.) 2" x 1⅝" brass spun tip butt hinges w/ screws
(2) ⅝"-dia. brass knobs
(2) Magnetic door catches w/ screw plates
(1) Brass 8-day chiming clock movement
(1) Standard pendulum w/ bob
(1 set) Weight fillings
(1 set) Black clock hands
(1) 11" x 11" clock face dial
(1 pc.) 19-gauge wire (14" long)
(1) 11¼" x 11¼" single-strength glass
(1) 9½" x 27⅝" single-strength glass
(1) 1½" x 1½" L-bracket

CUTTING DIAGRAM

ALSO NEED: SCRAP ¼" HARDBOARD FOR PART F AND ONE 48" x 48" SHEET OF ¼" PLYWOOD FOR PARTS M, N, O, AND EE

¾ x 4½ - 96 (3 Bd. Ft.)
A | W | W | Y | Y | G | H | H

¾ x 5½ - 96 (3.7 Bd. Ft.)
K | K | K | K | CC | Q

¾ x 4½ - 96 (3 Bd. Ft.)
T | B | E
B | C

¾ x 4½ - 96 (3 Bd. Ft.)
L | L | L

¾ x 5½ - 96 (3.7 Bd. Ft.)
D | C
D | E | R | R | BB | BB

¾ x 4½ - 96 (3 Bd. Ft.)
L | P | P | P | P

¾ x 5½ - 96 (3.7 Bd. Ft.)
J | J | J | V | X
V | X

¾ x 6 - 96 (4 Bd. Ft.)
S | DD | DD | DD | CC
U | Z | AA

FRAME MOLDINGS

What makes this clock an heirloom piece of furniture, and not simply a stack of frames and case sides? The exact fit between all of the parts. To achieve this fit, the solid-wood case sides must be perfectly flat and the frames must be square.

The clock has four distinct frames that separate the clock's sections. The five sections are the kickboard, lower case, pendulum case, head case, and crown

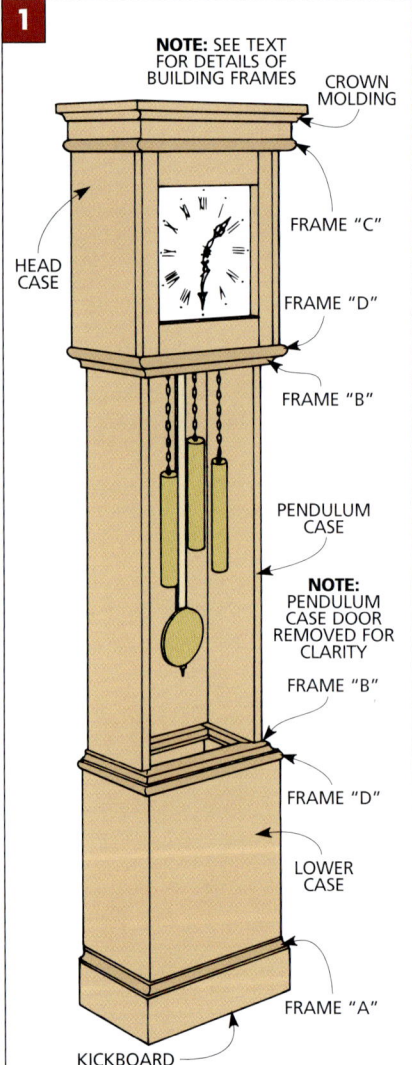

1

NOTE: SEE TEXT FOR DETAILS OF BUILDING FRAMES

CROWN MOLDING

HEAD CASE

FRAME "C"

FRAME "D"

FRAME "B"

PENDULUM CASE

NOTE: PENDULUM CASE DOOR REMOVED FOR CLARITY

FRAME "B"

FRAME "D"

LOWER CASE

FRAME "A"

KICKBOARD

molding *(Fig. 1)*. The frames are used to orient and align the cases. At the bottom of the clock is the kickboard frame "A." Frame "B" is at the top and bottom of the pendulum case. At the top of the head case, separating the head case from the crown molding, is frame "C." Lastly, frame "D" is located at both the bottom of the head case and the top of the lower case.

FRAME DESIGN

Frames for casework are typically joined in two ways. If the frame is mostly decorative, it's joined with 45° miters at all four corners. Structural frames, on the other hand, like the ones you find in a well-built chest of drawers, are usually made with mortise and tenon joints.

For the six frames on this clock, I actually borrowed from each of these designs *(Figs. 2 and 3)*. I wanted to hide the end grain on the front corners of the frames, so I decided to use miter joints.

Since the visible edges of the frames have routed profiles on them, and I wanted this profile to extend all the way to the back edges of the frames, I used butt joints on the back corners.

STRONG FRAME. And since the frames lie flat and are screwed to the main sections of the case, there's not much force pulling apart the mitered front corners. But the backs of the frames aren't attached to any other part of the case. So here I strengthened the butt joint with a mortise and a loose tenon. (You could use a regular mortise and tenon.)

FRAME PARTS

All six frames are built the same way. I found it most efficient to build them all at once, rather than one at a time as I needed them for the clock.

I cut the four sections needed for each frame to exact size before assembly.

Note: Refer to *Figs. 4, 5, 6, and 7* on the opposite page for the exact measurements of each frame.

FRONTS. The critical dimension on these frames is the length of the front piece. As you build each frame, start by cutting 45° miters on the front piece so the long-point to long-point measurement equals the dimensions given in the drawings on the next page.

SIDES. Next cut a miter on one end of each side piece. Then trim each of these pieces to length with a square cut across the back end.

BACKS. Now cut the back of the frame to finished width and length. The length of these back pieces should equal the short-point to short-point distance between the miters on the front pieces. (They will have to be 1½" longer if you use a mortise and tenon joint.)

MORTISE AND FLOATING SPLINES. To join the frame backs to the sides, I cut ¼"-wide mortises using a straight bit in a table-mounted router.

After cutting the mortises to size, cut the molding splines (F) *(Fig. 2)*. I used ¼" hardboard for these pieces and I rounded over the edges of the splines to fit the mortises.

ASSEMBLING THE FRAMES

To keep each frame square, flat, and flush across its joints, I clamped the frame, one section at a time, to a piece of plywood with square corners *(Fig. 3)*.

To do this, begin by gluing one of the side pieces to the front piece. Next, glue the back to this front and side assembly. Then add the last side piece.

To clamp the frame in place, I put one C-clamp on either side of the miter joints, then I placed a bar clamp to hold the back in place *(Fig. 3)*. I also put waxed paper under each joint so the frames wouldn't be glued to the plywood.

ROUT EDGES. When the glue has dried and the frame is complete, rout the decorative profile on the front edges. (Refer to the procedures shown on the facing page for routing these profiles.) Do not rout any of the frame backs.

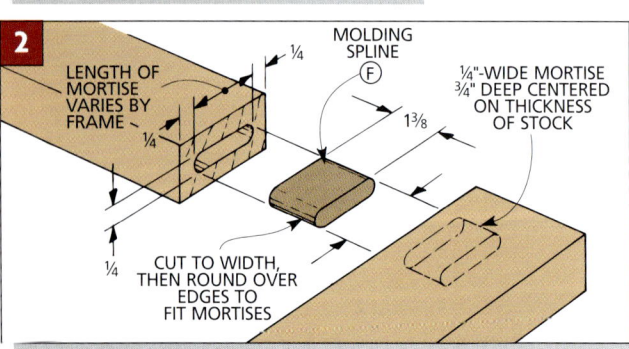

2

LENGTH OF MORTISE VARIES BY FRAME

¼

¼

¼

MOLDING SPLINE (F)

¼

¼"-WIDE MORTISE ¾" DEEP CENTERED ON THICKNESS OF STOCK

1⅜

CUT TO WIDTH, THEN ROUND OVER EDGES TO FIT MORTISES

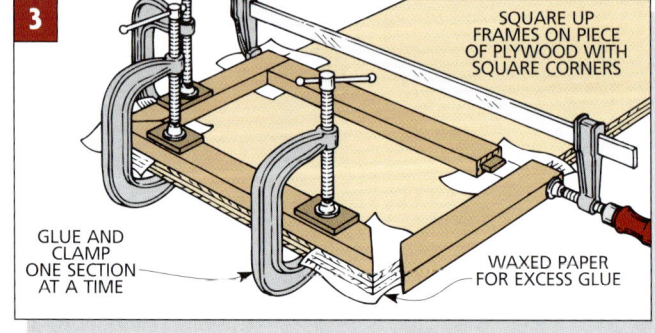

3

SQUARE UP FRAMES ON PIECE OF PLYWOOD WITH SQUARE CORNERS

GLUE AND CLAMP ONE SECTION AT A TIME

WAXED PAPER FOR EXCESS GLUE

FRAME "A"

Frame "A" separates the kickboard from the lower case (*Fig. 1*).

To rout the profile on the front and sides of the frame, first use a 1/2" cove bit (*Fig. 4a*). It's routed in several passes until there's a 1/4"-thick shoulder along the outside of the frame.

Complete the profile by forming a 1/8"-deep rabbet along the lower outside edge. I used a 3/4" straight bit and left a 1/8"-thick shoulder (*Fig. 4b*).

Drill seven countersunk shank holes on the rabbeted side of the frame (*Fig. 4*).

FRAME "B"

You will need two of these frames — one for the top of the pendulum case, and one for the bottom of the pendulum case.

Shape the molding for frame "B" by routing the 1/2" cove in several passes until there's a 1/8"-thick shoulder on the bottom edge (*Fig. 5a*).

Next rout a 1/8" deep, 1/8"-wide decorative rabbet above the cove (*Fig. 5b*). Finally, soften the upper inside edge with a 1/4" roundover bit (*Fig. 5*).

Then, drill just four counterbored shank holes (*Fig. 5*).

FRAME "C"

Frame "C" gets attached to the top of the head case. It separates the head case from the crown molding.

The "bullnose" profile is routed in two stages. First form a profile around the upper outside edge of the frame using a 1/2" roundover bit (*Fig. 6a*).

Second, complete the bullnose using a 1/4" roundover bit on the lower edge (*Fig. 6b*). Rout both roundovers to the full depth of cut of each bit

FRAME "D"

The front and sides of frame "D" also have a bullnose profile (*Fig. 7a*).

One of these frames is attached to the top of the lower case, and the other is turned over and attached to the bottom of the head case (*Fig. 1*).

Finally, you'll need to drill the fourteen countersunk holes on one of the frame "D" assemblies (*Fig. 7a*). For the other frame "D" assembly, you'll drill eight holes, but they're only on the side pieces — omitting the six shank holes on the front piece (*Fig. 7*).

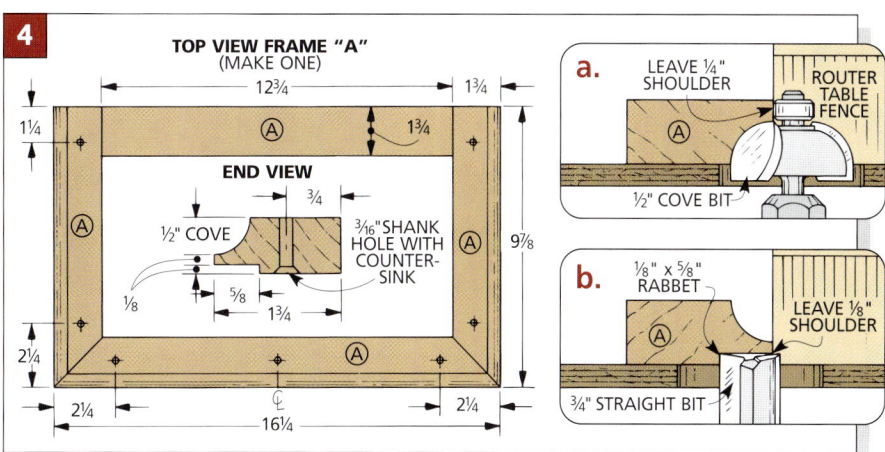

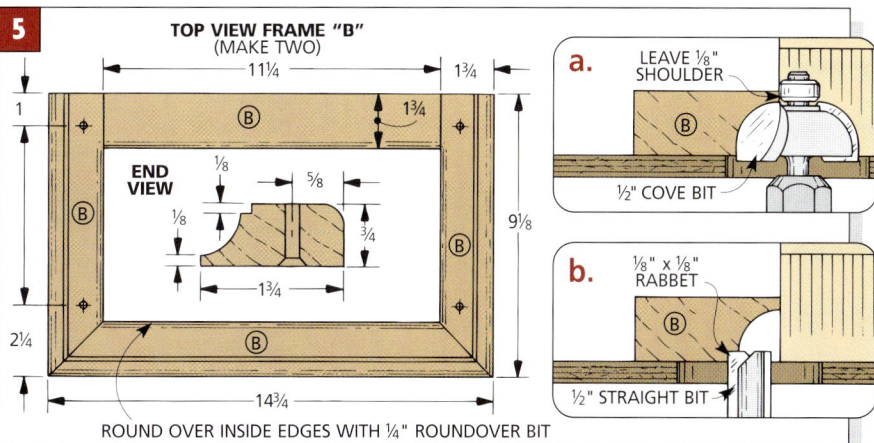

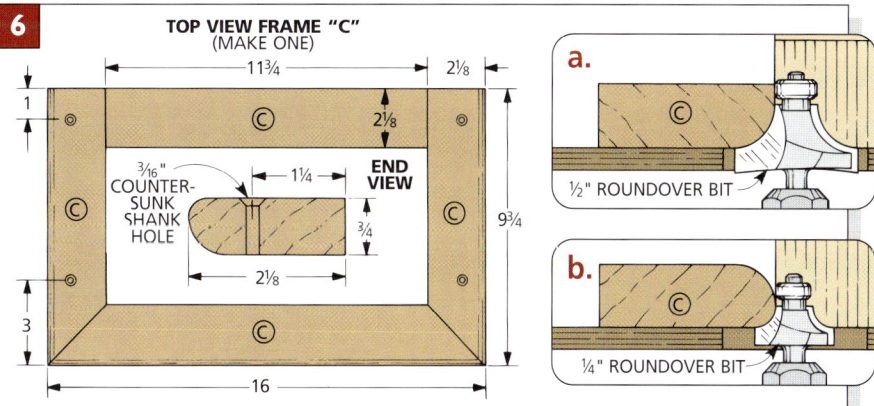

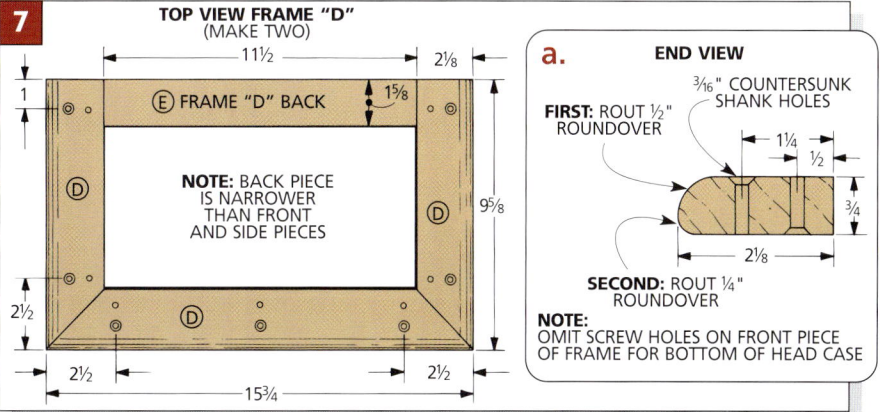

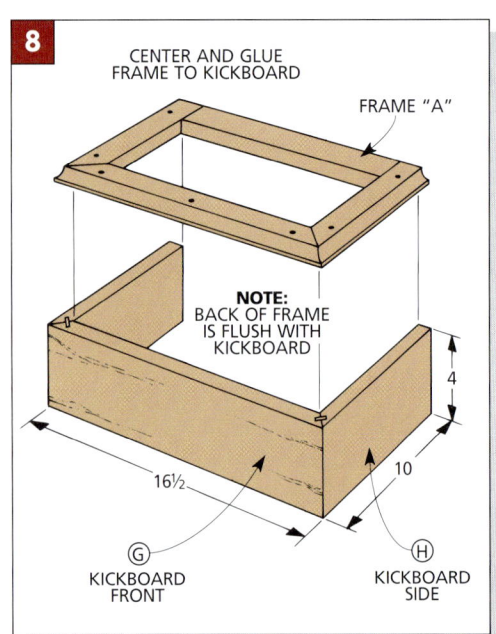

8 CENTER AND GLUE FRAME TO KICKBOARD

FRAME "A"

NOTE: BACK OF FRAME IS FLUSH WITH KICKBOARD

4

10

16½

G — KICKBOARD FRONT

H — KICKBOARD SIDE

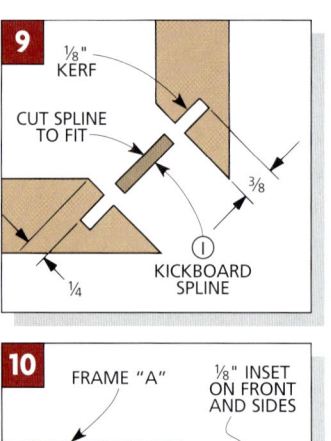

9 ⅛" KERF

CUT SPLINE TO FIT

⅜

¼

I — KICKBOARD SPLINE

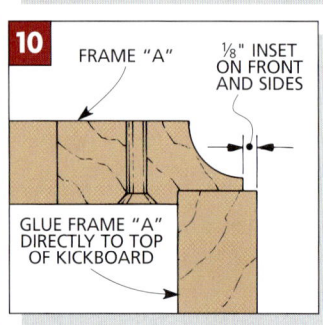

10 FRAME "A"

⅛" INSET ON FRONT AND SIDES

GLUE FRAME "A" DIRECTLY TO TOP OF KICKBOARD

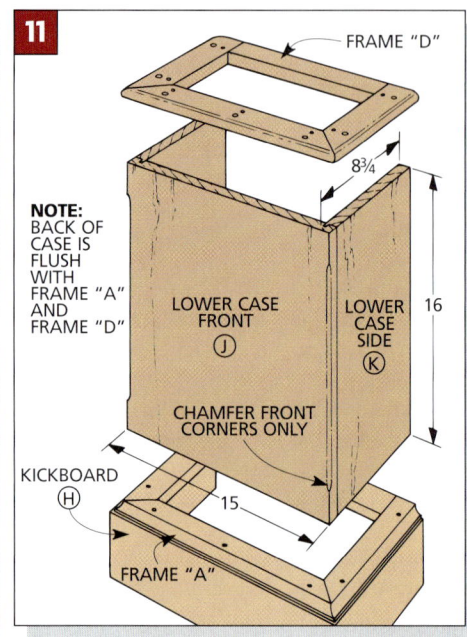

11 FRAME "D"

NOTE: BACK OF CASE IS FLUSH WITH FRAME "A" AND FRAME "D"

8¾

LOWER CASE FRONT J

LOWER CASE SIDE K

16

CHAMFER FRONT CORNERS ONLY

KICKBOARD H

15

FRAME "A"

KICKBOARD

After making all six frames I began work on the kickboard. The kickboard consists of a front and two sides. Start by ripping all three pieces to width (4") *(Fig. 8)*.

MITERS. Now, miter both ends of the kickboard front (G) so it's ¼" longer than the front of frame "A" (16½" from long point to long point). Then miter the fronts of both kickboard sides (H), and cut off the backs so they're ⅛" longer than the sides of frame "A" (10") *(Fig. 8)*.

KERF AND SPLINE. Next, cut a kerf along the mitered edges of each piece, and cut hardwood splines (I) to fit the kerfs *(Fig. 9)*. Then glue the splines in place and clamp the unit square.

ATTACH FRAME. To complete the kickboard, center frame "A" on top of the kickboard and glue it in place *(Fig. 10)*.

LOWER CASE

Now begin work on the lower case. Start by edge-gluing boards for the front (J) and side panels (K) *(Fig. 11)*. These panels will stand on the kickboard frame. Cut all the panels to final length (16"), then cut the front panel to final width (15") so it's inset ⅛" from the coved top edge on frame "A" *(Fig. 13)*.

The lower case side panels attach flush with the back of frame "A." And before cutting the pieces to width, you can cut a tongue and groove joint

along the front edges of each front and side panel *(Fig. 12)*.

ASSEMBLE CASE. Now the side panels can be cut to final width (8¾") *(Fig. 11)*. Then spread glue inside each groove, and slide the tongued side panels into the grooved front panel. Clamp the case with pipe clamps until the glue dries.

CHAMFER EDGES. After the glue dries, rout a decorative chamfer along the outside edges of the front piece *(Fig. 14)*. Stop the chamfers 2" from the top and bottom of the case.

ATTACH UPPER FRAME. Now screw a "D" frame (the one with 14 shank holes) onto the top of the case assembly,

centering the frame across the sides. This should result in a ⅜" overhang around the front and sides of the lower case *(Fig. 13)*. The frame should be flush at the back edge of the lower case.

INSTALL ONTO KICKBOARD. Finally, screw this entire sub-assembly to the top of frame "A" on the kickboard *(Fig. 13)*.

PENDULUM CASE

The pendulum case consists of two tall sides held in place between a pair of "B" frames *(Fig. 15)*. (A door is added later.)

SIDES. To make the case sides (L) start by edge-gluing blanks to width (8" rough)

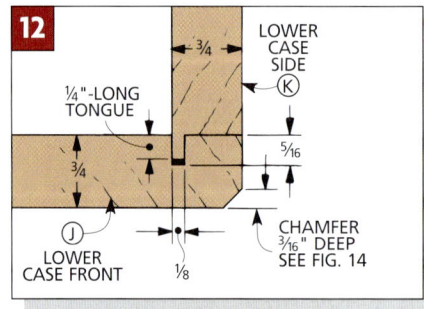

12 LOWER CASE SIDE K

¾

¼"-LONG TONGUE

¾

5/16

⅛

J — LOWER CASE FRONT

CHAMFER 3/16" DEEP SEE FIG. 14

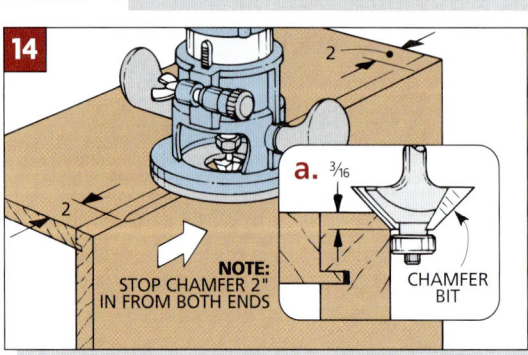

14 2

a. 3/16

2

NOTE: STOP CHAMFER 2" IN FROM BOTH ENDS

CHAMFER BIT

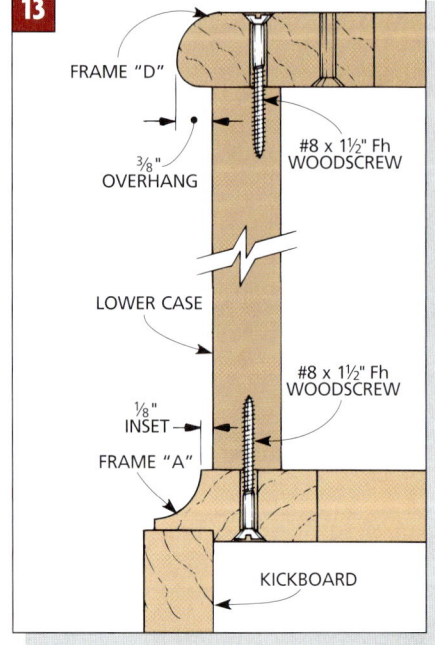

13 FRAME "D"

⅜" OVERHANG

#8 x 1½" Fh WOODSCREW

LOWER CASE

#8 x 1½" Fh WOODSCREW

⅛" INSET

FRAME "A"

KICKBOARD

and length (32" rough). Then trim each workpiece to a final width ½" less than the depth of frame "B" (7⅝" in my case), and 31½" long *(Fig. 15)*.

ALIGN FRAME TO CASE SIDE. With the side panels cut to size, the case can be assembled. To do this, stand one of the "B" frames on edge with its back (unshaped) section down. Then stand one of the case sides on edge. (This ensures that the back edges of both pieces are flush.) Now position the pendulum side piece ⅛" in from the shoulder of the frame *(Fig. 16)*.

PILOT HOLES. Using the pre-drilled shank holes in the frames as guides, drill pilot holes for both screws. Now screw the frame to the side piece with two No. 8 x 1½" Fh woodscrews. Then attach the other side piece.

The second "B" frame is attached to the other end of the case assembly in much the same manner. This frame mirrors the first frame — the cove-molded edges of both frames face toward each other and into the case opening *(Fig. 16)*.

INSTALL ONTO LOWER CASE. Now the pendulum case can be screwed in place onto the lower case with No. 8 x 1½" Fh woodscrews *(Fig. 17)*.

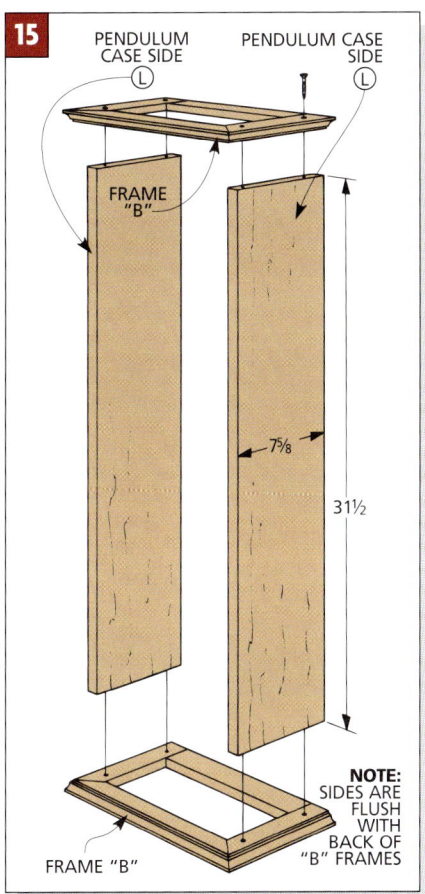

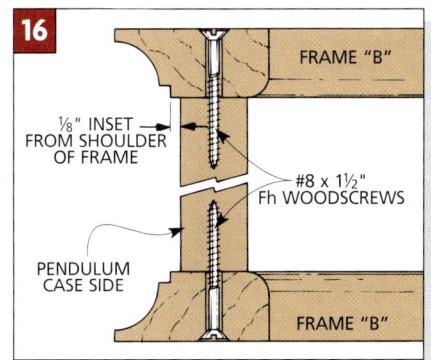

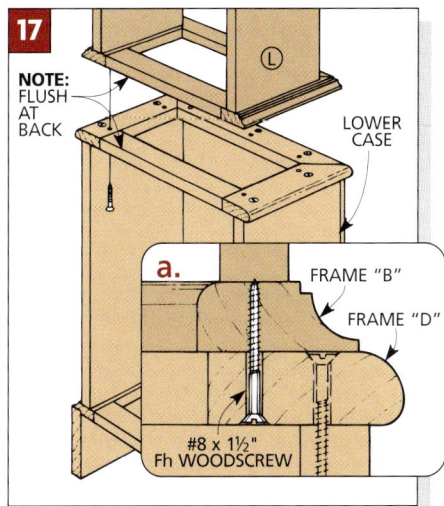

JOINERY .... *Splined Miter Joints*

For the kickboard of the Tall Case Clock I used a miter and spline joint. The miter joint hides the end grain. But I added a spline for a couple of reasons.

Note: A spline is just a thin piece of hardwood that runs across the joint.

ADVANTAGES. First, it provides more face grain glue surface. A miter joint is end grain to end grain, which is weak.

Second, miters tend to slide out of alignment as you clamp the joint together. A spline helps keep the pieces aligned.

KERFS. The hardwood spline fits into kerfs cut in both workpieces. After cutting the miters, lower the blade, but keep it tilted to 45°. Then move the rip fence to act as a stop *(Fig. 1)*.

The position of the rip fence will determine the location of the kerf *(Fig. 1a)*. I prefer to offset the kerf toward the heel rather than the point of the miter *(Fig. 2)*.

With the spline near the heel, the tip isn't as likely to crack off if the joint is stressed. By positioning it near the heel,

you can insert a longer spline to provide more glue surface.

SPLINE. Now cut the hardwood splines to fit the kerfs. These splines are exposed, so I cut them so the grain runs perpendicular to the joint line *(Fig. 3)*. (If the spline is not exposed you could use ⅛" hardboard instead.)

Also, to ensure that the spline won't prevent the miter from closing completely, I cut the spline a hair shorter than the total depth of both kerfs.

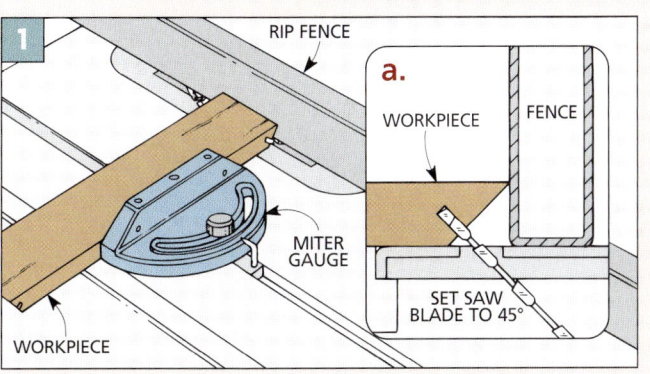

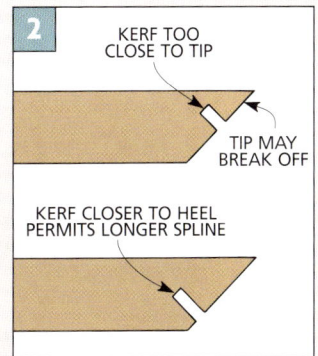

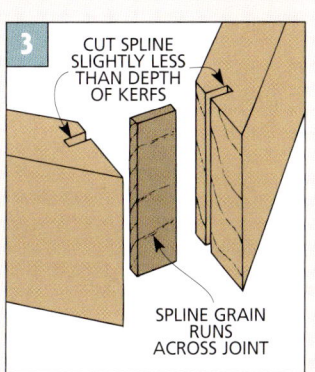

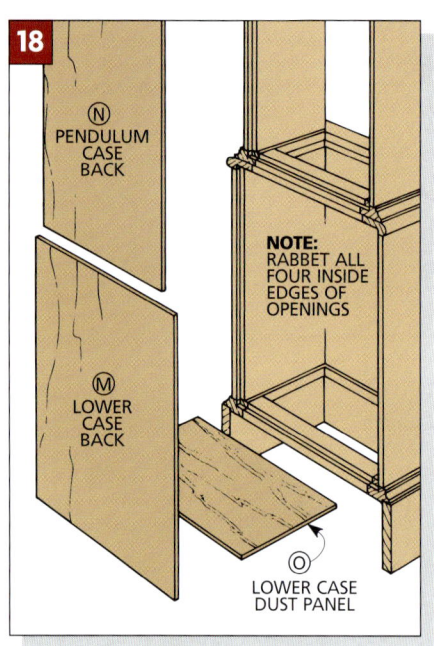

18

PENDULUM CASE BACK ⓝ

NOTE: RABBET ALL FOUR INSIDE EDGES OF OPENINGS

LOWER CASE BACK ⓜ

ⓞ LOWER CASE DUST PANEL

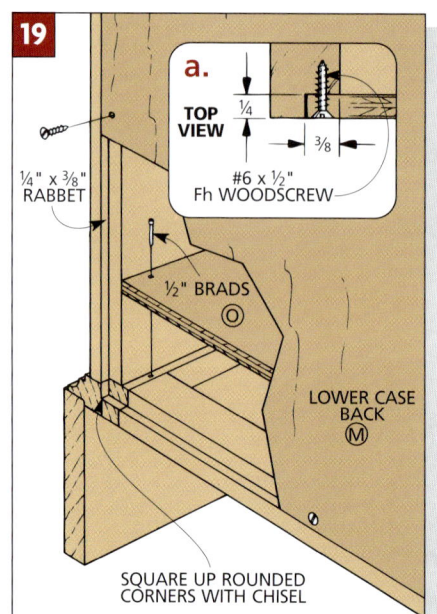

19

a.

TOP VIEW

¼

⅜

#6 x ½" Fh WOODSCREW

¼" x ⅜" RABBET

½" BRADS ⓞ

LOWER CASE BACK ⓜ

SQUARE UP ROUNDED CORNERS WITH CHISEL

The router will leave rounded corners in the rabbets, so I used a sharp chisel and a mallet to square up the corners of each of the rabbets.

CUT PANELS. Now measure the size of each of these openings and cut a lower case back (M) and a pendulum case back (N) to fit from ¼" plywood.

Also measure and cut a dust panel (O) to lay flat in the bottom of the kickboard assembly *(Fig. 18)*.

ATTACH PANELS. I installed the dust panel with ½"-long brads. But so the case backs can be removed later if needed, I attached them with No. 6 x ½" Fh wood-screws only *(Fig. 19)*.

HEAD CASE

The head case is made up of two solid wood sides that are separated by a dial frame, and attached to two bullnose frames *(Fig. 20)*. Frame "C" goes at the top of this case and frame "D" is located at the bottom.

HEAD SIDES. Begin by edge-gluing two head case sides (P), and trimming each to 1⅛" less than the depth of frame "D." (Mine was 8½" x 15⅛".)

DIAL FRAME. The frame that holds the clock face is built a little differently than

PLYWOOD BACKS

Once the pendulum case has been attached to the lower case, ¼" plywood backs can be screwed into rabbets routed around the back edges of both cases.

ROUT RABBETS. First, lay the entire assembly face down across a pair of saw-horses. Then rout a ¼" deep, ⅜"-wide

rabbet around all four inside edges of both case openings *(Fig. 19)*. I did this with a ⅜" rabbetting bit in my hand-held router. These are narrow boards, though, and holding the router steady can be a difficult task. So to make it easier, there are a couple of ways to handle this operation. (For more on routing on the edge of a workpiece, see the Shop Tip box below.)

SHOP TIP .... *Routing on an Edge*

When I began routing the rabbets for the back panels on the clock, I had trouble keeping the router level on the narrow edge of the case.

If you try to balance the router on the narrow edge, it will probably tip one way or the other and dig into the wood *(Fig. 1)*.

There are a couple of ways to solve this problem. If the box or case is constructed in such a way that

clamps will reach around it, clamp on a 2x4 block flush with the edge to be routed *(Fig. 2)*. This provides an extra 1½" of solid support for the router base.

The second method is to add an auxiliary base to the router *(Fig. 3)*. The base serves as a bridge across the case to the opposite side. I make this auxiliary base from a short piece of ¼" hardboard.

After drilling a hole in the hard-board platform for the bit to come through, I use double-sided carpet tape to stick the auxiliary platform to the plastic base on my router. (Or, you can remove your existing base and screw the new platform directly to your router.)

Then, you can rout as usual with the new base straddling over both edges of the case.

1

WORKPIECE EDGE TOO NARROW TO KEEP ROUTER STEADY

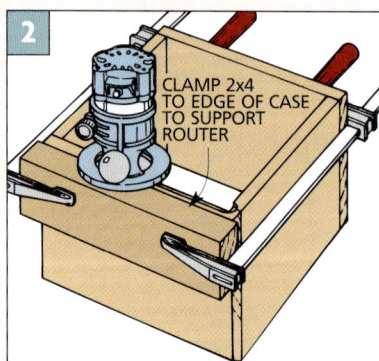

2

CLAMP 2x4 TO EDGE OF CASE TO SUPPORT ROUTER

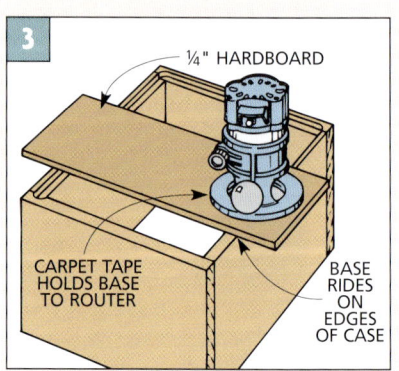

3

¼" HARDBOARD

CARPET TAPE HOLDS BASE TO ROUTER

BASE RIDES ON EDGES OF CASE

the case frames. It has mortise and loose tenon joints at each corner.

To make this frame, first cut two dial frame stiles (Q) to size ($1\frac{5}{8}$" wide x $15\frac{1}{8}$" long). Then cut two dial frame rails (R) to size ($2\frac{7}{16}$" x $10\frac{1}{4}$"). I used the same mortise and loose tenon joints on the dial frame as I did for the molding frames. The molding splines (F) I used here are $1\frac{15}{16}$" wide and $1\frac{3}{8}$" long .

Note: If you're using regular blind mortise and tenon joints instead, be sure to cut the dial frame rails $1\frac{1}{2}$" longer to allow for the tenons.

Then you can assemble the head case frame *(Figs. 20 and 20a)*.

Once it's assembled, you can soften the front inside edge of the dial frame with a $\frac{1}{8}$" roundover bit *(Fig. 20)*.

ASSEMBLY. Next, assemble the head case by gluing the dial frame between the two head case sides (P). Clamp the U-shaped sub-assembly so the pieces are flush at the top, bottom, and front.

After the glue dries, screw the bullnose frames ("C" and "D") to the top and bottom of this sub-assembly *(Fig. 20b)*.

Note: The bullnose profiles should be facing the same direction.

Position the frames so they're flush with the head case sides at the back, and centered from side to side. Then drill pilot holes through the shank holes in the frames, and screw the frames to the side pieces *(Fig. 20b)*.

SCREW TO PENDULUM CASE. Finally, screw the head case to the top of the pendulum case using four No. 8 x $1\frac{1}{2}$" Fh woodscrews *(Fig. 21)*.

CROWN MOLDING

The crown molding assembly is made up of three U-shaped frames. A field frame (S), made up of a front and two side pieces, stands on edge. The cove frame (T) front and sides are routed and lie flat, and the rim frame (U) front and sides lie flat on top of this *(Fig. 22)*.

Make these frames by first cutting three strips of $\frac{3}{4}$" stock to a rough length and finished width for each frame. Then, on the cove frame (T) pieces, rout the same cove and rabbet profile as on frame "B" (refer to *Fig. 5* on page 17).

Now miter the front section of each frame to final length *(Fig. 22)*. Then miter each side piece, and cut it to length. Finally, assemble the frames by gluing the front and then the sides onto the frame directly below it *(Fig. 22a)*.

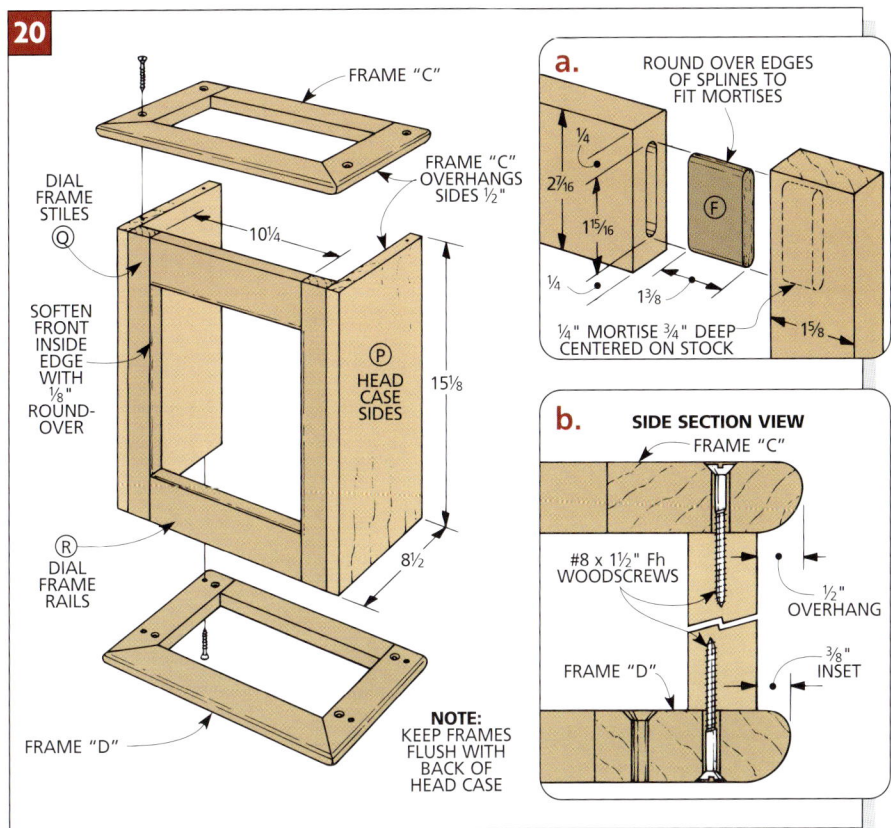

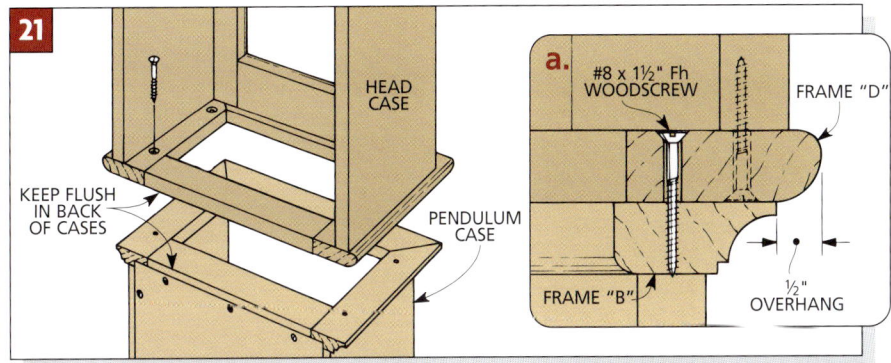

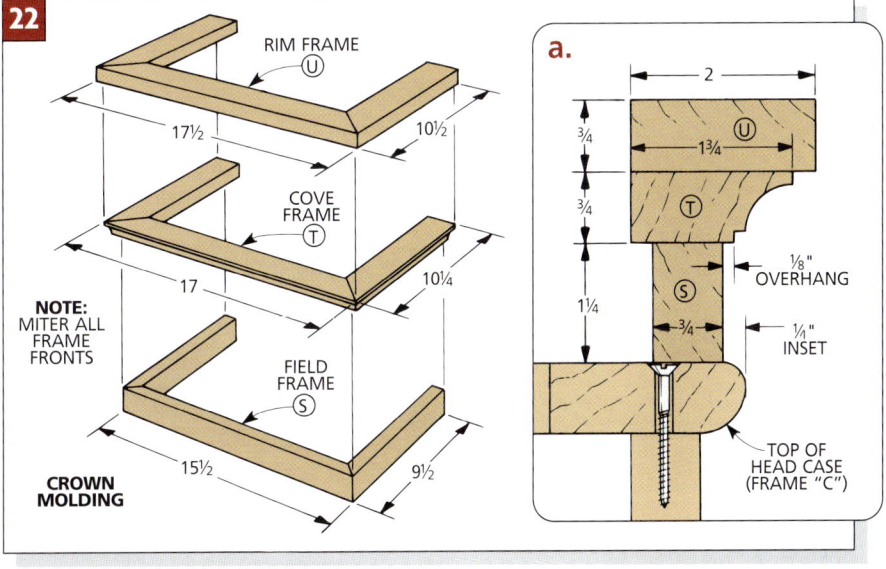

DOORS

The clock has two doors — a head case door that allows access to the clock hands, and a pendulum case door that allows access to the weights. The frames for both doors are made using mortise and loose tenon joinery, and both frames have a rabbet along the inside edge to accept a glass (or wood) panel.

STILES AND RAILS. The stiles and rails for both doors are all $2\frac{1}{4}$" wide. To determine the length of the door frame stiles, I subtracted $\frac{1}{8}$" from the height of the door openings. (This allows for $\frac{1}{16}$" clearance above and below the finished doors.) Now cut the two pendulum case door stiles (V) and two head case door stiles (W) to length *(Fig. 23)*.

To determine the length of the pendulum case door rails (X) and head case door rails (Y), subtract $4\frac{1}{2}$" from the width of the pendulum case and the head case *(Fig. 23)*. (Mine are $8\frac{3}{4}$" and $10\frac{1}{2}$".)

SPLINES AND MORTISES. To assemble the frames, first rout mortises on all the mating pieces on a router table. Then make the $1\frac{3}{4}$"-wide loose tenons for the two frames *(Fig. 23a)*. Now glue up the frames, clamping them flat and square with the loose tenons in place.

INSIDE RABBETS. With the door frames assembled, cut the rabbets that receive the glass panels. I used a $\frac{3}{8}$" rabbeting bit in the router *(Fig. 23b)*. Cut these rabbets $\frac{3}{8}$" deep. Then use a chisel and mallet to square up the round corners.

STOPPED CHAMFERS. The faces of both frames have a stopped chamfer routed along the inside edges of the rails, and both edges of the stiles *(Fig. 23)*. With a pencil, mark the stopping points for the inside chamfers $\frac{5}{8}$" from the corners *(Fig. 23c)*. Mark the stopping points for the outside chamfers 2" from the ends of the stiles *(Fig. 23c)*.

PANEL STOPS. After routing the stopped chamfers, cut the window stops (Z) from $\frac{1}{4}$"-thick stock *(Fig. 24)*.

Note: If you're building the pendulum case with a wood panel, the stop needs to be larger so it overlaps the rabbet and holds a screw *(Fig. 24)*.

The glass panels should be cut to fit the dimensions of the door openings (less $\frac{1}{8}$"). Then install the glass with the panel stops. They're mitered at their ends and then nailed in place.

HANGING THE DOORS. Each door has two 2" brass butt hinges. The hinges are positioned 2" from the top and bottom

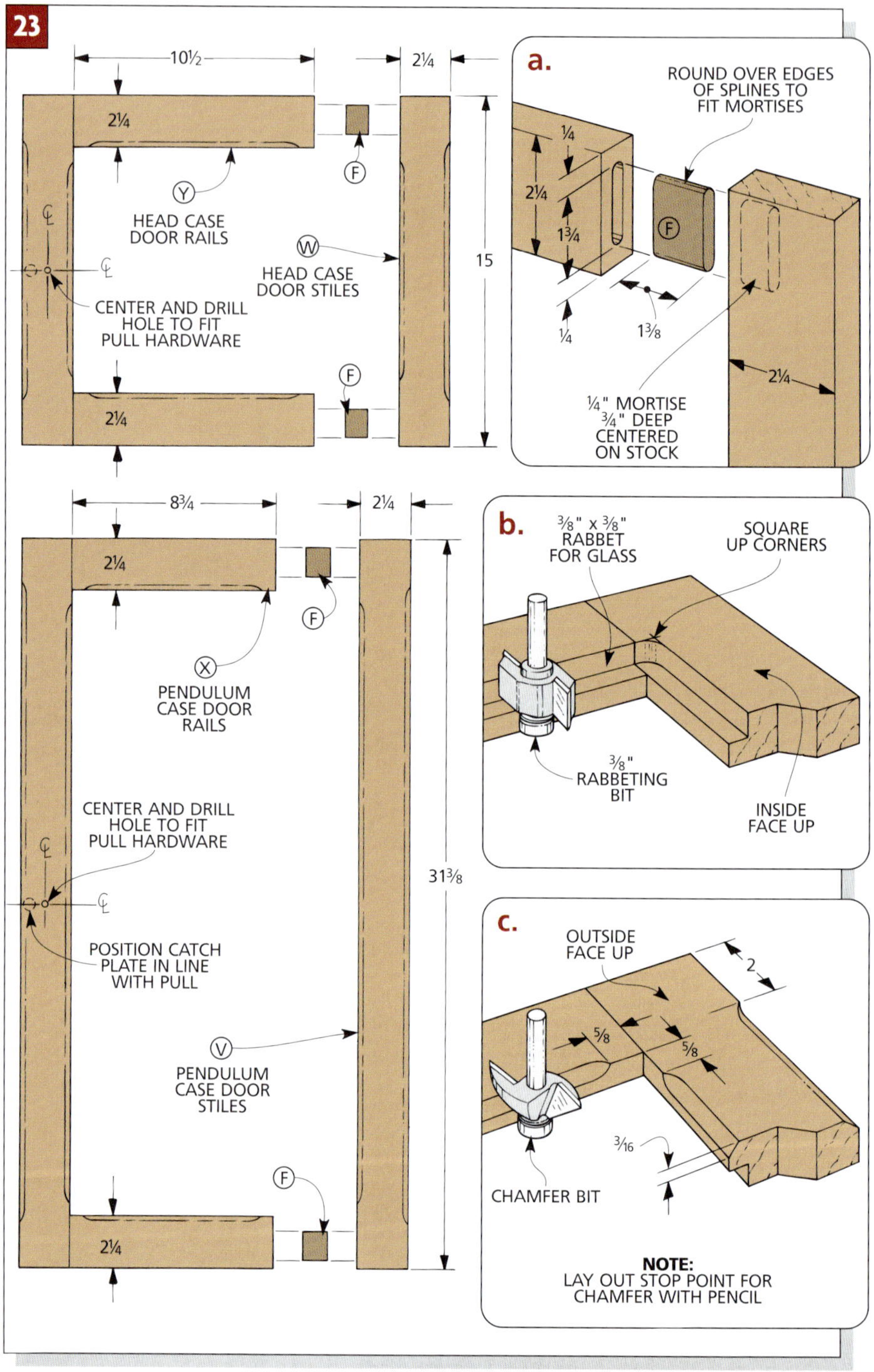

23
10½
2¼
2¼
2¼
F
Y
HEAD CASE DOOR RAILS
W
HEAD CASE DOOR STILES
15
CENTER AND DRILL HOLE TO FIT PULL HARDWARE
8¾
X
PENDULUM CASE DOOR RAILS
CENTER AND DRILL HOLE TO FIT PULL HARDWARE
31⅜
POSITION CATCH PLATE IN LINE WITH PULL
V
PENDULUM CASE DOOR STILES
a.
ROUND OVER EDGES OF SPLINES TO FIT MORTISES
¼
1¾
1⅜
¼" MORTISE ¾" DEEP CENTERED ON STOCK
b.
⅜" x ⅜" RABBET FOR GLASS
SQUARE UP CORNERS
⅜" RABBETING BIT
INSIDE FACE UP
c.
OUTSIDE FACE UP
2
⅝
⅝
³⁄₁₆
CHAMFER BIT
NOTE: LAY OUT STOP POINT FOR CHAMFER WITH PENCIL

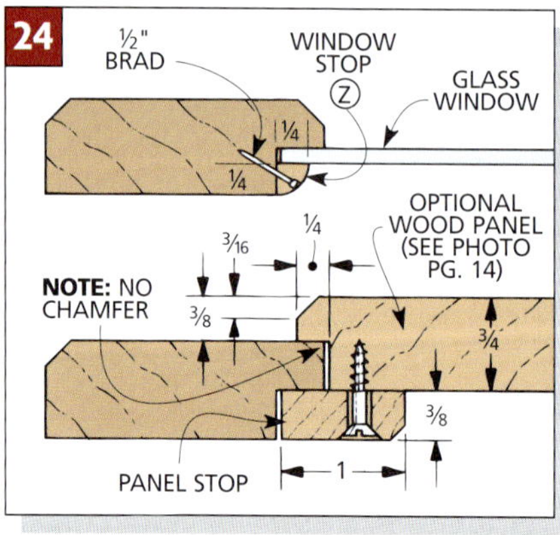

24
½" BRAD
WINDOW STOP
Z
GLASS WINDOW
¼
¼
OPTIONAL WOOD PANEL (SEE PHOTO PG. 14)
NOTE: NO CHAMFER
³⁄₁₆
⅜
¾
⅜
1
PANEL STOP

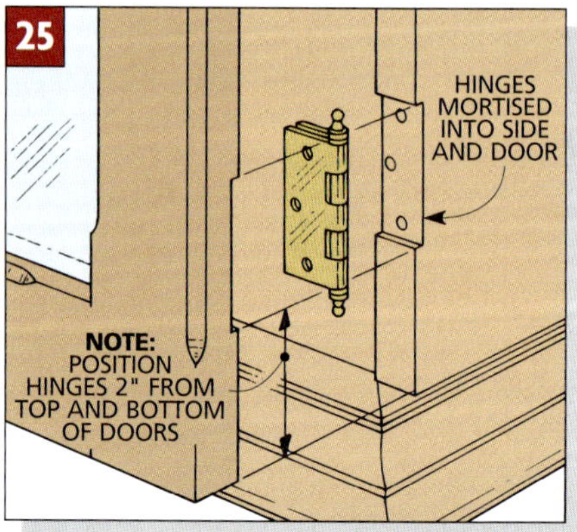

25
HINGES MORTISED INTO SIDE AND DOOR
NOTE: POSITION HINGES 2" FROM TOP AND BOTTOM OF DOORS

of each door, and are mortised into both the case and the frame stiles *(Fig. 25)*.

PULLS AND CATCHES. Now drill holes in both frames for door pulls *(Fig. 23)*. Then drill holes for two door catches. Finally, install the pulls and catches. (For more on catches, see the Shop Tip on page 13.)

CLOCKWORKS

The clockworks consists of two major components — the movement and the chime rods. Before you can install these, center and screw the clock dial face on the back of the dial frame *(Fig. 26)*.

Note: These steps are for the clockworks I used. You may need to alter these steps to fit your works.

SEATBOARD. The clock movement sits on a grooved seatboard (AA) that straddles two supports *(Fig. 26)*. To make the seatboard, rip a piece of $^3/_4$" stock to width *(Fig. 26a)*. Then cut it to length to fit between the head case sides.

Now cut two $^1/_8$" wide, $^1/_8$"-deep grooves along the length of the seatboard to accept the movement *(Fig. 26a)*.

The brass chains that support the weights hang through a slot centered between these grooves. To form the slot, first drill a pair of $^3/_4$" holes *(Fig. 26a)*. Then complete the slot by connecting these end holes with two jig saw cuts.

To mount the seatboard into the case, drill countersunk holes on each end of the seatboard *(Fig. 26a)*. Also, drill a hole near one end to allow the chime silencing wire to pass through.

SEATBOARD SUPPORTS. Next, to support the seatboard, cut two seatboard supports (BB) ($5^1/_2$" x $4^3/_4$"). Then screw the supports to the inside of the head case sides, and the seatboard across the top of the supports *(Fig. 26)*.

SCREW BLOCKS. In order to mount the chime board to the back of the head case, I attached screw blocks (CC) to the top inside of the case *(Fig. 26)*. These blocks are cut to match the length of the seatboard supports. Once they are cut to size, screw them in place *(Fig. 27)*.

CHIME BOARD. The chimes are screwed to a chime board (DD), which acts as a back panel for the head case. To make it, edge-glue a $^3/_4$"-thick panel and cut it to fit in the back of the case *(Fig. 26)*.

Next, bore $^3/_{16}$" holes through the back of the chime board to mount the chime

block *(Fig. 26)*. Counterbore the holes on the back side to accept the large washers that come with the chimes *(Fig. 27)*. Also drill countersunk shank holes to mount the chime board to the screw blocks and seatboard supports.

INSTALL WORKS. Now, set the clockworks on the seatboard. The handshaft should be centered in the dial hole. If it isn't, remove and re-cut the supports.

To add the chimes, first screw the chime block onto the chime board. Then screw the chime board to the screw blocks and seatboard supports *(Fig. 26)*.

DUST PANEL. After you've fine-tuned the movement and chimes, top the case with a dust panel (EE) *(Fig. 26)*. Cut it to fit and screw it in place (no glue). You may need to remove it to adjust the works. Finally, add a wall bracket *(Fig. 28)*. ∎

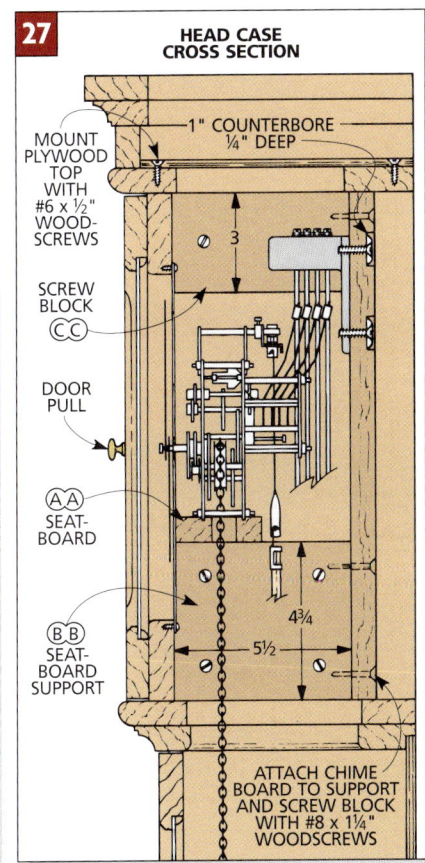

27 HEAD CASE CROSS SECTION

MOUNT PLYWOOD TOP WITH #6 x ½" WOOD-SCREWS

1" COUNTERBORE ¼" DEEP

SCREW BLOCK (CC)

DOOR PULL

(AA) SEAT-BOARD

(BB) SEAT-BOARD SUPPORT

4¾"

5½"

ATTACH CHIME BOARD TO SUPPORT AND SCREW BLOCK WITH #8 x 1¼" WOODSCREWS

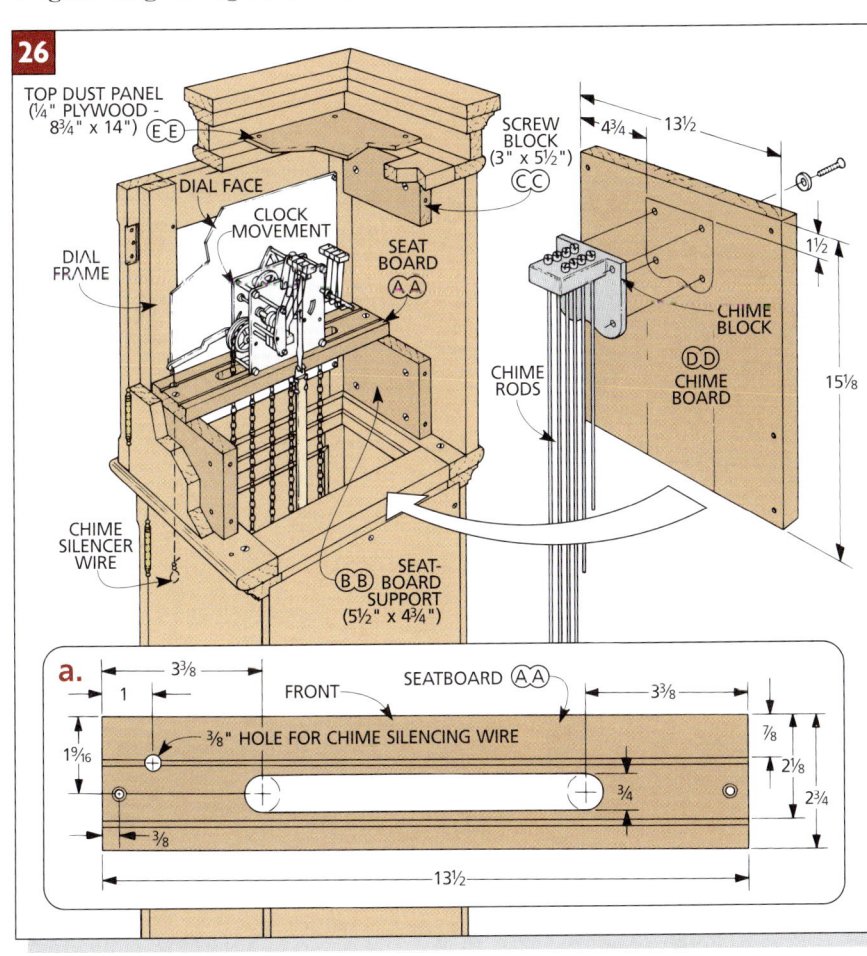

26

TOP DUST PANEL (¼" PLYWOOD - 8¾" x 14") (EE)

DIAL FACE

CLOCK MOVEMENT

DIAL FRAME

SEAT BOARD (AA)

SCREW BLOCK (3" x 5½") (CC)

4¾   13½

1½

CHIME BLOCK

CHIME RODS

(DD) CHIME BOARD

15⅛

CHIME SILENCER WIRE

(BB) SEAT-BOARD SUPPORT (5½" x 4¾")

a.

3⅜

1

FRONT   SEATBOARD (AA)

3⅜

⅜" HOLE FOR CHIME SILENCING WIRE

1⁹⁄₁₆

¾

⅜

13½

⅞

2⅛

2¾

28

SCREW L-BRACKET INTO HEAD FRAME

THEN SCREW BRACKET TO WALL STUD

DESIGNER'S NOTEBOOK

It's hard to improve on a classic design, but adding a gabled pediment will beautifully enhance this clock's looks — and it's easy to build. A scroll-sawn cutout in the kickboard front completes the change.

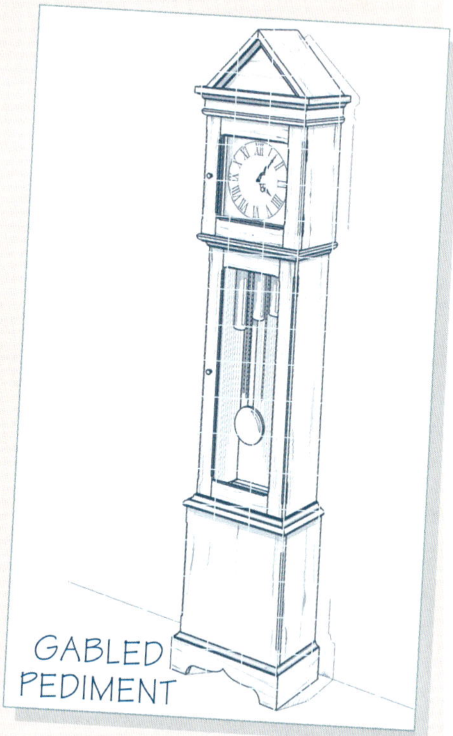

GABLED PEDIMENT

CONSTRUCTION NOTES:

■ Start construction by using the pattern below to lay out and cut the scroll-sawn design on the kickboard front (G) *(Fig. 1)*.

■ Then cut the pieces for the rest of the clock and assemble it as before.

■ Now you can begin building the gabled pediment. Start by cutting the front and rear gables (FF) to size *(Figs. 2 and 3)*.

■ The roof panels are attached to the gables with tongue and groove joints. Add the tongues by cutting a rabbet along the top edges of the panels *(Figs. 2 and 3)*.

■ Next drill a pair of countersunk shank holes on the outside face of the rear gable *(Fig. 3)*. Then later, you'll be able to attach the pediment assembly to a cleat.

■ Once the gables are complete, set them aside and work on the roof panels (GG). Start by cutting them to size *(Fig. 4)*. Then cut the grooves for the tongues on the gables. Be sure to note the different locations of these grooves.

■ To complete the roof panels, miter one end of each at 45° and then miter the opposite end to its final length *(Fig. 4)*.

■ Now you're ready to assemble the pediment. To do this, glue the roof panels to the gables, making sure to have the countersunk holes in the back gable facing toward the back of the clock.

■ Then you can screw the assembly to the clock. But first, you'll need to make a cleat (HH) *(Fig. 5)*. It's just a ¾"-square piece of hardwood with two sets of holes. One set lets you attach it to the case, the other is for screwing the rear gable in place.

■ Finally, add the pediment molding. To make it, first rout a profile identical to frame "B" on a 1¾"-wide blank (refer to *Fig. 5* on page 17). After you rout the molding to final shape, rip the blank to ¾" wide and cut it to rough length.

■ Now miter one end of the blank. To determine its final length, hold the piece beneath the roof panel and mark its length. It should be cut flush with the miter at the bottom of the panel *(Fig. 5)*.

■ Repeat this procedure, mitering the second piece of molding to length, before gluing them both in place.

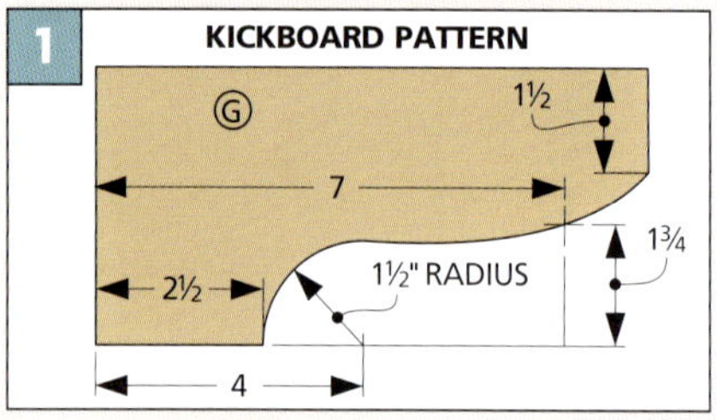

1
KICKBOARD PATTERN
G
1½
7
1¾
2½
1½" RADIUS
4

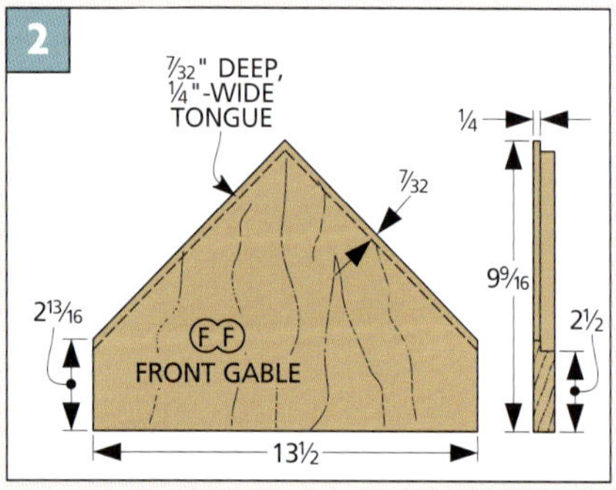

2
7⁄32" DEEP, ¼"-WIDE TONGUE
7⁄32
¼
9 9⁄16
2 13⁄16
FF
FRONT GABLE
2½
13½

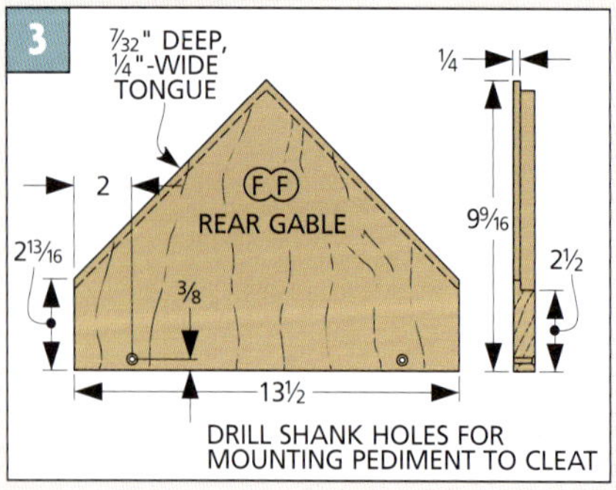

3
7⁄32" DEEP, ¼"-WIDE TONGUE
¼
2
FF
REAR GABLE
9 9⁄16
2 13⁄16
⅜
2½
13½
DRILL SHANK HOLES FOR MOUNTING PEDIMENT TO CLEAT

MATERIALS LIST

NEW PARTS

FF	Front/Rear Gables (2)	¾ x 9 9⁄16 - 13½
GG	Roof Panels (2)	¾ x 9½ - 11 1⁄16
HH	Cleat (1)	¾ x ¾ - 13½
II	Pediment Molding	¾ x ¾ - 20 rgh.

HARDWARE SUPPLIES

(4) No. 8 x 1¼" Fh woodscrews

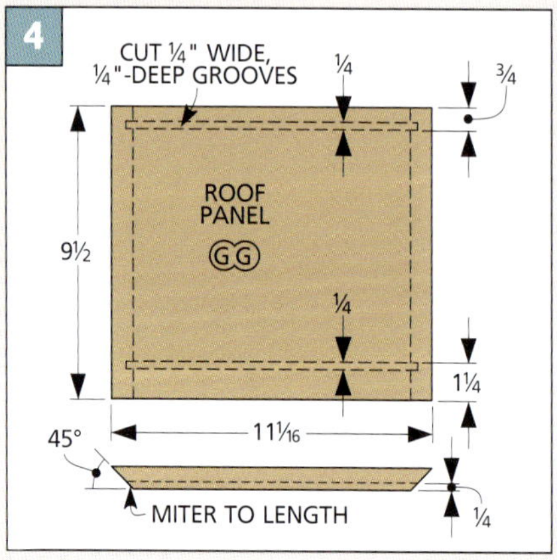

4
CUT ¼" WIDE, ¼"-DEEP GROOVES
¼
¾
ROOF PANEL
GG
9½
¼
1¼
45°
11 1⁄16
MITER TO LENGTH
¼

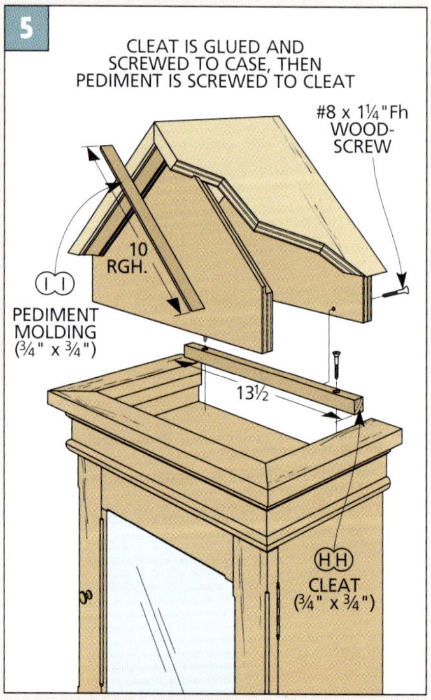

5
CLEAT IS GLUED AND SCREWED TO CASE, THEN PEDIMENT IS SCREWED TO CLEAT
#8 x 1¼"Fh WOOD-SCREW
10 RGH.
II
PEDIMENT MOLDING (¾" x ¾")
13½
HH
CLEAT (¾" x ¾")

Tambour Clock

The double curve gives this clock a classic, graceful shape. Building it involves a couple of techniques you may not have used before — kerf bending and routing shapes with a template.

The first time you see this Tambour Clock, you might think it's made from a solid block of walnut. But a closer look will show that only the base is solid wood. When you open the door on the back and look inside, you see that the front and back of the case are plywood. And the top is hardboard covered with walnut veneer on the outside.

So that begs the question, why all the different materials to make a clock that looks like a solid piece of wood? Couldn't you just make it out of a solid block?

Of course, you could. And it would probably be a more straightforward way to make the clock. But an arched-topped case made from solid wood just wouldn't be as attractive, mostly because there would be a lot of end grain exposed. End grain mixed with face grain looks like Morse code — a series of lines interrupted by a bunch of dots.

TECHNIQUES. Making the clock mostly from plywood presents several challenges. First of all, ³/₄" plywood doesn't bend around curves as tight as those on this clock. Instead, I kerf-bent a piece of ¹/₄" hardboard and covered it with veneer resawn from plywood. (This is covered in a Technique article on page 28.)

Another challenge with this clock was cutting the curved front and back pieces to identical shape. There's a simple trick for doing this, though. It involves cutting the parts to rough shape first, then routing them to final shape using a template and a flush trim bit.

CLOCKWORKS AND KIT. Before you build the clock, it's best to have all the clock parts in hand. I used an inexpensive quartz movement that's battery powered (see Sources on page 126).

DESIGN OPTIONS. I thought it would be nice to add decorative trim pieces to the front of the clock. All that's needed is to cut two pieces of veneer and apply them to the clock face. For more on this, see the Designer's Notebook on page 31.

EXPLODED VIEW

OVERALL DIMENSIONS:
18¼W x 4½D x 8⁵⁄₁₆H

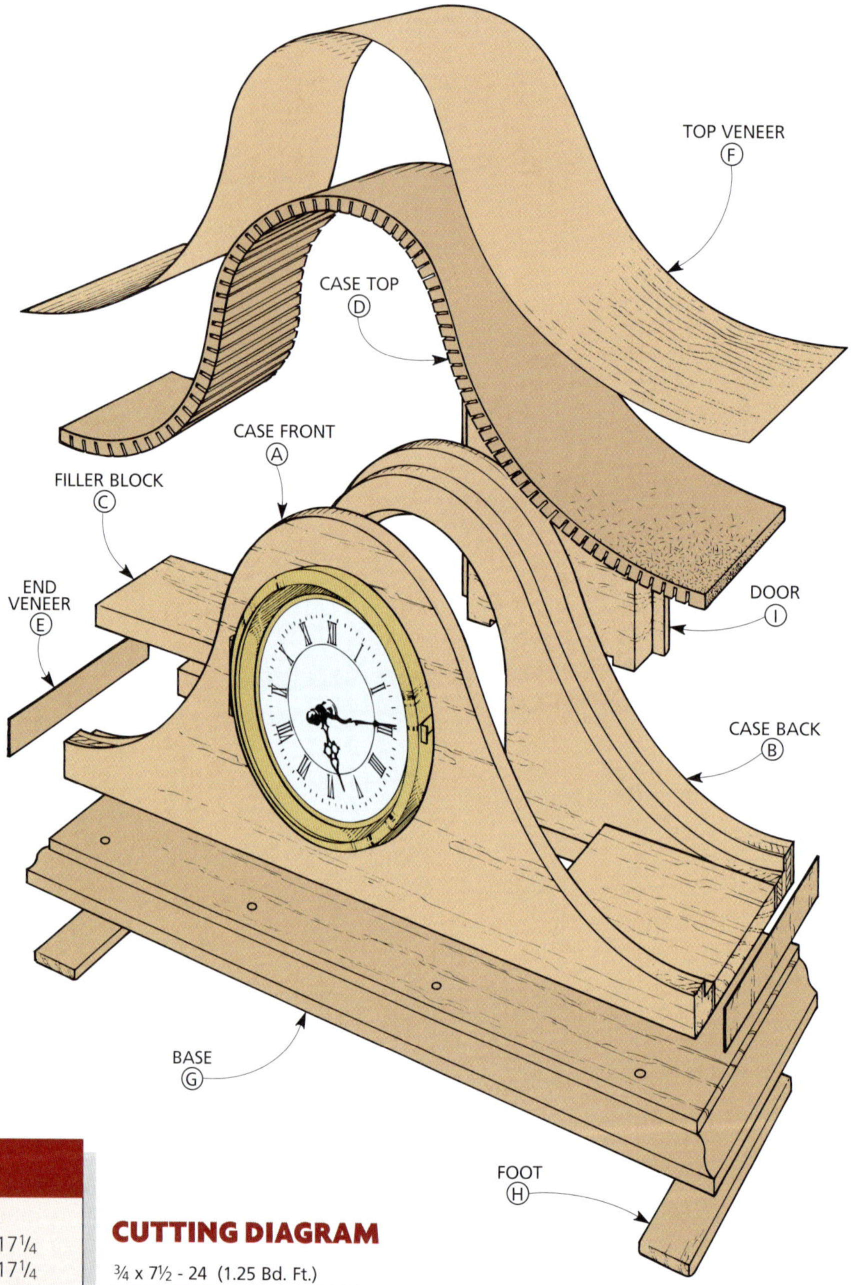

TOP VENEER
F

CASE TOP
D

CASE FRONT
A

FILLER BLOCK
C

END VENEER
E

DOOR
I

CASE BACK
B

BASE
G

FOOT
H

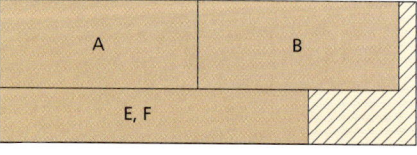

MATERIALS LIST

WOOD

A	Case Front (1)	¾ ply - 7⁵⁄₁₆ x 17¼
B	Case Back (1)	¾ ply - 7⁵⁄₁₆ x 17¼
C	Filler Blocks (2)	½ x 2½ - 4
D	Case Top (1)	¼ hdbd. - 3¼ x 24 rgh.
E	End Veneer (2)	¹⁄₁₆ x 4¼ - ¾ rough
F	Top Veneer (1)	¹⁄₁₆ x 4¼ - 24 rough
G	Base (1)	¾ x 4½ - 18¼
H	Feet (2)	¼ x 1¼ - 4¼
I	Door (1)	¾ x 6½ - 5

HARDWARE SUPPLIES

(8) No. 6 x 1½" Fh woodscrews
(1) No. 4 x ½" Rh woodscrew
(50) ¾" wire brads
(2) 1" x 1" brass hinges
(1) Brass door pull
(1) Bullet catch
(1) Clock movement (⁵⁄₁₆" x ¹⁵⁄₁₆" shaft)
(1) 5⅛" punched clock face w/ bezel
(1 pr.) 2¹⁄₁₆"-long serpentine hands

CUTTING DIAGRAM

¾ x 7½ - 24 (1.25 Bd. Ft.)

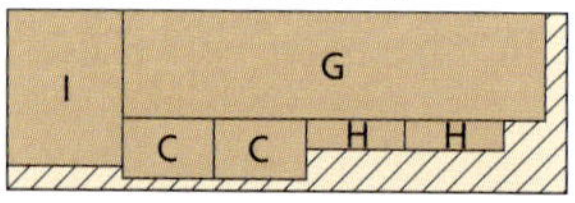

I		G	
	C C	H H	

¼" HARDBOARD - 12 x 24

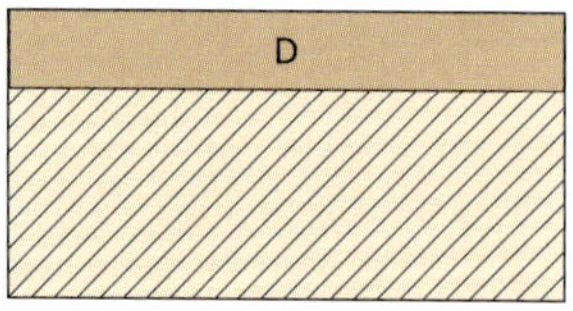

D

¾" PLYWOOD - 12 x 36

A	B
E, F	

NOTE: VENEER (E, F) IS RESAWN FROM ¾" PLYWOOD

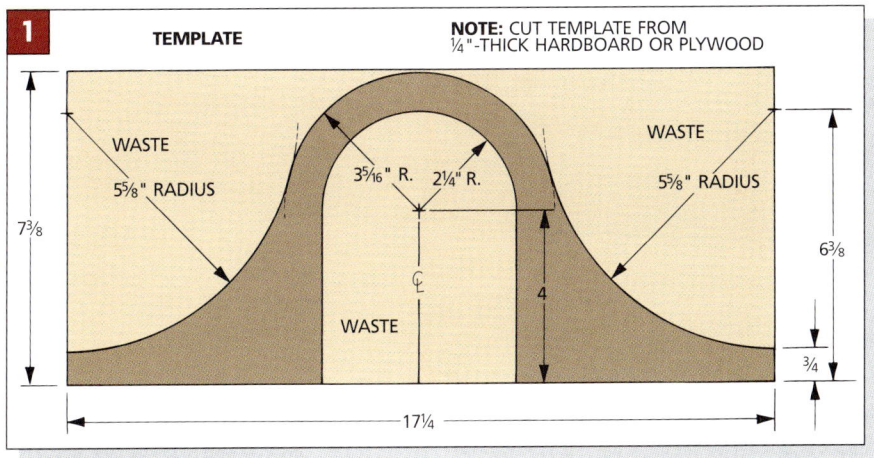

1 TEMPLATE

NOTE: CUT TEMPLATE FROM ¼"-THICK HARDBOARD OR PLYWOOD

WASTE
5⅝" RADIUS

3⁵⁄₁₆" R. 2¼" R.

WASTE
5⅝" RADIUS

7⅜

4

6⅜

¾

WASTE

17¼

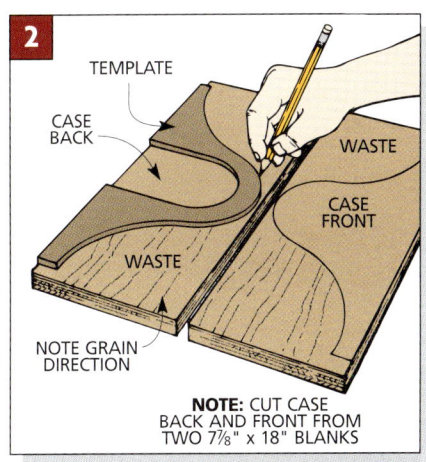

2 TEMPLATE

CASE BACK

WASTE

CASE FRONT

WASTE

NOTE GRAIN DIRECTION

NOTE: CUT CASE BACK AND FRONT FROM TWO 7⅞" x 18" BLANKS

CASE FRONT & BACK

This whole clock is based on a double-curved shape for the front, back, and top. And that shape begins with a template.

TEMPLATE. I made a template as a guide so I could use a router to cut the front and back pieces to exact shape.

I began making the template by drawing the curved shape on a piece of ¼"-thick hardboard *(Fig. 1)*. The dia-gram includes the outline of the double-curved shape, and also the shape of the opening for the back door.

After the outline is drawn, cut the template to shape slightly oversize with a jig saw or band saw. Also rough-cut the opening for the back door.

Now, very carefully file or sand up to the pencil lines to produce a smooth, curved shape. (You want the template as perfect as you can get it.)

CUT PLYWOOD. Next you'll want to cut two blanks of plywood for the case front and case back. Also cut one piece of ply-wood (4½" x 25½") for the top and end veneer strips. (More on these strips later.)

Note: To get the best color and grain match, I cut all three blanks from the same piece of plywood.

To make the case front (A) and case back (B), first draw the shape of the tem-plate onto both plywood blanks *(Fig. 2)*. Then cut the pieces to rough shape. But don't cut out the door opening yet.

FLUSH TRIM SMOOTH. Now use the template with a router table and a flush trim bit to get the exact shape.

To do this, attach the template to one of the blanks using carpet tape *(Fig. 3)*. Then using a flush trim bit, rout around the profile of the shape *(Fig. 3a)*. Do this on both the case front and case back.

DOOR OPENING. On the piece for the back, also rough-cut the opening for the door *(Fig. 3)*. Then use the template and a flush trim bit once again to smooth the opening to shape.

RABBET EDGES. The next step is to pro-vide a way to mount the case top. (The case top is a piece of ¼" hardboard that's covered with veneer.)

I used a rabbeting bit on the router table to cut a ¼"-deep rabbet along the curved edge of each piece *(Fig. 4)*.

CASE ENDS. After rabbeting the edges, the next step is to cut two filler blocks (C) to fit between the case front and back *(Fig. 5)*. These blocks hold the front and back together, and provide a surface to mount the veneer on the end of the case.

ASSEMBLE CASE. Now the case front and back can be assembled as a unit, with the filler blocks glued between them *(Fig. 5)*. (I also used a temporary spacer at the top of the clock case while clamping the front and back to the filler blocks.)

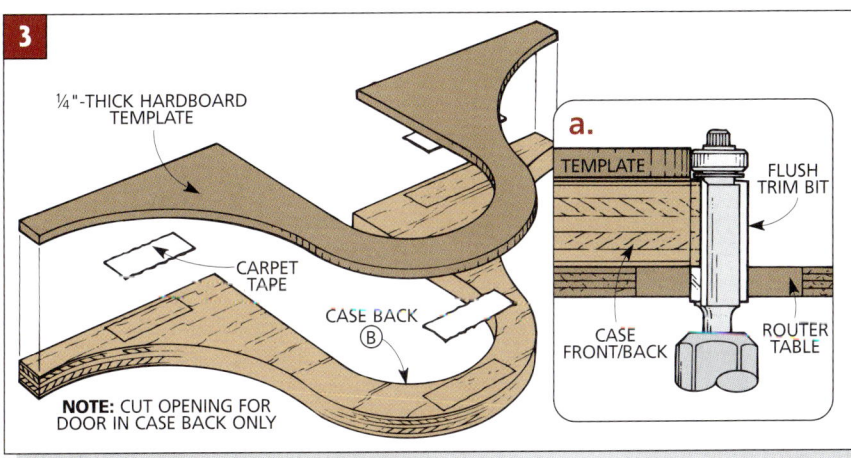

3 ¼"-THICK HARDBOARD TEMPLATE

CARPET TAPE

CASE BACK (B)

a. TEMPLATE

FLUSH TRIM BIT

CASE FRONT/BACK

ROUTER TABLE

NOTE: CUT OPENING FOR DOOR IN CASE BACK ONLY

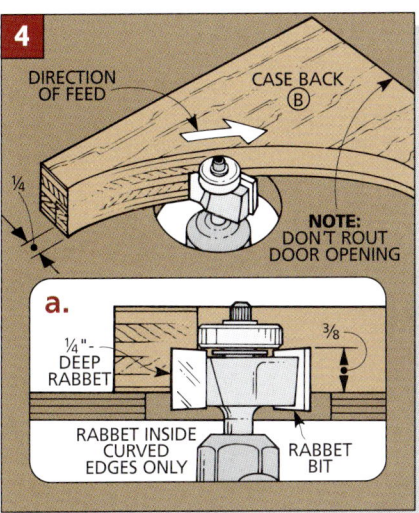

4 DIRECTION OF FEED

CASE BACK (B)

¼

NOTE: DON'T ROUT DOOR OPENING

a. ¼"-DEEP RABBET

⅜

RABBET INSIDE CURVED EDGES ONLY

RABBET BIT

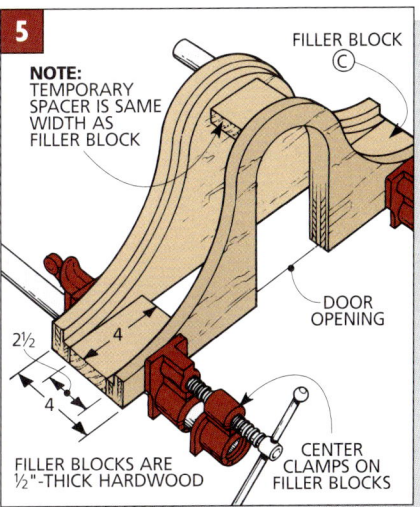

5 FILLER BLOCK (C)

NOTE: TEMPORARY SPACER IS SAME WIDTH AS FILLER BLOCK

2½

4

4

DOOR OPENING

CENTER CLAMPS ON FILLER BLOCKS

FILLER BLOCKS ARE ½"-THICK HARDWOOD

The curved top of the clock has two parts: a kerf-bent piece of hardboard and a strip of veneer that covers it.

TOP. To make the bent case top (D), begin by ripping a strip of ¼" hardboard to width to fit between the rabbets on the case front and back. The strip should be about 24" long.

KERF CUTS. To get the hardboard to curve around the shape of the front and back pieces, I cut a series of narrow kerfs on this piece *(Fig. 6a)*.

ATTACH TO CASE. After the piece is kerfed, mount it into the rabbet in the case. First spread a bead of glue in the rabbets. Then tack the top in place, spacing the brads along the curve *(Fig. 6b)*.

SAND SMOOTH. When the top is attached, sand it so the surface is smooth, and so that both the hardboard and plywood are flush *(Fig. 7)*.

VENEER. You could use a piece of flexible veneer to cover the top and ends.

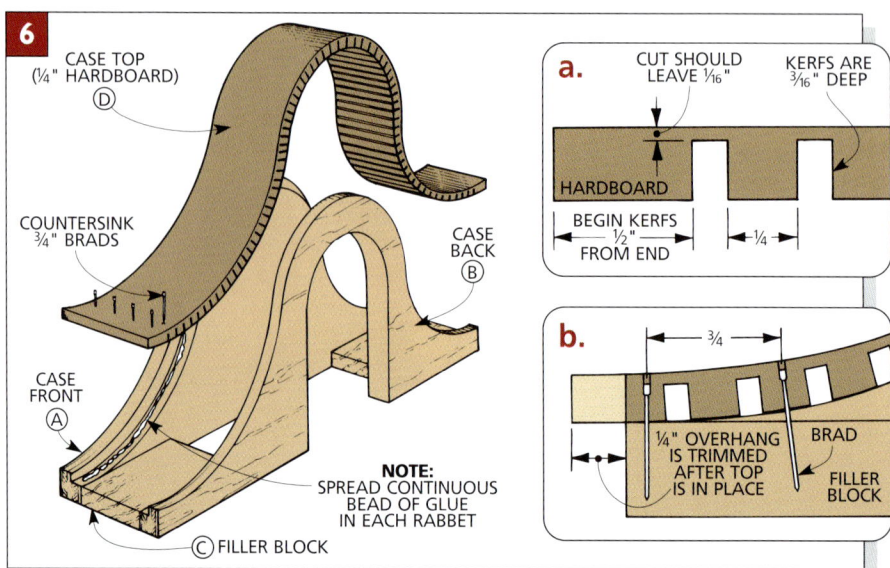

But for the best color match with the rest of the case, I sliced a strip of veneer off the same plywood I used for the case front and back. (See the Technique box below to learn how I did this.)

ATTACH VENEER. To attach the veneer, first cut two end veneer strips (E), and glue them to the ends of the case (with contact cement) *(Fig. 8)*. Note that the grain runs vertically on these strips.

TECHNIQUE .... *Resawing Veneer from Plywood*

The wood on the top and the ends of the Tambour Clock needed to match the plywood I used for the front and back of the clock. But a piece of ¾" plywood can't be kerf-bent to follow the curve of the clock, so I had come up with another solution.

What I did was resaw the veneer from a piece of the same sheet of plywood that I used for the front and back of the case. Then I cut a series of narrow kerfs in a piece of ¼" hardboard to act as a base for the veneer. A couple of filler blocks back up the veneer on the ends.

PLYWOOD BLANK. To make the matching veneer sheets, start with a piece of plywood slightly wider and a little longer than the finished size you'll need for the project. (For the clock top and end strips I used one piece 4½" wide and 25½" long.)

SAW SET-UP. Though veneer could be resawn off the plywood with a band saw, I decided to use the table saw with a combination blade. Start by raising the blade so it's slightly higher than half the width of the plywood.

Now comes the tricky part of the process — setting the rip fence so a thin layer can be cut off the waste side of the blade *(Fig. 1)*. What you're trying to do is cut off the face veneer plus a little bit of the plywood just beneath it. This second layer (called a crossband) gives the face veneer some support.

TRIM OFF THE VENEER. Once the rip fence is set, turn on the saw and slowly run the plywood over the blade. Then flip the piece end for end and make a second cut to remove the veneer and a thin layer of the crossband *(Fig. 1)*.

CLEAN OFF CROSSBAND. The next step is to remove the crossband from the back. You need to do this for two reasons. First, the thickness of the veneer has to be consistent so it will glue down smooth and flat. Also, you don't want any of the crossband layer to show along the edge of the veneer once it's glued in place.

To do this, I used a portable belt sander *(Fig. 2)*. Start by using a 120 grit sanding belt. Be careful not to sand through or to sand the edges of the veneer too thin.

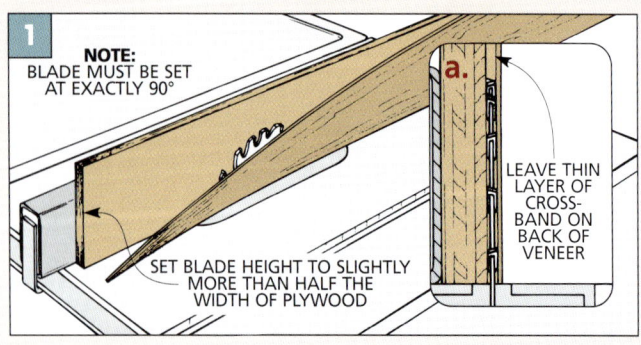

Then glue on the top veneer (F), starting at the top of the arch and working down the flared sides. Now trim the sides and ends flush with the case with a utility knife, and lightly sand all the edges.

BASE

The double-curved case is mounted to a $1/4$"-thick hardwood base. The base is screwed to the case from the bottom. For the best appearance, I once again tried to match the color of the base to the color of the case front.

CUT TO SIZE. To make the clock base (G) first cut a piece of solid wood to finished size so it's 1" longer and $1/2$" wider than the bottom of the clock case *(Fig. 9)*.

ROUT OGEE. Next, to give the clock a more finished appearance, I routed a Roman ogee around the front and ends (but not the back) of the clock base with a $5/32$" Roman ogee bit *(Fig. 9)*.

FEET. The clock base rests on a pair of hardwood feet (H) *(Fig. 10)*. Glue these in place to the bottom of the base, insetting them $1/4$" from the side and front, but flush to the back edge.

ATTACH BASE TO CASE. When the feet are attached, the base can be screwed from below to the case. To do this, first drill countersunk shank holes into the bottom of the base *(Fig. 9)*. Then, temporarily clamp the base to the case and drill pilot holes into the case using the shank holes as guides *(Fig. 11a)*. Now screw the base to the case.

ACCESS DOOR

In order to have access to the clockworks, I added a door to the back of the case. The opening in the back of the case has already been cut to shape. So now the door has to be cut to fit.

CUT TO SIZE. To make the door (I), first measure the size of the door opening. Then cut a blank $1/4$" wider than the height of the opening and $1/2$" longer than the width of the opening *(Fig. 12)*. This will orient the grain of the door horizontally — the same direction as the case back. This size also allows for a lip on the sides and top of the door. (There's no lip on the bottom edge of the door blank.)

LAY OUT ARC. After the door blank is trimmed to size, the next step is to lay out the arc on the top of the door *(Fig. 12)*. You can rough-cut this arc to shape with a jig saw (or band saw), and then just file or sand it smooth.

RABBET. To prevent the door from falling into the case, I added a lip around the edge. This lip is formed by routing a $3/8$" rabbet around the sides and top on the inside face of the door *(Fig. 12a)*.

Note: Be sure not to rabbet the bottom edge of the door.

ROUND OVER EDGES. Finally, to soften the outside edges of the access door, sand a slight roundover around the sides and top *(Fig. 12a)*.

Note: Once again, don't round over the bottom edge, or any of the inside edges of the door.

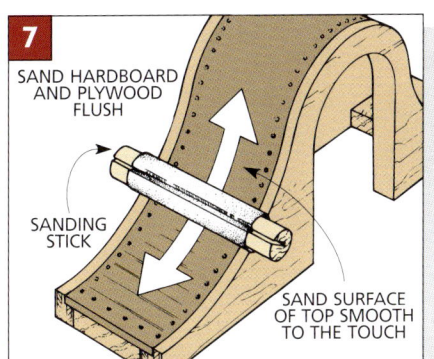

7 SAND HARDBOARD AND PLYWOOD FLUSH

SANDING STICK

SAND SURFACE OF TOP SMOOTH TO THE TOUCH

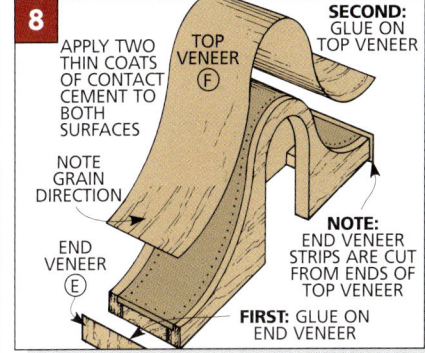

8 APPLY TWO THIN COATS OF CONTACT CEMENT TO BOTH SURFACES

NOTE GRAIN DIRECTION

END VENEER (E)

TOP VENEER (F)

SECOND: GLUE ON TOP VENEER

NOTE: END VENEER STRIPS ARE CUT FROM ENDS OF TOP VENEER

FIRST: GLUE ON END VENEER

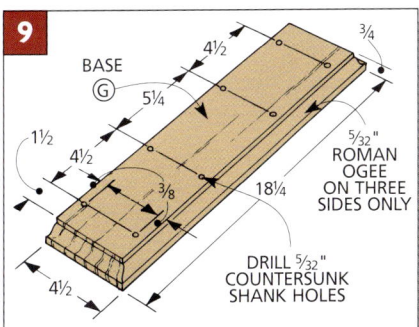

9 BASE (G)

$4\frac{1}{2}$ $3/4$

$5\frac{1}{4}$

$5/32$" ROMAN OGEE ON THREE SIDES ONLY

$1\frac{1}{2}$

$4\frac{1}{2}$

$3/8$

$18\frac{1}{4}$

$4\frac{1}{2}$

DRILL $5/32$" COUNTERSUNK SHANK HOLES

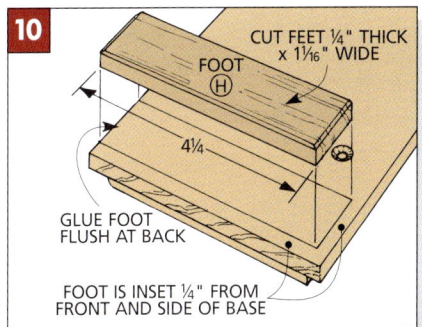

10 CUT FEET $1/4$" THICK x $1\frac{1}{16}$" WIDE

FOOT (H)

$4\frac{1}{4}$

GLUE FOOT FLUSH AT BACK

FOOT IS INSET $1/4$" FROM FRONT AND SIDE OF BASE

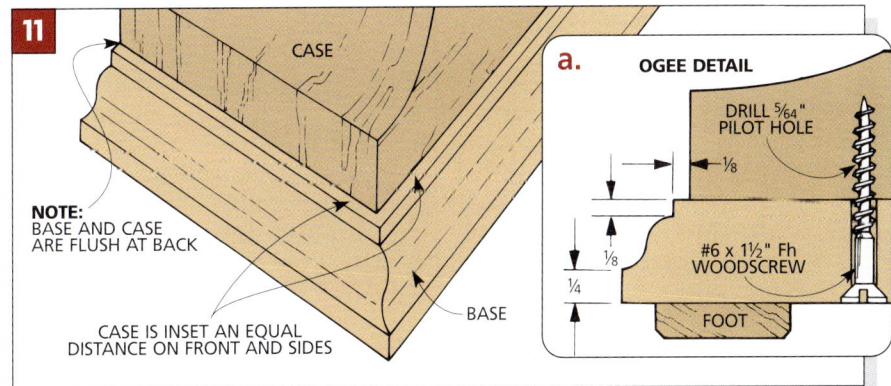

11 CASE

a. OGEE DETAIL

DRILL $5/64$" PILOT HOLE

$1/8$

$1/8$

$1/4$

#6 x $1\frac{1}{2}$" Fh WOODSCREW

FOOT

NOTE: BASE AND CASE ARE FLUSH AT BACK

CASE IS INSET AN EQUAL DISTANCE ON FRONT AND SIDES

BASE

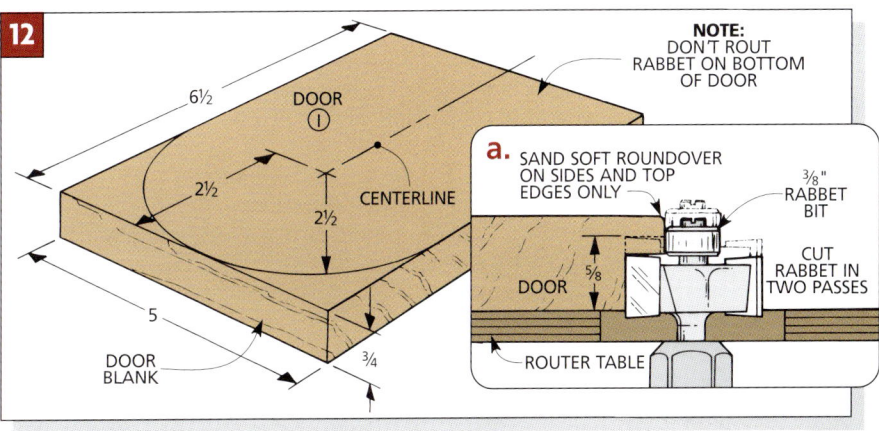

12

$6\frac{1}{2}$

DOOR (I)

$2\frac{1}{2}$

$2\frac{1}{2}$

CENTERLINE

5

DOOR BLANK

$3/4$

NOTE: DON'T ROUT RABBET ON BOTTOM OF DOOR

a. SAND SOFT ROUNDOVER ON SIDES AND TOP EDGES ONLY

$3/8$" RABBET BIT

CUT RABBET IN TWO PASSES

DOOR

$5/8$

ROUTER TABLE

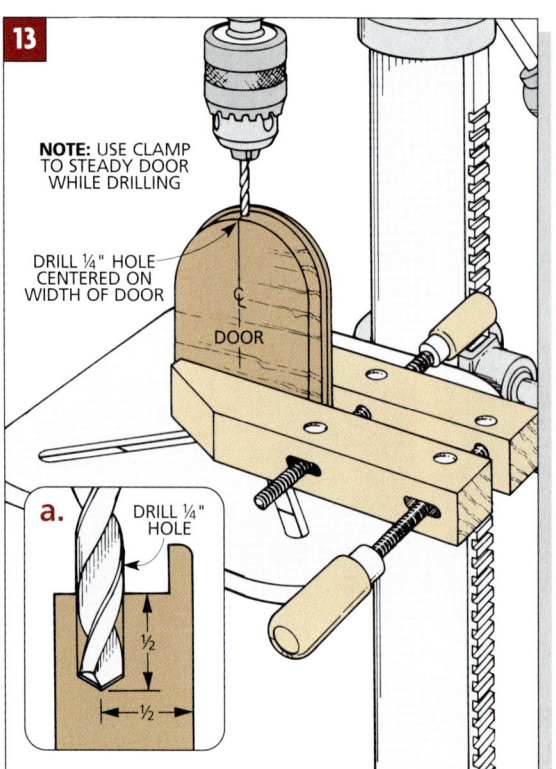

13 NOTE: USE CLAMP TO STEADY DOOR WHILE DRILLING

DRILL ¼" HOLE CENTERED ON WIDTH OF DOOR

DOOR

a. DRILL ¼" HOLE

½

½

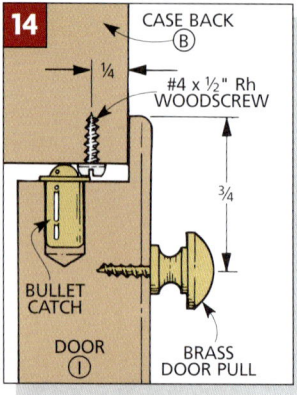

14 CASE BACK (B)

¼

#4 x ½" Rh WOODSCREW

¾

BULLET CATCH

DOOR (I)

BRASS DOOR PULL

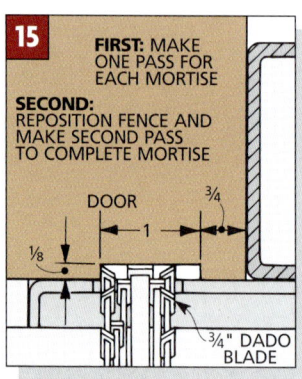

15 FIRST: MAKE ONE PASS FOR EACH MORTISE

SECOND: REPOSITION FENCE AND MAKE SECOND PASS TO COMPLETE MORTISE

DOOR

¾

1

⅛

¾" DADO BLADE

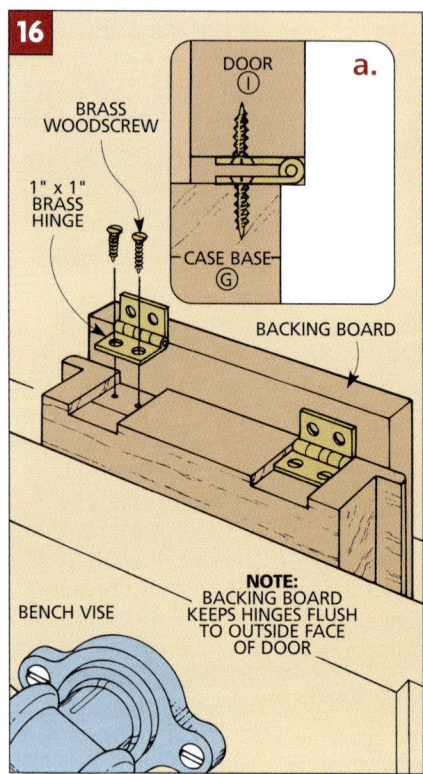

16 DOOR (I) **a.**

BRASS WOODSCREW

1" x 1" BRASS HINGE

CASE BASE (G)

BACKING BOARD

BENCH VISE

NOTE: BACKING BOARD KEEPS HINGES FLUSH TO OUTSIDE FACE OF DOOR

DOOR CATCH. Now, to keep the door from dropping open, I installed a bullet catch in the door. To do this, drill a hole into the top of the door, centered on the door's width *(Fig. 13)*. Then tap the bullet catch into the hole *(Fig. 14)*.

Next, add a small roundhead screw in the top edge of the door opening to act as a trap for the bullet catch *(Fig. 14)*.

DOOR PULL. Before installing the access door hinges, screw the door pull to the door *(Fig. 14)*. (You'll need to get it open when testing the hinges.) The pull is centered on the door's width, ¾" down from the top edge.

HINGE MORTISES. To attach the door, I used a dado blade to cut mortises for the hinges across the door's bottom edge *(Fig. 15)*. Set the dado blade to the same height as the diameter of the barrel of the hinge. This way, both leaves of the hinge are mortised into the door — not into the clock base *(Fig. 16a)*.

ATTACH DOOR. With the mortises cut, the hinges can be screwed to the door.

Note: The barrel of each hinge should be attached flush to the outside face of the door *(Fig. 16)*. To do this, I used a backing board to keep the hinges flush to the outside of the door face. With one leaf of each hinge attached to the door, screw the other leaf to the base (G) *(Fig. 16a)*.

APPLY FINISH. It's easiest to apply the finish to the clock before installing the clockworks. I used a coat of sealer and two coats of a satin finish.

Note: If you want to add trim to the case to give it the look of a faux raised field, see the Designer's Notebook on the next page. Then glue the trim on before applying the finish.

CLOCKWORKS

I used a battery-operated quartz movement in my Tambour Clock *(Fig. 17)* (see Sources, page 126). There's also a dial with a hinged glass bezel attached to the front of the clock.

INSTALL WORKS AND DIAL. To install the clockworks, first locate the center-point of the case front and drill a ½" hole at this point for the shaft of the movement (that holds the hands) *(Fig. 17)*. Then place the movement inside the case with the shaft protruding through the hole.

Now slide the clock dial onto the front of the case, over the hand shaft. Both the dial and the works are held to the case by threading the mounting washer and nut onto the shaft from the front.

NAIL DIAL. Before completely tightening the nut, adjust the position of the dial on the front so there's an equal space around the sides and top of the dial.

Also, to make sure the dial is oriented properly, place a square on the clock base and line up the 12 and 6 markings on the

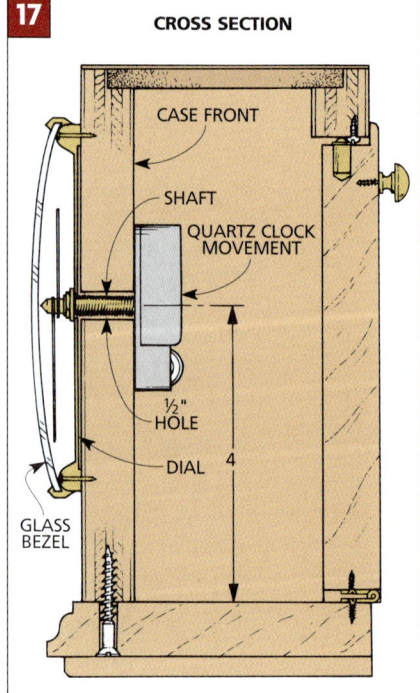

17 CROSS SECTION

CASE FRONT

SHAFT

QUARTZ CLOCK MOVEMENT

½" HOLE

DIAL

4

GLASS BEZEL

dial. Now tighten the mounting nut, and tack the dial in place with brass nails.

HANDS. With the dial nailed in place, slip the hour and minute hands onto the shaft. Then secure the hands to the shaft with the knurled hand nut. Finally, mount the bezel hinge to the dial face with some small brass screws, install the battery, and adjust the clock. ◼

DESIGNER'S NOTEBOOK

Adding veneer trim to the front of the Tambour Clock is a simple way to give it the look of a much more detailed piece. The face of the clock looks like it's been raised from a piece of solid walnut burl.

CONSTRUCTION NOTES:

■ After building the clock, it occured to me that it would be easy to add two decorative trim pieces to its front. The trim pieces give the clock front the look of raised fields. All that's involved is cutting two triangular pieces from a contrasting (or complementary) piece of veneer.

■ Since my clock case was walnut plywood, I made the trim pieces out of walnut burl veneer (see Sources, page 126). If the veneer is mounted to a thin base piece, it looks thicker and stands out a little further from the front of the case.

■ To make the base pieces for the veneer, start by cutting two rectangular blanks from a thin piece of ⅛"-thick hardwood *(Fig. 2)*. (I cut these from scrap left over from the clock base.)

■ Now, cut two pieces of veneer the same size as the hardwood pieces. Then glue a piece of veneer to one side only of each piece of hardwood.

■ Two things make the trim pieces look good on the front of the clock. First, the curves on the trim match the curves on the case front. Also, the spacing between

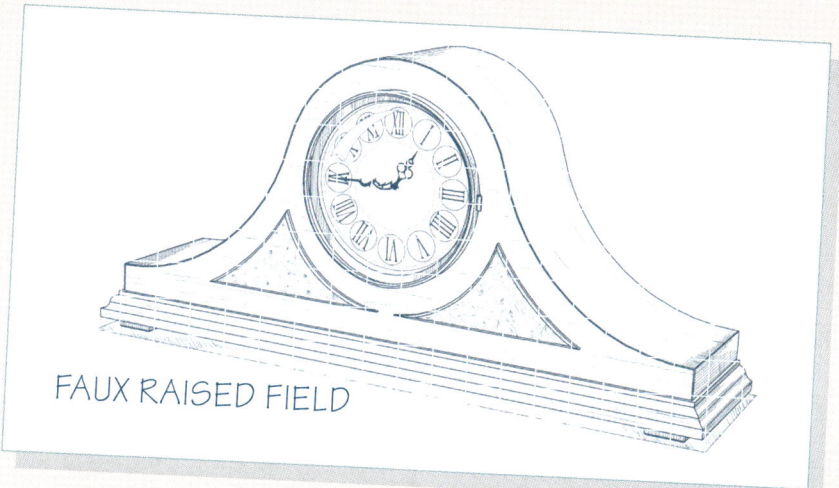

FAUX RAISED FIELD

the trim piece and bezel is equal to the spacing between the trim and the curved edge of the case.

■ To get the pieces the proper shape and size, make a template from cardboard using the faux raised field pattern *(Fig. 1)*.

■ Next, draw an outline of the template on each of the trim blanks. Then cut out the triangular trim with the band saw *(Fig. 3)*.

■ Once the workpieces have been cut and sanded to shape, sand a bevel on all

three top edges. This helps to blend the trim pieces into the clock face. I did this with a short length of dowel wrapped with sandpaper *(Fig. 4)*.

■ To get the trim pieces aligned properly, first position them temporarily on the case front. Then draw a light pencil reference mark around each piece.

■ Now apply glue to the back of the trim pieces and press them in place with hand pressure, using the marks to line them up.

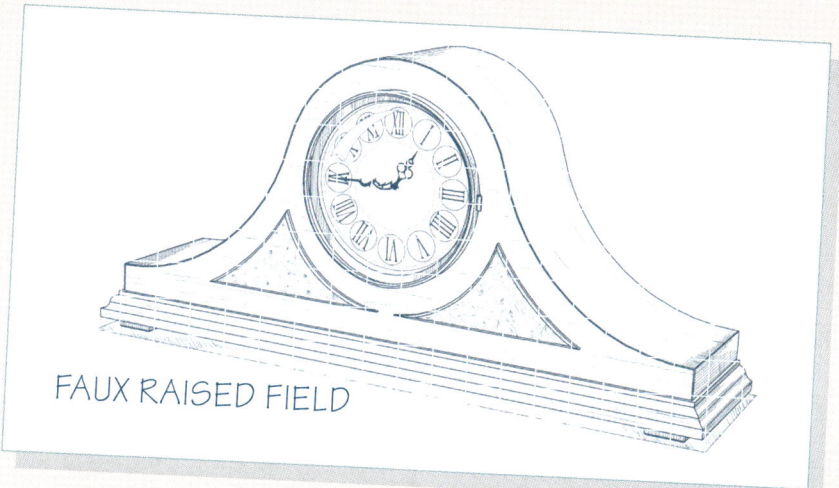

1 FAUX RAISED FIELD PATTERN

6⅜
5⅝
3⁵⁄₁₆
3⁵⁄₁₆
8⅝

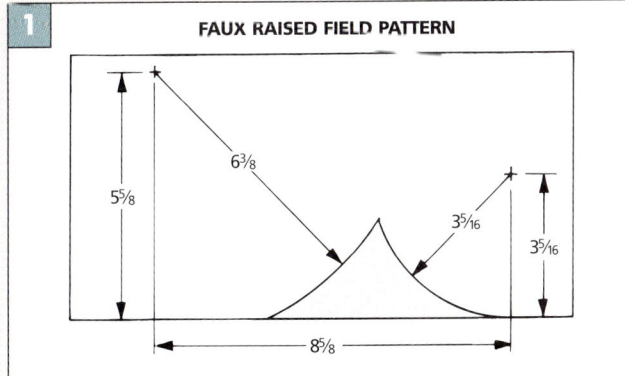

2 GLUE VENEER TO HARDWOOD BASE — CARDBOARD PATTERN — BURL VENEER — DRAW PATTERN ON TRIM BLANK — ⅛" HARDWOOD BASE

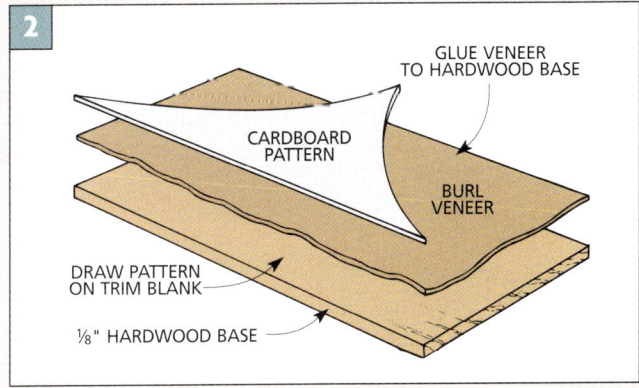

3 BAND SAW — CUT OUT TRIM PIECE, THEN SAND EDGES SMOOTH — TRIANGULAR TRIM PIECE

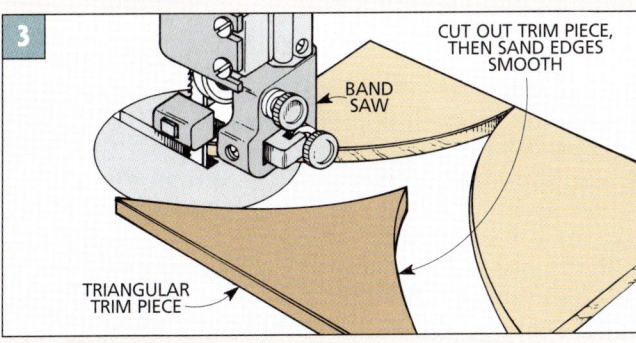

4 CHAMFER ALL EDGES OF TRIM PIECE WITH SANDING STICK — TRIM PIECE

Mantel Clock

You'd expect a large project to attract a lot of attention. But sometimes, a small weekend project, like this Mantel Clock, can surprise you. The beauty of the curly maple "pops out" with just the right finish.

There's no doubt about it — what grabs your attention right away is the wood. It's curly maple. This figured wood is so dramatic that you might be inclined to think "the wood makes the clock." But while I appreciate the beauty of the curly maple, there are a couple of other reasons why I like this Mantel Clock as much as I do.

DESIGN. First of all, there's the design. This Mantel Clock has a traditional look to it. Its clean lines and simple moldings at the top and bottom would look great no matter what wood you used.

CONSTRUCTION. In spite of its elegance, there's nothing very difficult about building this Mantel Clock. Everything is held together with simple joinery, and it houses a quartz movement, which is readily available and easy to install. In fact, a quartz movement and a hardware kit is available from *Woodsmith Project Supplies*. For more information about what's included in this kit, see page 126.

QUARTER-ROUND MOLDING. The clock face is protected by a framed piece of glass. To hold the glass in place inside the frame, I used small strips of quarter-round. And even though they're small, I came up with an easy way to make them with a table-mounted router and the table saw. (See the Shop Tip on page 36.)

WOOD. As I mentioned earlier, I used a piece of curly maple to make the clock case. If you've never stained figured maple, you're in for a bit of a surprise. Unlike regular maple (where you want to apply an even stain), figured maple requires an uneven stain. The goal is to bring out the "waves" — not hide them. (See the Finishing Tip on page 37.)

EXPLODED VIEW

OVERALL DIMENSIONS:
10W x 4D x 12¼H

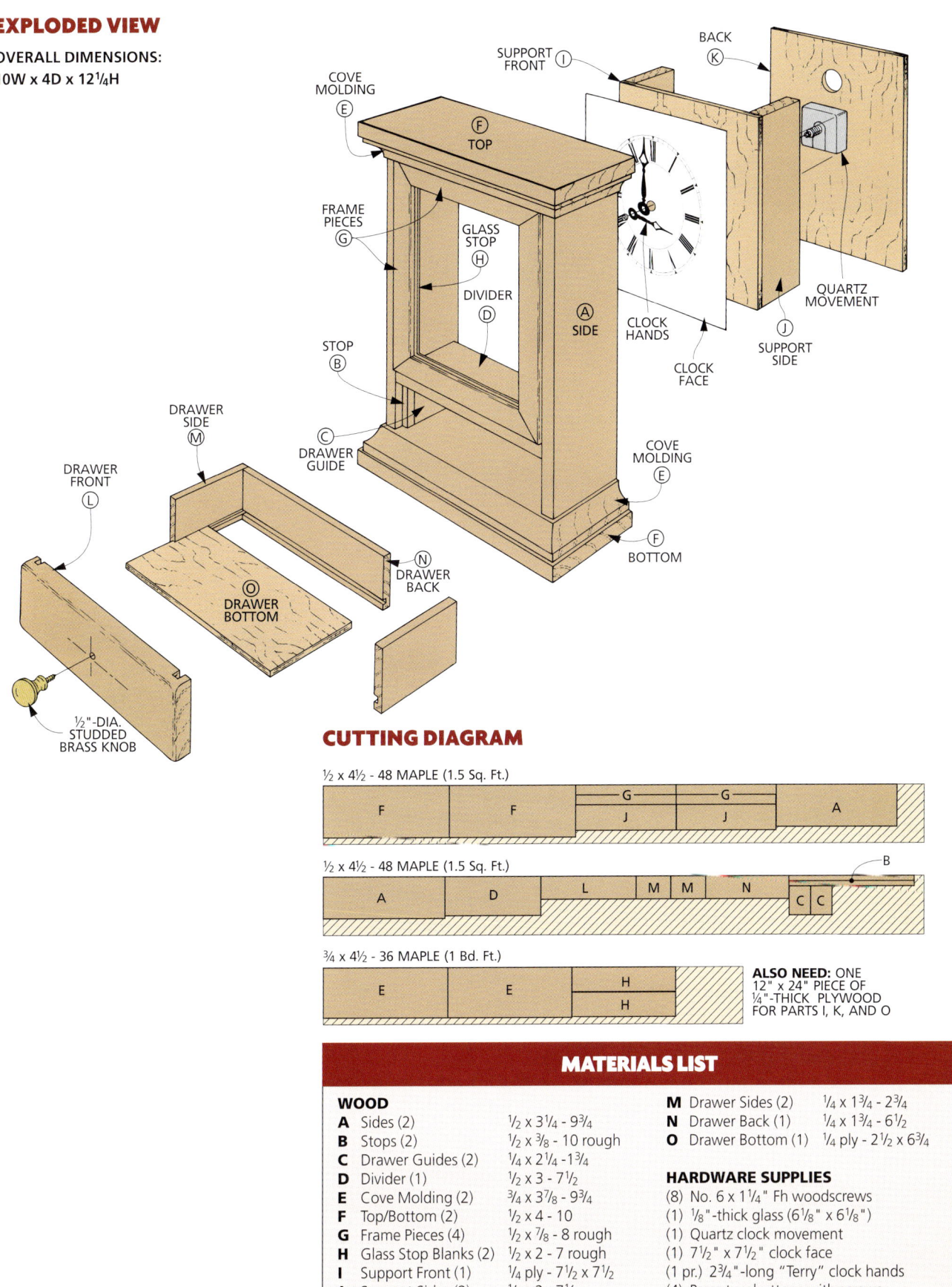

COVE MOLDING Ⓔ

SUPPORT FRONT Ⓘ

BACK

ⓀK

Ⓕ TOP

FRAME PIECES ⒼG

GLASS STOP ⒽH

DIVIDER ⒹD

Ⓐ SIDE

CLOCK HANDS

CLOCK FACE

QUARTZ MOVEMENT

ⒿJ

SUPPORT SIDE

STOP ⒷB

ⒸC DRAWER GUIDE

COVE MOLDING Ⓔ

Ⓕ BOTTOM

DRAWER SIDE ⓂM

DRAWER FRONT ⒧L

DRAWER BACK ⓃN

Ⓞ DRAWER BOTTOM

½"-DIA. STUDDED BRASS KNOB

CUTTING DIAGRAM

½ x 4½ - 48 MAPLE (1.5 Sq. Ft.)

| F | F | G G / J J | A |

½ x 4½ - 48 MAPLE (1.5 Sq. Ft.)

B

| A | D | L M M N | C C |

¾ x 4½ - 36 MAPLE (1 Bd. Ft.)

| E | E | H / H |

ALSO NEED: ONE 12" x 24" PIECE OF ¼"-THICK PLYWOOD FOR PARTS I, K, AND O

MATERIALS LIST

WOOD
A	Sides (2)	½ x 3¼ - 9¾
B	Stops (2)	½ x ⅜ - 10 rough
C	Drawer Guides (2)	¼ x 2¼ - 1¾
D	Divider (1)	½ x 3 - 7½
E	Cove Molding (2)	¾ x 3⅞ - 9¾
F	Top/Bottom (2)	½ x 4 - 10
G	Frame Pieces (4)	½ x ⅞ - 8 rough
H	Glass Stop Blanks (2)	½ x 2 - 7 rough
I	Support Front (1)	¼ ply - 7½ x 7½
J	Support Sides (2)	½ x 2 - 7½
K	Back (1)	¼ ply - 7½ x 9¾
L	Drawer Front (1)	½ x 1¾ - 7½
M	Drawer Sides (2)	¼ x 1¾ - 2¾
N	Drawer Back (1)	¼ x 1¾ - 6½
O	Drawer Bottom (1)	¼ ply - 2½ x 6¾

HARDWARE SUPPLIES
(8) No. 6 x 1¼" Fh woodscrews
(1) ⅛"-thick glass (6⅛" x 6⅛")
(1) Quartz clock movement
(1) 7½" x 7½" clock face
(1 pr.) 2¾"-long "Terry" clock hands
(4) Brass turnbuttons with screws
(1) ½"-dia. studded brass knob
(4) ¾"-dia. felt pads

The body of this clock is quite simple. It starts out as an H-shaped frame that's sandwiched between a layer of molding and a top and bottom.

SIDES. I began work on the clock body by cutting the two sides (A) to size from $\frac{1}{2}$"-thick stock *(Fig. 1)*.

With the sides cut, next I cut a $\frac{3}{8}$"-wide groove, $\frac{1}{4}$" deep in each side. Then I glued two stops (B) in each groove *(Figs. 1 and 1a)*. The lower stop is for a drawer. The upper stop will position both the clock face and a face frame.

There's a $\frac{1}{2}$"-wide gap between the upper and lower stops *(Fig. 1b)*. This is for a divider that's added later *(Fig. 2)*. One easy way to create this gap is to use a $\frac{1}{2}$"-thick scrap piece as a temporary spacer between the stops.

The last step for the sides is to glue a drawer guide (C) behind each lower stop *(Fig. 1b)*. These guides fit flush with the outside face of the lower stop and should stop $\frac{1}{4}$" short of the back edges of the sides to allow for a plywood back.

DIVIDERS. Next, to create the H-shaped frame, I connected the two sides with a divider (D) *(Fig. 2)*. This piece is sized so it's flush with the sides in front and the drawer guides in back *(Fig. 2a)*.

The trick to gluing the divider and sides is to keep the assembly square. So I used a spacer that matched the length of the divider *(Fig. 3)*.

TOP AND BOTTOM. With these pieces assembled, I added a layer of molding to each end and then added the top and bottom pieces *(Fig. 2)*. To do this, I cut two cove molding (E) pieces to size from $\frac{3}{4}$"-thick stock *(Fig. 2)*. These pieces are sized to create a $\frac{1}{8}$" lip at the sides and front *(Fig. 5a)*. (Mine were $3\frac{7}{8}$" x $9\frac{3}{4}$".)

Then to shape the molding, I routed the ends and front with a $\frac{1}{2}$" cove bit *(Figs. 4 and 4a)*. Next I screwed them to the sides, flush with the back *(Fig. 5)*.

Now, with the cove molding in place, I added the top and bottom (F) pieces *(Fig. 2)*. They overhang the cove molding $\frac{1}{8}$" on the front and sides *(Fig. 6a)*. But there is no profile routed on their edges and they're simply glued in place.

FRAME

With the basic body of the clock complete, I turned my attention to the frame that holds the glass *(Fig. 7)*. This is a simple mitered frame that fits the opening

above the divider and stands a little proud of the sides ($\frac{1}{8}$").

FRAME PIECES. To begin, I ripped the frame pieces (G) to width from $\frac{1}{2}$"-thick stock *(Fig. 7)*. But before mitering the frame pieces to final length, there's a little

decorative shaping that needs to be done to them on the router table.

First, to soften the edges on the front of the frame pieces, I routed a $\frac{1}{8}$" roundover along both outside edges of each frame piece *(Fig. 7a)*.

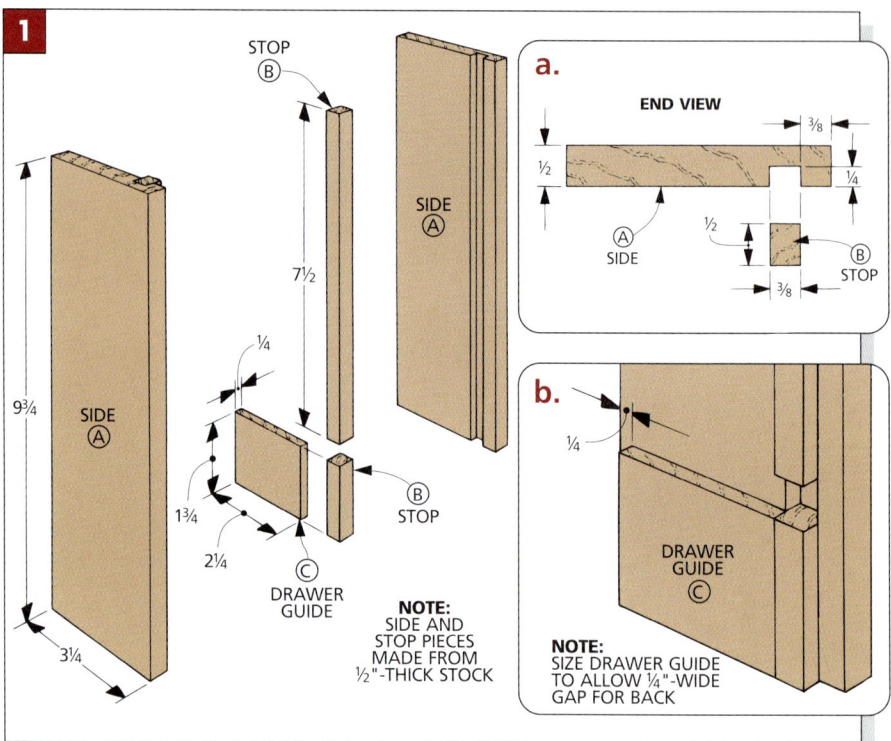

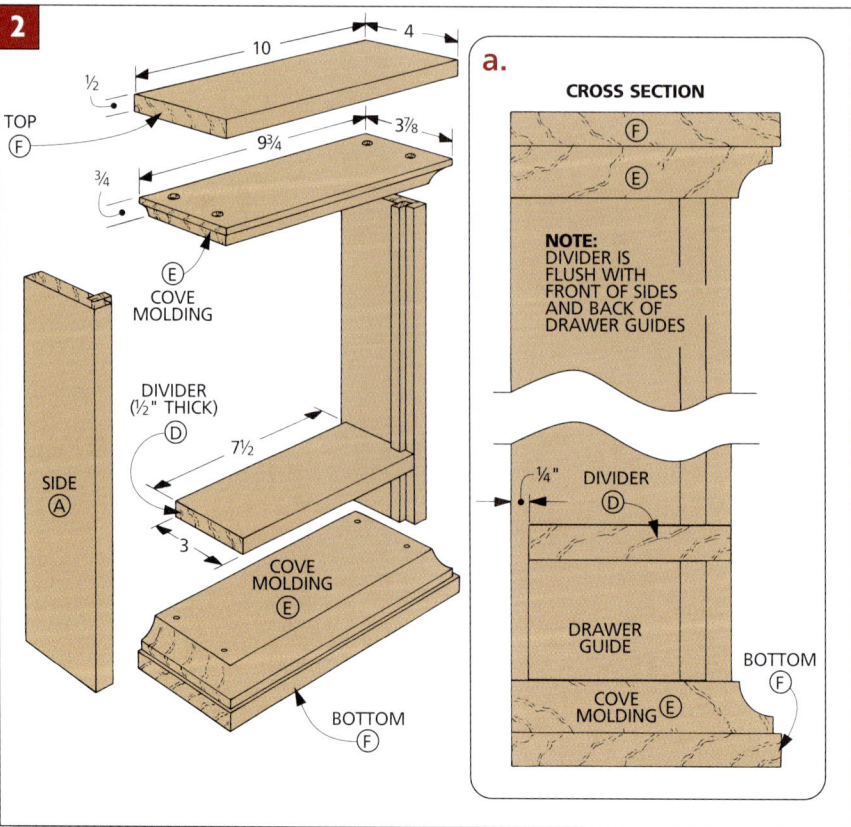

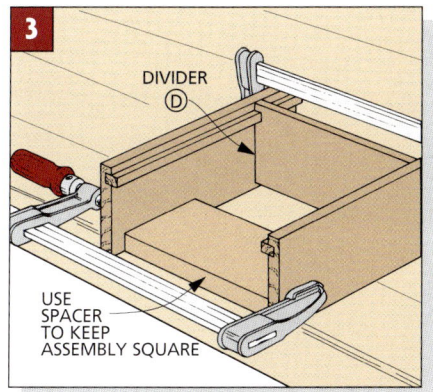

3

DIVIDER Ⓓ

USE SPACER TO KEEP ASSEMBLY SQUARE

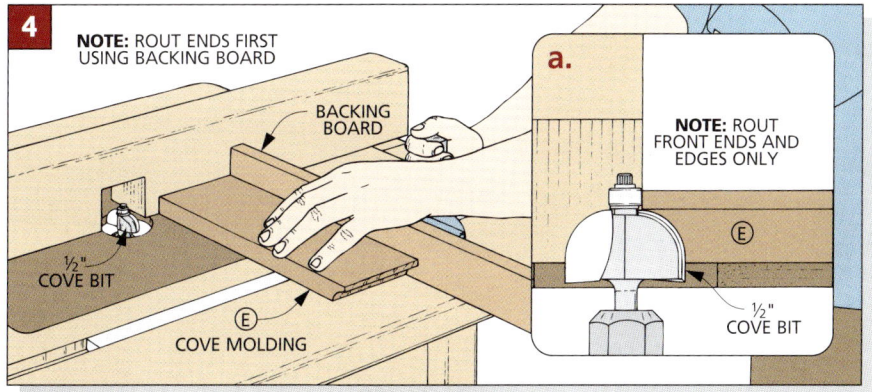

4

NOTE: ROUT ENDS FIRST USING BACKING BOARD

BACKING BOARD

½" COVE BIT

COVE MOLDING

Ⓔ

a.

NOTE: ROUT FRONT ENDS AND EDGES ONLY

Ⓔ

½" COVE BIT

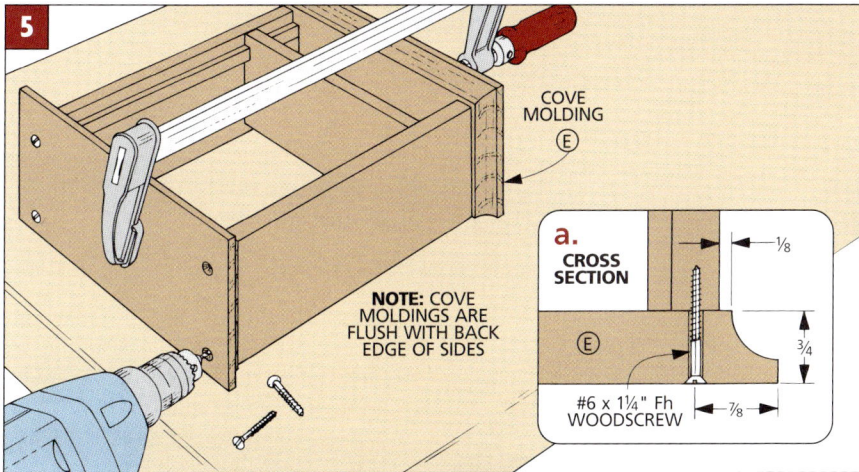

5

COVE MOLDING Ⓔ

a.

CROSS SECTION

Ⓔ

⅛

¾

#6 x 1¼" Fh WOODSCREW

⅞

NOTE: COVE MOLDINGS ARE FLUSH WITH BACK EDGE OF SIDES

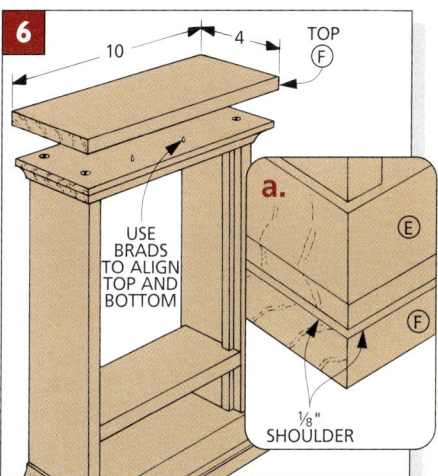

6

10   4   TOP Ⓕ

USE BRADS TO ALIGN TOP AND BOTTOM

a.

Ⓔ

Ⓕ

⅛" SHOULDER

Next, to hold the ⅛"-thick piece of glass and the ¼"-thick glass stops, I routed a rabbet along the inside edge of each frame piece *(Fig. 7b)*.

Note: Because this rabbet is ⅜" deep, I'd recommend routing it in two passes.

When the rabbet is routed, the frame pieces can be mitered to length *(Fig. 7)*. Here, you want a snug fit. So after mitering one end of each, I cut each piece to length so it fit the opening exactly.

ASSEMBLY. Now the frame can be glued together. But don't glue it into the body just yet. It's easier to make and fit the glass stops before the frame is in place.

GLASS STOPS. The glass stops (H) are simply ¼"-thick quarter-round strips *(Fig. 7)*. But making these strips can be dangerous. Because they're so small, they can get hung up in the insert plate on the table saw, and there's potential for kickback. So I started with oversize blanks and used a zero-clearance insert in the table saw. (For more on this, refer to the Shop Tip on page 36.)

With the stops routed and cut to size, they can be mitered to fit the rabbet in the frame. But I didn't glue the frame or tack in the stops quite yet. Instead, I waited until after the clock had been stained.

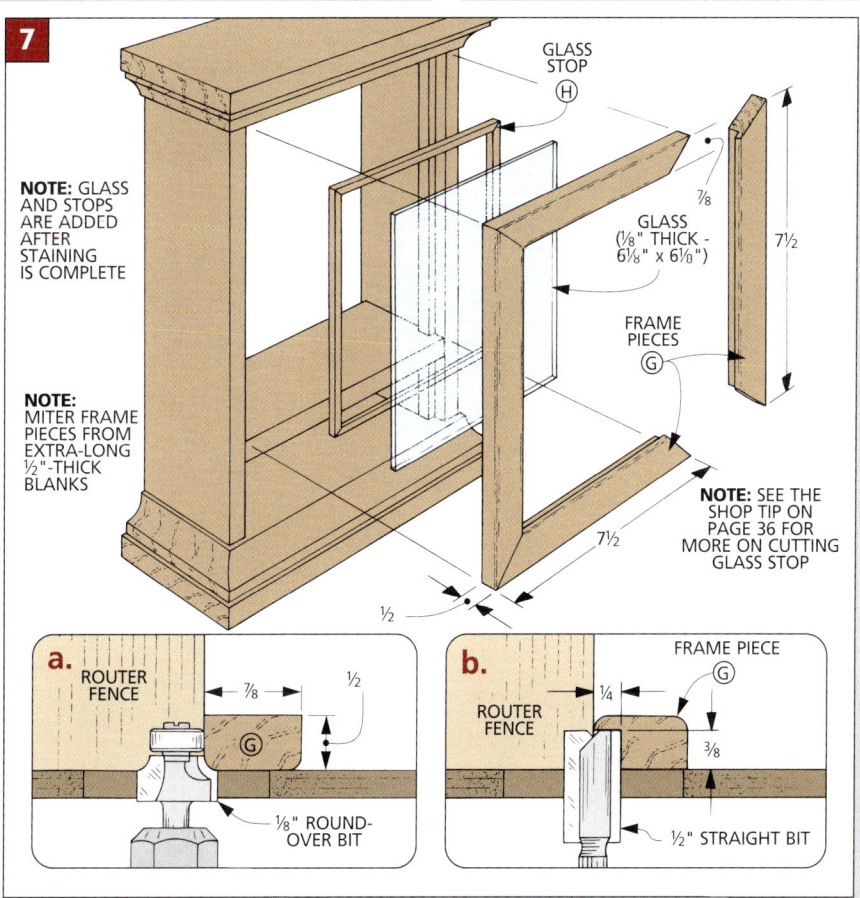

7

GLASS STOP Ⓗ

NOTE: GLASS AND STOPS ARE ADDED AFTER STAINING IS COMPLETE

NOTE: MITER FRAME PIECES FROM EXTRA-LONG ½"-THICK BLANKS

⅞

GLASS (⅛" THICK - 6⅛" x 6⅛")

7½

FRAME PIECES Ⓖ

NOTE: SEE THE SHOP TIP ON PAGE 36 FOR MORE ON CUTTING GLASS STOP

7½

½

a.

ROUTER FENCE

⅞   ½

Ⓖ

⅛" ROUND-OVER BIT

b.

FRAME PIECE Ⓖ

ROUTER FENCE

¼

⅜

½" STRAIGHT BIT

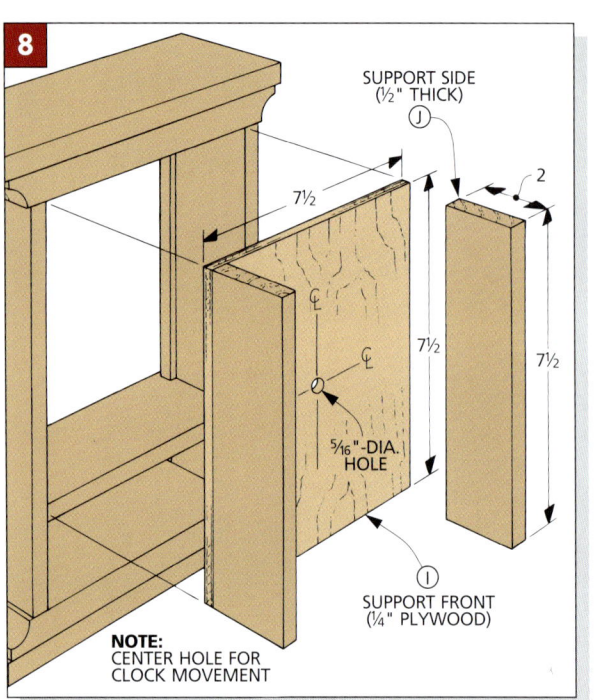

8

SUPPORT SIDE
(½" THICK)
Ⓙ

7½

2

7½

7½

5/16"-DIA.
HOLE

SUPPORT FRONT
(¼" PLYWOOD)
Ⓘ

NOTE:
CENTER HOLE FOR
CLOCK MOVEMENT

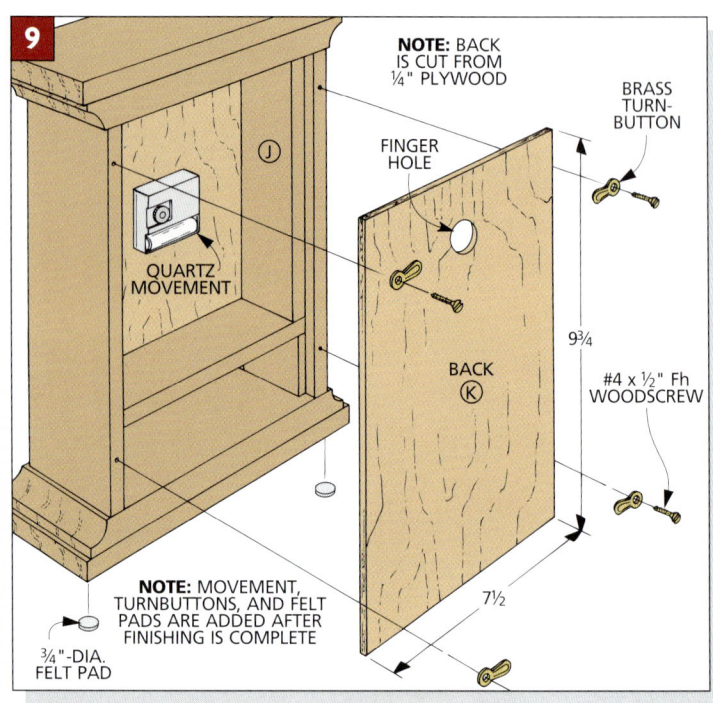

9

NOTE: BACK
IS CUT FROM
¼" PLYWOOD

BRASS
TURN-
BUTTON

FINGER
HOLE

Ⓙ

QUARTZ
MOVEMENT

BACK
Ⓚ

9¾

#4 x ½" Fh
WOODSCREW

NOTE: MOVEMENT,
TURNBUTTONS, AND FELT
PADS ARE ADDED AFTER
FINISHING IS COMPLETE

7½

¾"-DIA.
FELT PAD

MOVEMENT SUPPORT

At this point, the body of the clock is almost complete. All that's left is to mount the movement and add a back. I used a slightly unusual method for doing this.

The clock face and quartz movement are attached to a support that slides into the opening in the back *(Fig. 8)*.

This way, if you ever need to get at the hands in front, all you have to do is slide the movement support out through the back of the clock body.

FRONT. To make it, I cut a support front (I) from ¼"-thick plywood *(Fig. 8)*. The size of this piece depends on the opening inside the clock. I cut my front to fit the opening exactly (7½" x 7½"), and then trimmed it slightly so it would slide in without too much trouble.

The only thing that needs to be done to this front piece is to drill a 5/16"-dia. hole in the center. This is for mounting the quartz movement *(Fig. 8)*.

SIDES. Next, cut two support sides (J) to match the height of the front *(Fig. 8)*.

Cut them to width so they're flush with the back of the divider when the support is slid inside the clock *(Fig. 9)*. (Mine were 2" wide. This may vary depending on the thickness of your clock face.)

With the sides glued to the front, the support is complete. But wait to add the quartz movement until after the clock has been finished. At that point, adding the movement is just a matter of feeding its post through the front and the clock face and securing them with a nut. Then the hands can be attached to the post.

SHOP TIP .... *Quarter-Round Molding*

To hold a piece of glass in a frame, I often use small strips of quarter-round (see photo below). But routing and ripping thin strips can be dangerous, so I use oversize blanks.

First, I cut a ¾"-thick blank roughly 2" wide. Then to create the quarter-round profile, I rout two edges with a ¼" roundover bit *(Fig. 1)*. Now, cutting the strips

from the blank is a simple, two-step process. First, 3/8"-deep kerfs are cut ¼" from the rounded edges *(Fig. 2)*. (To be safe, use a zero-clearance insert plate in your table saw.)

Next, the blank can be stood on edge, and the quarter-round molding strips can be cut away. Just make sure that the strips fall to the waste side of the blade *(Fig. 3)*.

1

BLANK FOR
GLASS STOPS

ROUTER
FENCE

¾

¼" ROUND-
OVER BIT

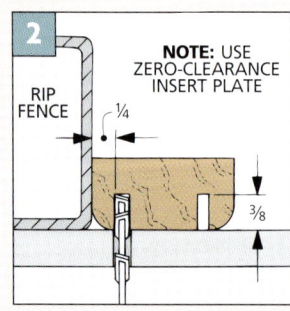

2

RIP
FENCE

¼

NOTE: USE
ZERO-CLEARANCE
INSERT PLATE

3/8

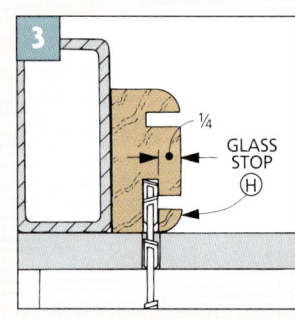

3

¼

GLASS
STOP
Ⓗ

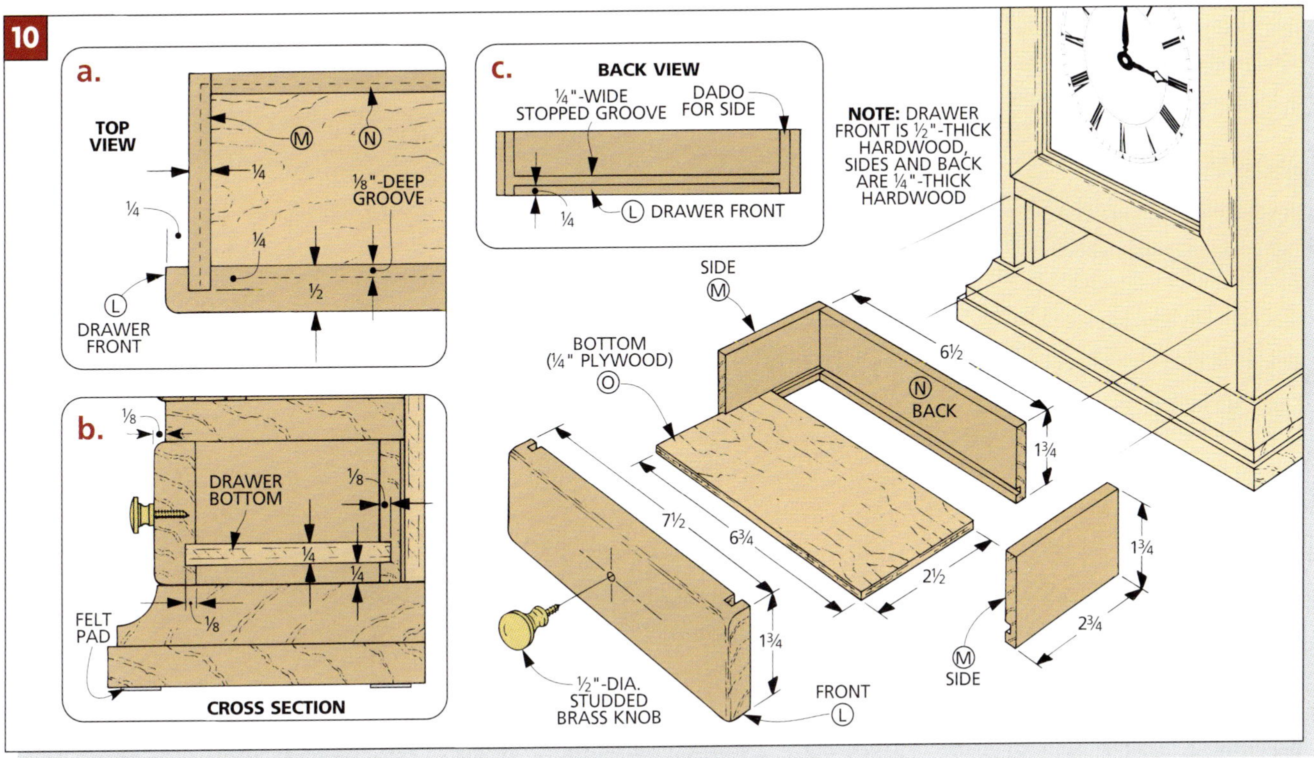
10
a.
TOP VIEW
M
N
1/4
1/8"-DEEP GROOVE
1/4
1/4
1/2
L
DRAWER FRONT
b.
1/8
DRAWER BOTTOM
1/8
1/4
1/4
1/8
FELT PAD
CROSS SECTION
c.
BACK VIEW
1/4"-WIDE STOPPED GROOVE
DADO FOR SIDE
1/4
L DRAWER FRONT
NOTE: DRAWER FRONT IS 1/2"-THICK HARDWOOD, SIDES AND BACK ARE 1/4"-THICK HARDWOOD
SIDE M
BOTTOM (1/4" PLYWOOD) O
N BACK
6 1/2
1 3/4
7 1/2
6 3/4
2 1/2
1 3/4
1 3/4
2 3/4
M SIDE
1/2"-DIA. STUDDED BRASS KNOB
FRONT L

BACK. The last piece to add to the body is a back (K) *(Fig. 9)*. This 1/4" plywood piece is cut to fit the opening in back. To make it easy to remove, drill a finger hole.

To hold the back in place, turnbuttons are screwed to the back of the sides.

DRAWER

Now the only thing left to add is a small drawer *(Fig. 10)*. It also stands slightly proud of the front (1/8") *(Fig. 10b)*.

Note: This drawer is shallow, so there won't be much binding as it's slid in. Because of this, I built the drawer to fit the opening exactly. Then I sanded it down until it slid in smoothly.

FRONT. Start by cutting the 1/2"-thick front (L) to fit the opening *(Fig. 10)*.

Next, to hold the sides, I cut two 1/4"-deep dadoes on the back face of the front *(Fig. 10a)*. (These are inset so the sides fit inside the clock.) Then after the dadoes were cut, I routed a 1/8" roundover on the outside edges of the front.

SIDES AND BACK. Now, the 1/4"-thick stock used for the drawer sides (M) and back (N) can be cut to size *(Fig. 10)*.

To hold the bottom, I added a 1/8"-deep groove in all the pieces *(Fig. 10b)*. But you don't want this groove visible on the ends of the drawer front, so rout a stopped groove between the dadoes *(Fig. 10c)*.

Now the drawer bottom (O) can be cut to size, and the drawer can be glued together. The back simply butts between the sides *(Fig. 10a)*.

FINISH. To finish the clock, I decided to highlight the curly maple with a water-based aniline dye and apply a wipe-on finish (see the Finishing Tip below).

Now, all that's left to complete the clock is to add a 1/2"-dia. brass knob to the drawer front and add four felt pads to the bottom of the case *(Figs. 9 and 10b)*. ■

FINISHING TIP .... *Staining Figured Maple*

If you've ever stained regular maple, then the first thing to know about figured maple is that it's a completely different ball game. With regular maple, you want an even stain, which is hard because maple's grain is so wavy. With figured maple, you want an uneven stain. This may sound disastrous, but the goal is to highlight the waves — not hide them.

This increases the depth and contrast and brings out the figure (see photo).

To stain figured maple, I use an aniline dye. (An off-the-shelf pigment stain tends to hide the figure.) And I typically choose water-based dyes because they're the easiest to work with and the most lightfast. (For sources, see page 126.) But the water will raise the grain. So I wet the wood and lightly sand off the whiskers before applying the stain.

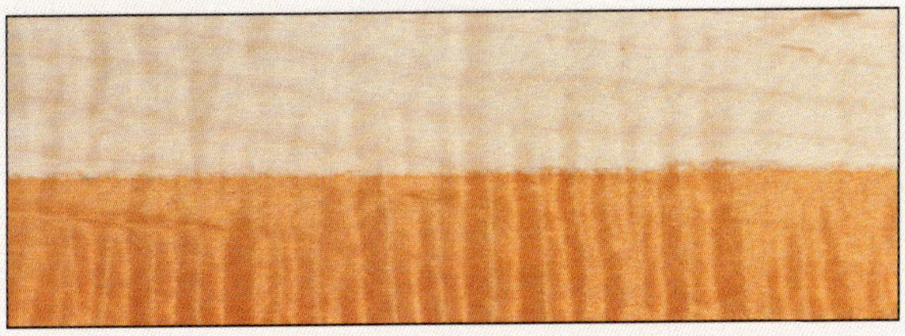

Then, I keep a wet edge. If the stain dries before you're finished, it could leave lap marks.

Aniline dyes look great when they're wet, but flat and dull when dry. Don't worry. Applying a top coat brings back the satin look.

CHESTS & TRUNKS

Stripped down to their essence, the projects in this section are just boxes. What's intriguing is the different ways a box can be built.

The steamer trunk is a plywood box with splined miters at the corners. Like the trunk, the bedside chest uses plywood panels for the case. But here, web frames hold the panels together and create the openings for the drawers.

For the jewelry chest, frames and panels are assembled with stub tenon and groove joinery. Then the frames are glued to each other to form the case.

The solid-wood construction of the blanket chest allows for dovetailed corners on the main case and the drawers, with splined miters used again on the drawer carcase.

Steamer Trunk

This trunk is a handsome — and lasting — piece of transportable furniture. Authentic hardware provides a period look, while mitered corners hide the modern plywood construction.

Not that many years ago, I would spend hours in my grandfather's attic rummaging through his old steamer trunk. It was exciting to think about where the stuff had come from and where the trunk had been.

That trunk is now in my attic. And if it weren't so beat up and musty-smelling, I'd bring it down and show it off. But instead of trying to renew the antique family heirloom, I decided to build a brand new Steamer Trunk of my own.

WOOD. Old trunks were built light so they could be carried. Since I don't plan on carrying this trunk around much, I used $3/4$" oak plywood for the case. But I used $1/4$" plywood for the lid panels so the lid would be easier to lift.

CONSTRUCTION. To avoid exposing the plywood edges of the trunk case, the sides are joined with a splined miter joint. The lid panels are reinforced with oak slats that are tenoned into frames.

FINISH. A dark-colored stain makes the oak trunk look like an heirloom. But finding just the right shade proved to be a challenge. After a bit of experimenting, I came up with a mixture of artists' oil and boiled linseed oil that produced a rich, translucent brown — like aged oak.

HARDWARE. In the past, trunks needed their corners and edges reinforced with hardware to prevent damage during travel. But today, trunks like this have extra hardware mostly for decoration. *Woodsmith Project Supplies* offers a kit with all the necessary hardware for this trunk. Ordering information and other sources can be found on page 126.

TRUNK LINING AND TRAY. If you decide to use the trunk as a sweater or blanket chest, you can line the inside with aromatic cedar to keep the contents smelling fresh and keep moths away. You could also build a removable tray to sit inside the top of the trunk. I'll explain how to add these features in the Designer's Notebook starting on page 48.

EXPLODED VIEW

OVERALL DIMENSIONS:
36½W x 20½D x 16⅛H

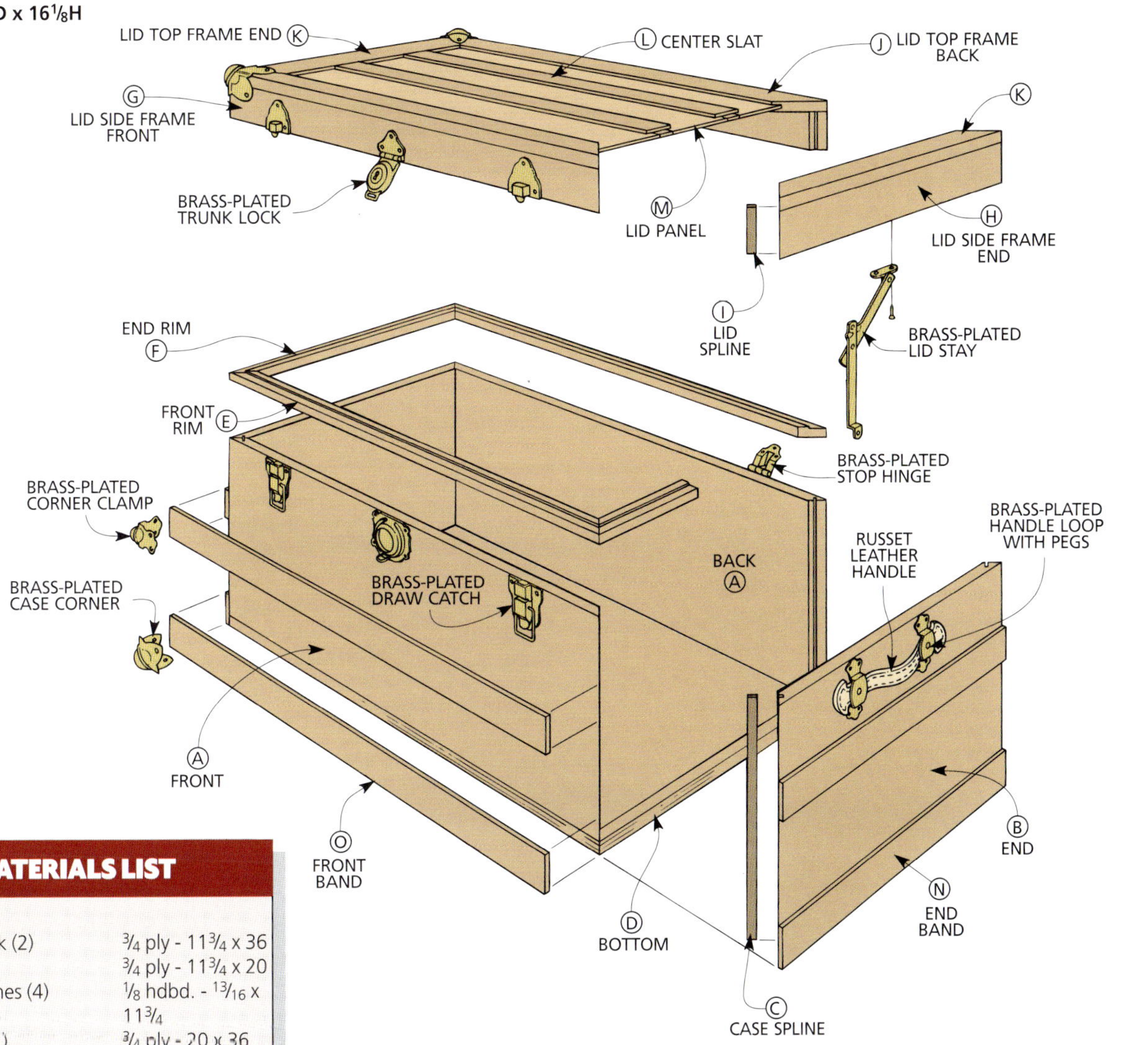

LID TOP FRAME END Ⓚ

Ⓖ LID SIDE FRAME FRONT

Ⓛ CENTER SLAT

Ⓙ LID TOP FRAME BACK

Ⓚ

BRASS-PLATED TRUNK LOCK

Ⓜ LID PANEL

Ⓗ LID SIDE FRAME END

Ⓘ LID SPLINE

BRASS-PLATED LID STAY

END RIM Ⓕ

FRONT RIM Ⓔ

BRASS-PLATED STOP HINGE

BRASS-PLATED CORNER CLAMP

BRASS-PLATED HANDLE LOOP WITH PEGS

BRASS-PLATED CASE CORNER

BRASS-PLATED DRAW CATCH

RUSSET LEATHER HANDLE

BACK Ⓐ

Ⓐ FRONT

Ⓑ END

Ⓞ FRONT BAND

Ⓝ END BAND

Ⓓ BOTTOM

Ⓒ CASE SPLINE

MATERIALS LIST

WOOD

A	Front/Back (2)	¾ ply - 11¾ x 36
B	Ends (2)	¾ ply - 11¾ x 20
C	Case Splines (4)	⅛ hdbd. - ¹³/₁₆ x 11¾
D	Bottom (1)	¾ ply - 20 x 36
E	Rim Front/Back (2)	¾ x 1¼ - 36
F	Rim Ends (2)	¾ x 1¼ - 20
G	Lid Side Frame Fr./Bk.(2)	¾ x 2¼ - 36
H	Lid Side Frame Ends (2)	¾ x 2¼ - 20
I	Lid Splines (4)	⅛ x 2¼ - ¹³/₁₆
J	Lid Top Frame Fr./Bk. (2)	¾ x 2 - 36
K	Lid Top Frame Ends (2)	¾ x 2 - 20
L	Center Slats (2)	¾ x 2 - 32½
M	Lid Panels (3)	¼ ply - 4½ x 32½
N	End Bands (4)	¼ x 2 - 20
O	Front/Back Bands (4)	¼ x 2 - 36½

HARDWARE SUPPLIES

(95) No. 6 x ½" Rh brass woodscrews
(36) 4d finish nails
(8) Brass-plated case corners
(4) Brass-plated corner clamps
(2) Brass-plated stop hinges
(4) Brass-plated handle loops w/ pegs
(2) Brass-plated draw catches
(1) Brass-plated trunk lock
(2) Brass-plated lid stays
(2) Russet leather handles

CUTTING DIAGRAM

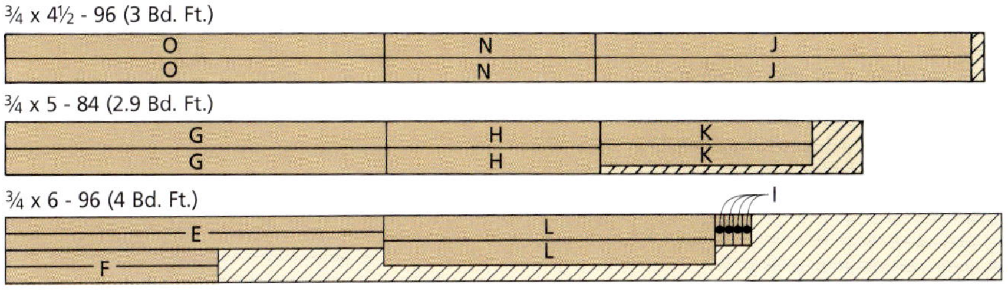

¾ x 4½ - 96 (3 Bd. Ft.)

O	N	J
O	N	J

¾ x 5 - 84 (2.9 Bd. Ft.)

G	H	K
G	H	K

¾ x 6 - 96 (4 Bd. Ft.)

E		L	I
F		L	

ALSO NEED: 24" x 96" PIECE OF ¾" PLYWOOD, 24" x 48" PIECE OF ¼" PLYWOOD AND SCRAP ⅛" HARDBOARD

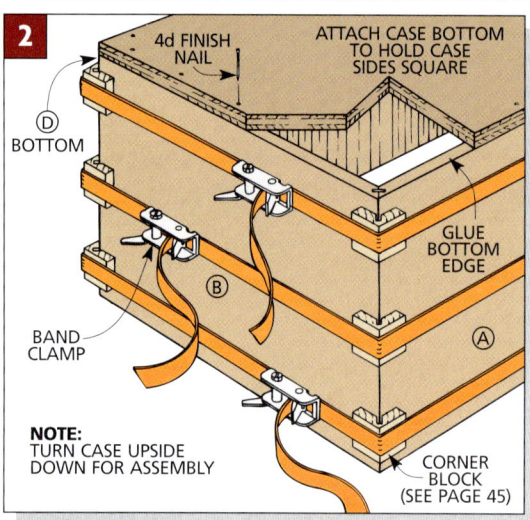

1

BACK Ⓐ

END Ⓑ

FRONT Ⓐ

11¾

20

36

24

Ⓐ Ⓐ

Ⓑ Ⓑ

Ⓓ

96

NOTE: CUT ALL FIVE PIECES FROM ONE 24" x 96" PIECE OF ¾" PLYWOOD

Ⓓ BOTTOM

2

4d FINISH NAIL

ATTACH CASE BOTTOM TO HOLD CASE SIDES SQUARE

Ⓓ BOTTOM

BAND CLAMP

Ⓑ

Ⓐ

GLUE BOTTOM EDGE

NOTE: TURN CASE UPSIDE DOWN FOR ASSEMBLY

CORNER BLOCK (SEE PAGE 45)

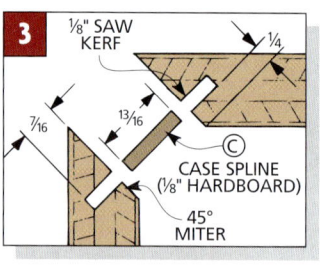

3

⅛" SAW KERF

¼

⁷⁄₁₆

¹³⁄₁₆

Ⓒ CASE SPLINE (⅛" HARDBOARD)

45° MITER

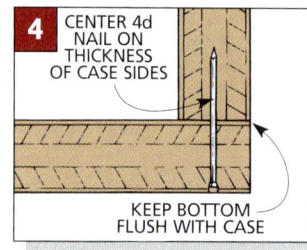

4

CENTER 4d NAIL ON THICKNESS OF CASE SIDES

KEEP BOTTOM FLUSH WITH CASE

TRUNK CASE

The main parts of the trunk case are cut from a half sheet (lengthwise) of ¾" plywood *(Fig. 1)*. I wanted continuous grain on the front, back, and sides.

CUT PIECES. Begin by cutting a double-wide piece for the case front and back (A) to rough size (24" x 37") *(Fig. 1)*.

Then cut another double-wide piece for the two ends (B) (24" x 21"). Now rip the double-wide pieces to get four pieces 11¾" wide *(Fig. 1)*.

MITER TO LENGTH. The case pieces are joined with splined miters *(Fig. 3)*.

To cut the joint, tilt the blade to 45° and cut a miter (bevel) on one end of all four pieces. Then miter the four pieces to finished lengths of 20" and 36" *(Fig. 1)*.

SPLINES. Next, cut a kerf for the splines in each mitered end *(Fig. 3* and the Shop Tip below). Then rip 11¾"-long splines (C) from ⅛" hardboard to fit in the kerfs. (For more on splined miters, see the Joinery article on page 44.)

ASSEMBLE CASE. Before gluing the case together, dry-assemble the sides to check the corners for square and to make sure they have tight joints *(Fig. 2)*. When you're satisfied the box is square, cut the bottom (D) to fit *(Fig. 1)*.

Note: The bottom fits *onto* the case sides *(Fig. 2)*. (The exposed edges are covered later with hardwood bands, refer to *Fig. 10a* on page 47.)

Now glue all four corners with splines in place, clamping the case square.

SHOP TIP . . . . *Miter Block*

To protect the mitered edges of a workpiece when cutting slots for splines, I use a block with an angled slot in it *(Fig. 1)*. The mitered edge fits in the slot so the wood fibers won't get crushed.

To make the block, rip a 2x4 to 1⅛" wide and cut it to length to match the mitered piece. Then, set the rip fence ⁹⁄₁₆" from the blade and cut a slot on the narrow edge of the block *(Fig. 2)*. Next, tilt the blade

to 45° and reset the rip fence *(Fig. 3)*. Run the block across the blade with the same face against the fence.

To cut the spline kerf in a mitered workpiece, fit the sharp edge of the mitered

panel into the slot *(Fig. 1)*. Then, put the block against the fence and set the blade to the right height *(Fig. 1a)*.

Next, cut the kerf in the workpiece with the block riding along the fence.

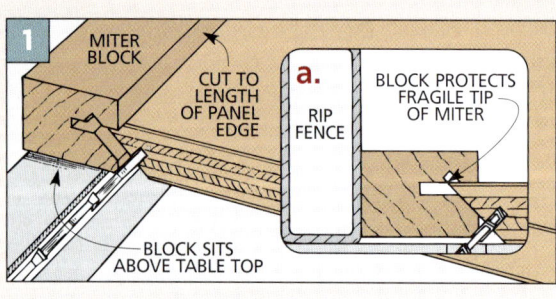

1 MITER BLOCK

CUT TO LENGTH OF PANEL EDGE

a. BLOCK PROTECTS FRAGILE TIP OF MITER

RIP FENCE

BLOCK SITS ABOVE TABLE TOP

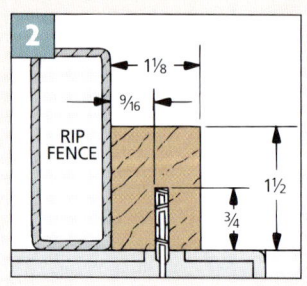

2 1⅛

⁹⁄₁₆

RIP FENCE

1½

¾

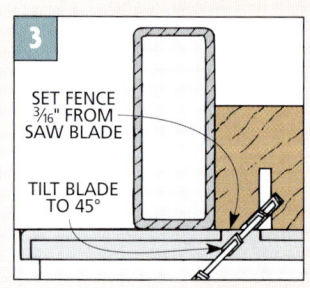

3 SET FENCE ³⁄₁₆" FROM SAW BLADE

TILT BLADE TO 45°

ATTACH BOTTOM. To keep the case square while the glue dries, attach the bottom with 4d (1½") finish nails *(Fig. 4)*.

CASE RIM

In order to hide the top edges of the plywood, the trunk case has a solid-wood rim *(Fig. 5)*. The trunk lid fits into a shallow rabbet cut around the upper outside edge of this rim *(Fig. 5b)*.

RIP AND RABBET. Begin making the case rim by ripping the rim front (E), back (E), and ends (F) to width (1¼") from ¾"-thick hardwood stock.

Before cutting the pieces to finished length, I cut the rabbets. To do this, first set the blade ¾" high *(Fig. 5a)*. Then set the rip fence to cut a ⅛"-deep rabbet on the outside face of each workpiece.

MITERS. When the pieces have been ripped to width and rabbeted, miter the

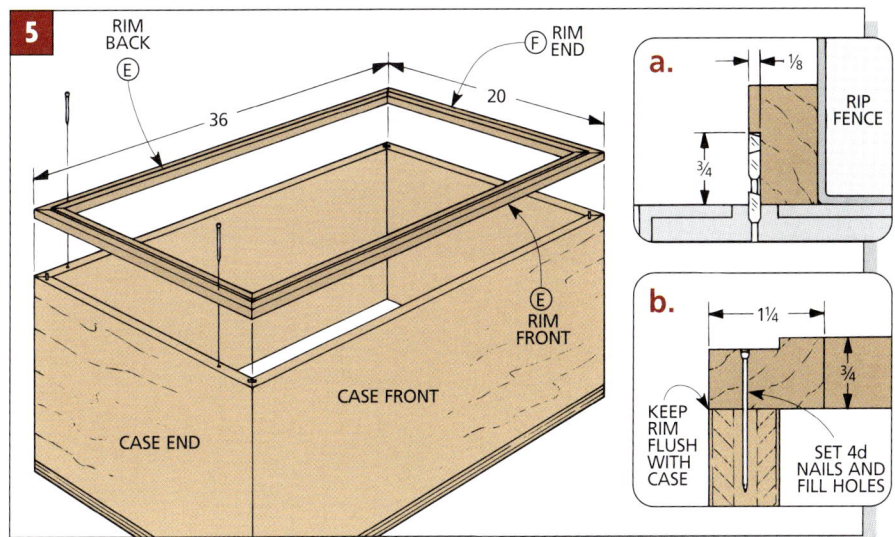

rim pieces to the same lengths as the case front, back, and ends *(Fig. 5)*. If it's not exact, you'll need to trim it flush later.

ATTACH RIM. Now glue and nail each piece of the rim onto the trunk case with the rabbet facing up and out *(Fig. 5b)*.

TECHNIQUE . . . . *Clean Cuts in Plywood*

When cross-cutting a plywood panel, the bottom layer of veneer often splinters out along the cut line. But there are some steps you can take to prevent it.

PLYWOOD BLADE. Perhaps the easiest way to avoid splintering is to use a blade that's made just for cutting plywood.

COMBINATION BLADE. But if you only have a combination blade, there are a few tricks you can use to get a clean cut. First, if the blade is crusted with sawdust or pitch, clean it thoroughly. Sometimes, however, even a clean combination blade will splinter the veneer.

There are two reasons for this. First, a combination blade has fewer teeth than a plywood blade, so it won't cut as cleanly.

The other reason is that the cutting edge of the teeth may be pushing the veneer down rather than slicing it off.

BLADE HEIGHT. One way to avoid this is to change the cutting angle of the teeth by raising or lowering the blade. If your panel is splintering on the bottom, lower the blade. If it's splintering on the top of the plywood, raise the blade.

SCORING CUT. The most common way to get a clean cut is to score the panel along the cut line before making the cut *(Fig. 1)*. To do this, cut through the veneer layer with a utility or other sharp knife. While this method works, it's sometimes difficult to line up the saw blade with the scored line.

SCORING ON THE SAW. An easier way to score the panel is to use the saw blade itself. The trick is to make the cut in two passes. On the first pass, set the blade just high enough to cut through the veneer *(Fig. 2a)*. Then raise the blade and finish the cut on the second pass. To keep the workpiece aligned with the blade during both cuts, clamp an extension fence with a stop block to your miter gauge *(Fig. 2)*.

BACKER BOARD. Another way to keep the veneer from splintering is to use a backer board *(Fig. 3)*. This is a piece of plywood or hardboard that's placed below the workpiece when making the cut. This way the veneer layer is supported and can be cut cleanly.

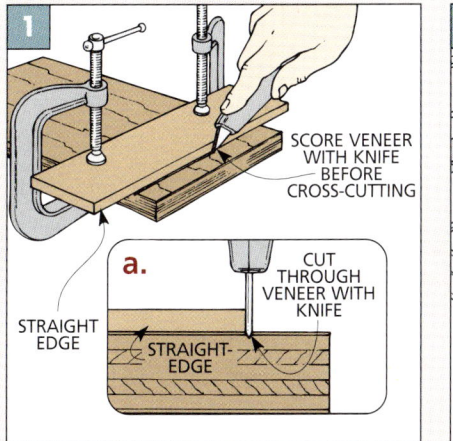

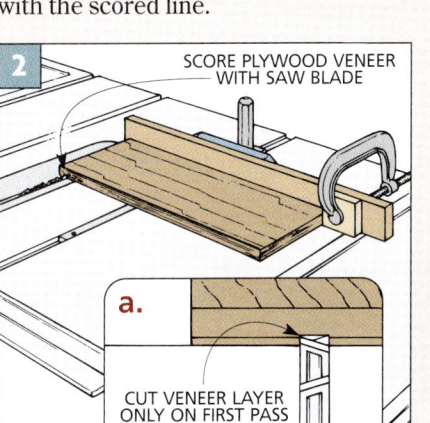

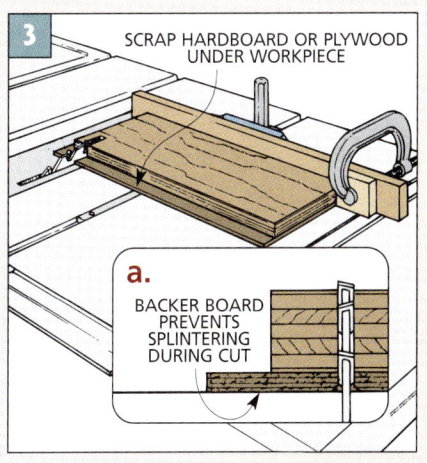

JOINERY . . . . *Splined Miters*

Miter joints are great for hiding end grain. When building the Steamer Trunk, miter joints are used on the four corners of the case to hide the ends of the plywood, and again on the side frame of the lid to hide the end grain.

Miter joints look like they're easy to cut and join together. Unfortunately, looks can be deceiving — especially when you're working with wide pieces.

CUTTING. The challenge is when you have to cut a perfect 45° bevel on the end of a wide workpiece. The trick here is to make a cut that's 90° to the adjacent edge over its entire length.

SPLINE. Cutting the miter is only half the problem. The other half is assembling it. I've found the best way to keep the miter from slipping out of alignment during clamping is to use splines.

A spline is a thin piece of wood that runs across the joint. It fits into kerfs cut in both sides of the joint (see photo).

There are a few advantages to adding a spline to a miter joint. First, since miters tend to slip out of alignment as you clamp

them together, the spline locks the joint while the clamps are tightened.

Second, the spline strengthens the joint by providing more glue surface.

PREPARATIONS. There are some steps to take before cutting a splined miter joint. First, to help support the workpiece over its entire length, and to have a surface to clamp a stop block to, I screw an auxiliary fence to my miter gauge and make sure the gauge is set 90° to the blade.

STOP BLOCK. Next, cut a stop block from a piece of scrap, and miter one end of the block at the same angle as the miter on the workpiece (*Step 3*).

ROUGH LENGTH. Before cutting a miter on large workpieces, it's a good idea to rough-cut all pieces so they're only about 1" longer than needed. This gives you enough length to work with, but reduces the workpiece to a manageable size.

TEST CUTS. Before making any finish cuts, it's a good idea to check the blade angle with test cuts on two pieces of scrap (*Step 1*). Check the angle with a try square (*Step 2*). If the joint is open at the heel (*detail 'a' in Step 2*), raise the angle of the blade (up from the table). If the joint is open at the point, lower the angle of the blade (down toward the table).

CUT PIECES. After the saw is set up, first cut miters on one end of all the workpieces using the auxiliary fence and stop block to hold the pieces in place (*Step 3*).

To cut the other end, mark the final length you want on the edge of one piece. Then align this mark with the blade.

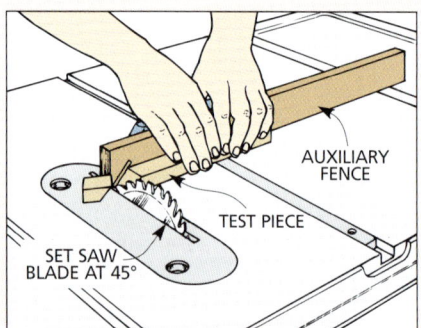

1 *Start with a test cut. Screw an auxiliary fence to the miter gauge and tilt the saw blade to 45°. Then make test cuts on the end of two pieces of scrap.*

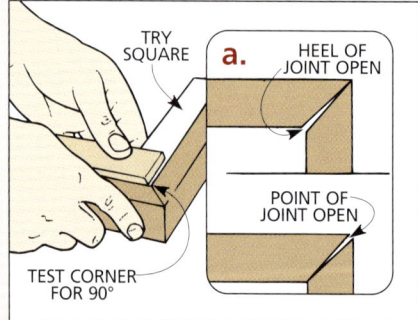

2 *To check the angle of the test cut, hold the pieces together around a try square. If heel is open, raise angle of blade from table. If point is open, lower the angle.*

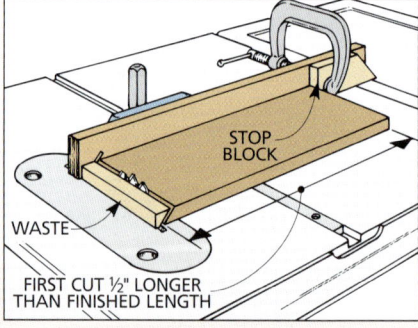

3 *After blade is set to cut exactly 45°, clamp a stop block to an auxiliary fence so blade will cut ½" longer than finished length. Then trim one end off workpiece.*

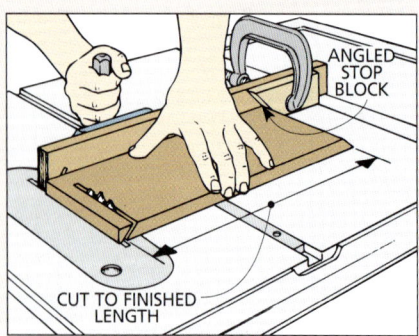

4 *Turn workpiece end for end and reset stop block to finished length. (Note angled stop block, see text.) Then trim off workpiece to finished length.*

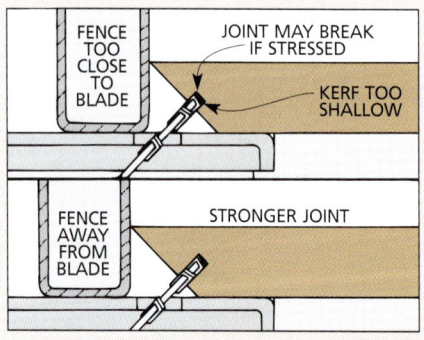

5 *To cut kerf for spline, lower the blade until it sticks above table about half the thickness of wood. Then position fence to act as a stop. Cut kerf near heel.*

6 *Make cut in test piece to check position and depth of kerf. Then cut kerf in all of the miters with the workpiece tight against the miter gauge and rip fence.*

Adjust the mitered end of the stop block at the end of the workpiece *(Step 4)*. (I always use a stop block rather than aligning to pencil marks to ensure the pieces are exactly the same length.)

KERFS. Once all of the miters are cut, the blade can be lowered (but left at the same angle) to cut kerfs for the splines. I use a carbide-tipped saw blade to cut $\frac{1}{8}$"-wide kerfs. Later, I cut the splines to fit.

When cutting the kerfs for the splines, use the rip fence as a stop for the mitered

end of the workpiece *(Step 5)*. Since you're not cutting through the piece, it's okay to use the rip fence and the miter gauge together. (To protect the point of the miter, see the Shop Tip on page 42.)

You want to position the fence so the kerf will be cut near the "heel" of the miter. In this position, the kerf can be deeper and thus the spline can be longer for more glue surface.

After the rip fence is positioned, cut a kerf in each miter *(Step 6)*.

MAKE SPLINES. Now splines are cut to fit the kerfs. If you cut the splines from solid wood, the grain must run perpendicular to the joint line *(Steps 7 and 8)*.

Cut the splines $\frac{1}{16}$" less than the depth of both kerfs, so the spline won't keep the joint from closing *(detail 'a' in Step 8)*.

And if the workpieces are wide (like the plywood for the trunk case), I use $\frac{1}{8}$" hardboard for the splines. If the ends of the splines will be exposed, they can be capped with solid-wood splines *(Step 9)*.

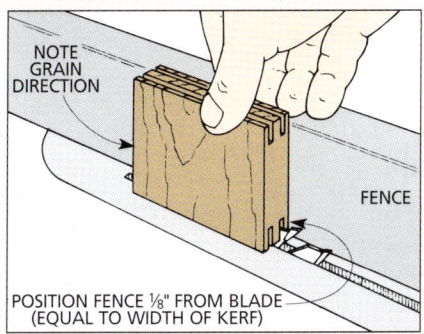

7 *To cut splines, first make four cuts in the edges of a wide piece of $\frac{3}{4}$" stock. Set distance from the blade to rip fence to equal width of kerf in the miter cut.*

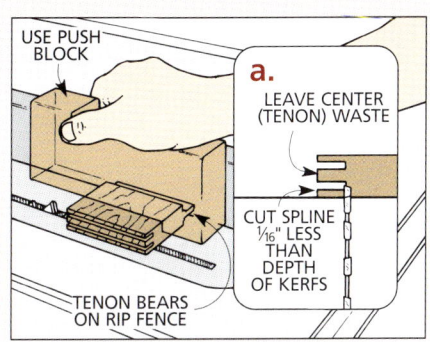

8 *Lower the blade until it's just high enough to cut off a spline. Now trim a spline off the waste side of the workpiece. Then cut splines off other three edges.*

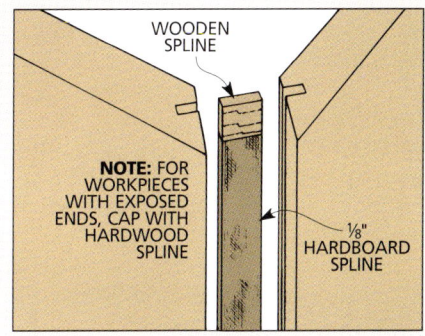

9 *For wide joints, $\frac{1}{8}$" hardboard can be used to make splines. Cut the hardboard shorter than the length of the miter and fill in ends with wooden splines.*

CLAMPING SPLINED MITER JOINTS

When I assembled the Steamer Trunk with splined miters, I used band clamps to clamp all four corners at the same time. A band clamp puts uniform pressure on all four joints, and directs the pressure toward the center of the project.

CLAMPING BLOCK. The only problem with band clamps is that they tend to round over the corners of the miter joint as the clamps are tightened. To prevent this, I use clamping blocks *(Fig. 1)*.

To make the clamping blocks, cut a large rabbet from one corner of a piece of 2x2 stock *(Steps 1 and 2 in Fig. 2)*. Then cut a relief kerf on the inside corner of the rabbet *(Step 3 in Fig. 2)*. This kerf allows space for glue squeeze-out so the blocks don't get glued to the project. It also protects the points of the miters when clamping *(Figs. 3 and 3a)*.

Finally, cut several $1\frac{1}{2}$"-long clamping blocks off the 2x2.

CARPET TAPE. It will seem like you need at least three hands to hold the clamping blocks in place while you try to tighten the band clamps. Since I only have two hands, I stuck a piece of carpet tape in the rabbet of the blocks to hold them in position over the corners before I positioned the band clamps *(Fig. 3a)*.

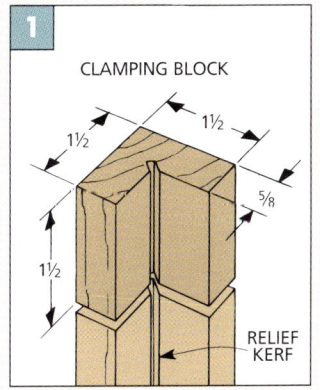

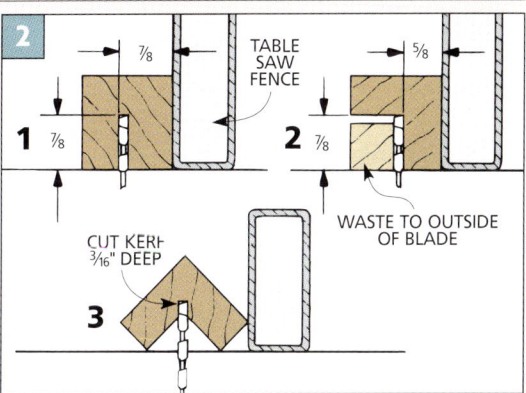

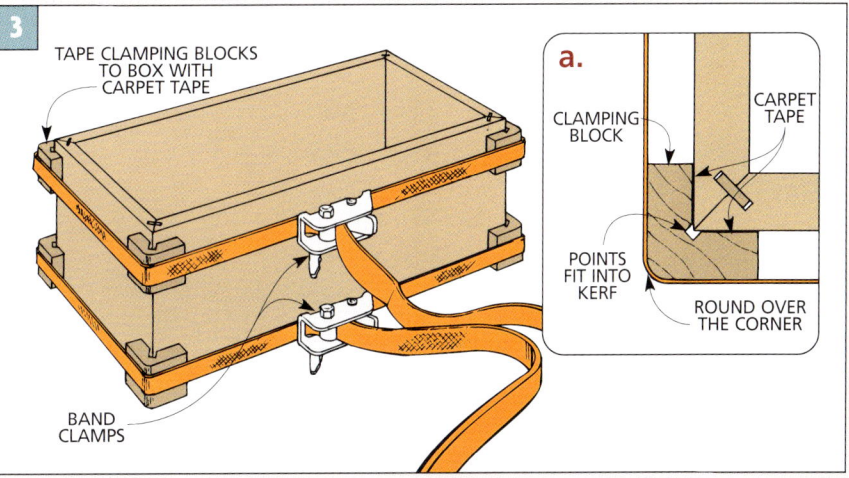

The lid of the trunk consists of a top frame that fits on a side frame *(Fig. 9)*.

SIDE FRAME. To make the lid, start by cutting the side frame pieces from ³/₄" solid wood to a width of 2¹/₄" *(Fig. 6b)*. Then cut the front and back pieces (G) to a rough length of 37", and the ends (H) to a rough length of 21".

Next miter all four pieces to the same size as the case rim *(Fig. 6)*. Then cut kerfs for the lid splines (I), and assemble the frame like you did the case *(Fig. 6a)*. (To keep the frame square, I assembled it on top of the case.)

TOP FRAME. The miters on the top frame are a little different than the miters on the rest of the trunk — they're cut across the face of the pieces *(Fig. 7)*. And because the mitered pieces will be glued down to the lid side frame, the joints don't need splines.

To make the top frame, first rip the top frame front (J), back (J), and ends (K) to a width of 2". Then miter them to length so they fit on top of the side frame pieces.

CENTER SLATS. Next, work on the center slats. These slats have ¹/₄"-long stub tenons that fit into grooves on the inside edges of the top frame *(Fig. 8)*.

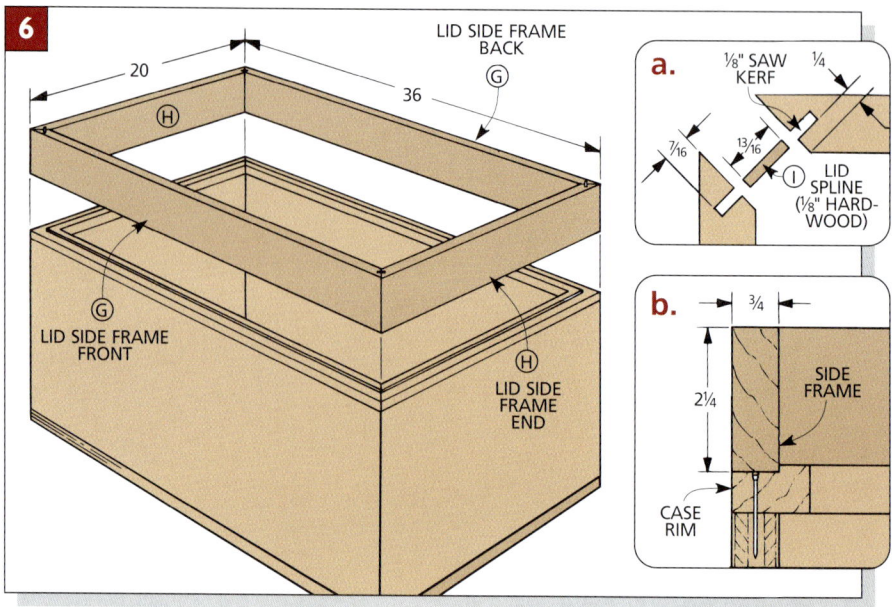

To make the slats (L), first cut two pieces of ³/₄" stock to a width of 2", and ¹/₂" longer than the inside length of the top frame (to allow for the tenons) *(Fig. 7)*.

GROOVES AND TENONS. Now cut grooves centered on the inside edge of each top frame section, and on both edges of the two center slats *(Fig. 7)*. Cut these grooves to width to hold the ¹/₄" plywood you will be using for the lid panels (M).

Note: Your plywood may be less than ¹/₄" thick. Be sure to cut the grooves to the exact thickness of your plywood.

Next, cut tenons on the ends of the center slats to fit in the groove *(Fig. 7a)*.

ROUNDOVER. To soften the edges of all the pieces, I routed ¹/₈" roundovers on three edges of the top frame pieces *(Fig. 8)*. Then round over all the edges of the center slats, and file over the ends.

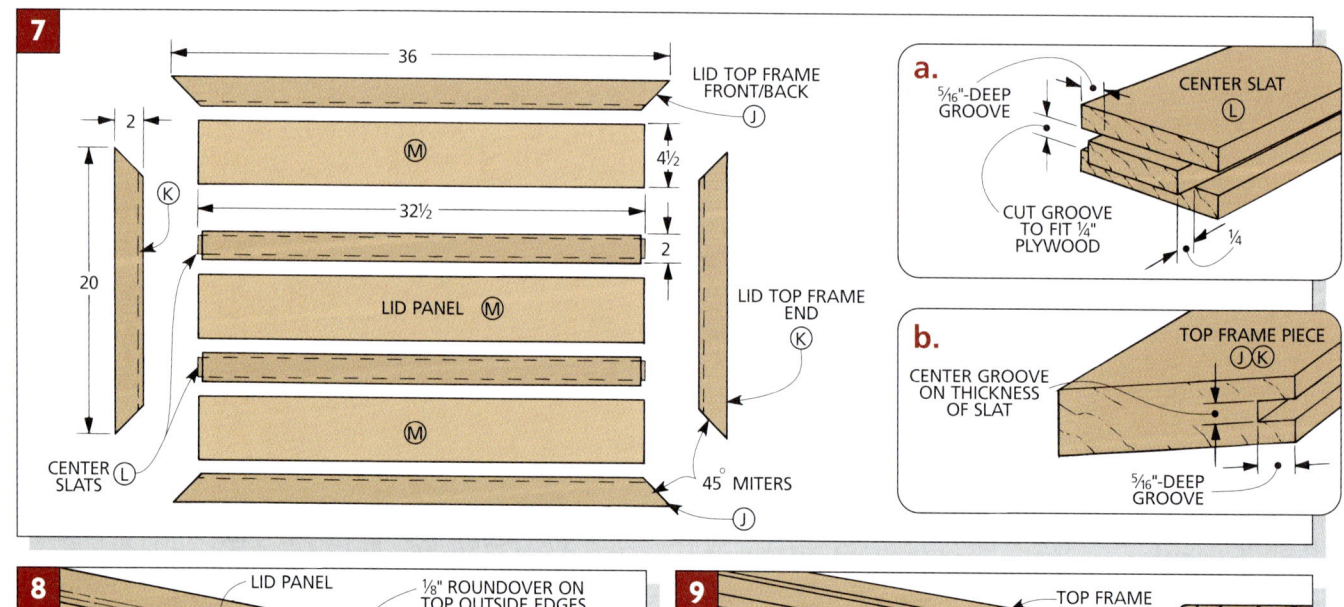

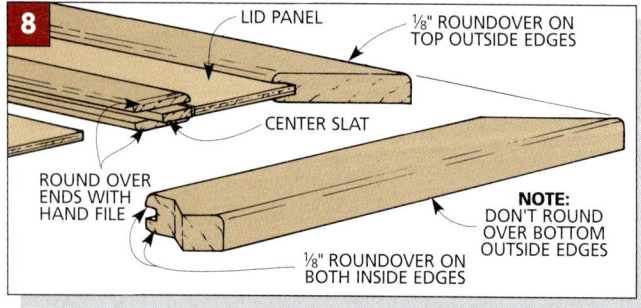

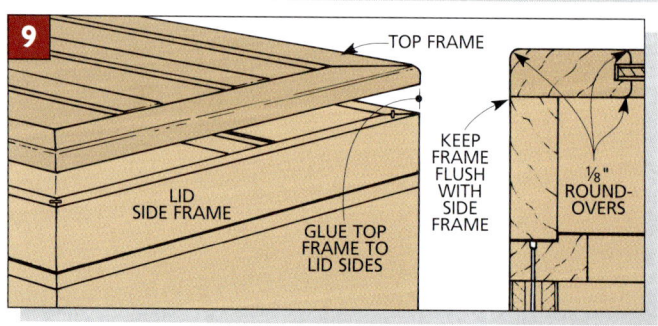

LID PANELS. Next you need to determine the width of the lid panels (M). To do this, first dry-assemble the top frame with the center slats in place. Then rip a piece of $\frac{1}{4}$" plywood into three equal-width panels ($4\frac{1}{2}$" for my lid). After that, cut the panels the same length as the center slats ($32\frac{1}{2}$") *(Fig. 7)*.

ASSEMBLY. With the panels cut to size, you're ready to glue and clamp together the top frame assembly. And once the glue dries, you can glue the top frame onto the side frame *(Fig. 9)*.

CASE BANDS

To make the trunk look more authentic, I added hardwood bands around the case *(Fig. 10)*. The $\frac{1}{4}$"-thick bands are resawn and planed from $\frac{3}{4}$" stock *(Fig. 10a)*.

CUT TO SIZE. To make the case bands, cut four 2"-wide end bands (N) the same length as the end of the case (20"). Then cut four 2"-wide front and back bands (O) so that they're $\frac{1}{2}$" longer than the trunk case ($36\frac{1}{2}$"). They should overlap the end bands *(Fig. 10)*.

ROUND OVER. After cutting the bands to length, rout the outside edges of each band with a $\frac{1}{8}$" roundover bit.

Note: Do not round over the inside edges, nor the ends of the end bands.

GLUE TO CASE. Now glue and clamp the bands to the trunk case, starting with the end bands *(Fig. 10)*. To make clamping the upper set of bands possible, I made a set of clamp extension blocks to use with my C-clamps *(Figs. 11 and 11a)*.

FINISH & HARDWARE

I applied a finish to the trunk before attaching the hardware.

Note: If you're building the cedar lining described in the Designer's Notebook on page 48, don't finish the cedar or you'll mask its aroma.

FINISH. To get the look of an "aged oak" trunk, I mixed a custom stain. To do this, start with a pint of boiled linseed oil, then add $1\frac{1}{2}$ tbsp. of burnt umber artists' oil color from an art supply store. (The two oils need a thorough mixing for the oil color to dissolve.)

After applying the stain, I wiped on two coats of a satin urethane finish.

HARDWARE. When the finish has completely dried, you can begin installing the trunk hardware.

Note: A complete hardware kit is available from *Woodsmith Project Supplies*.

See page 126 for ordering information and other sources.

The lid is attached with two stop hinges *(Fig. 12)*. On the front of the case, a pair of draw catches pull the lid tightly closed *(Fig. 13)*. You'll install the two-piece lock in the same manner as the draw catches *(Fig. 14)*.

A pair of lid stays inside the trunk prevents the lid from opening too far. I screwed one end of each stay to the inside of the case rim, and the other end to the top frame *(Fig. 15)*.

Attach the russet-colored leather handles to the ends of the trunk using two handle loops *(Fig. 16)*.

Finally, to complete the trunk's traditional look, I screwed on case corners and corner clamps *(Fig. 16)*. ■

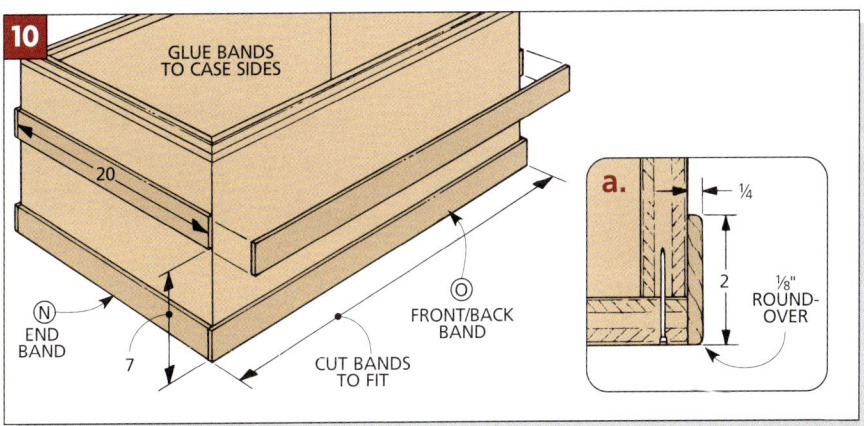

10 GLUE BANDS TO CASE SIDES

20

(N) END BAND

(O) FRONT/BACK BAND

7

CUT BANDS TO FIT

a. $\frac{1}{4}$ 2 $\frac{1}{8}$" ROUND-OVER

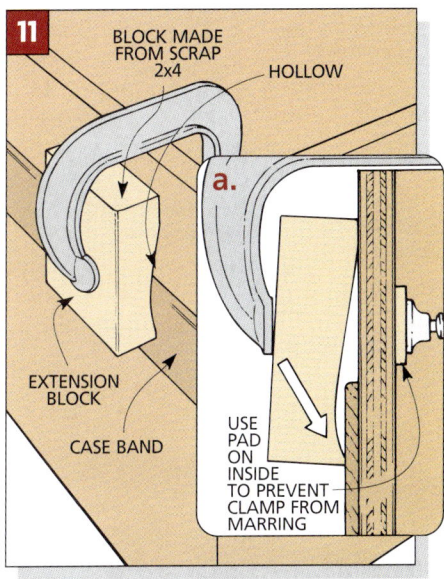

11 BLOCK MADE FROM SCRAP 2x4 HOLLOW

EXTENSION BLOCK

CASE BAND

a. USE PAD ON INSIDE TO PREVENT CLAMP FROM MARRING

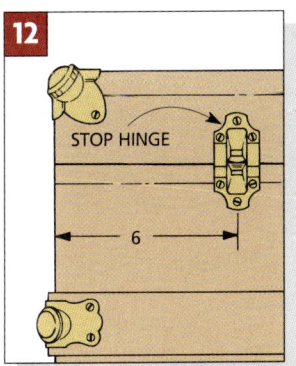

12 STOP HINGE

6

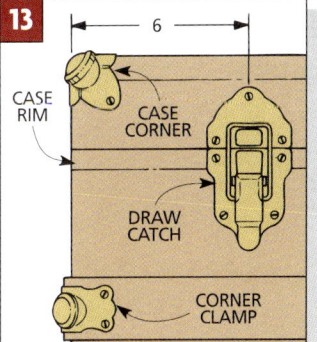

13 6

CASE RIM

CASE CORNER

DRAW CATCH

CORNER CLAMP

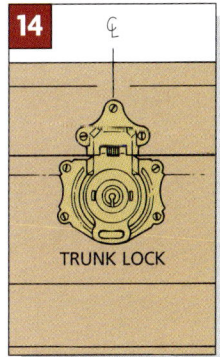

14 ℄

TRUNK LOCK

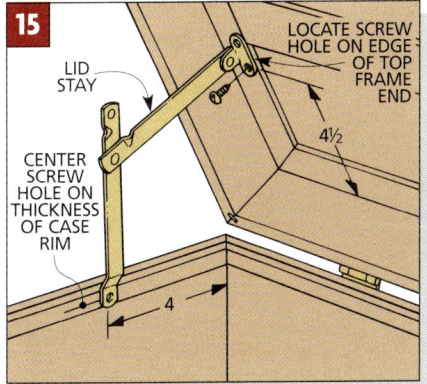

15 LID STAY

LOCATE SCREW HOLE ON EDGE OF TOP FRAME END

$4\frac{1}{2}$

CENTER SCREW HOLE ON THICKNESS OF CASE RIM

4

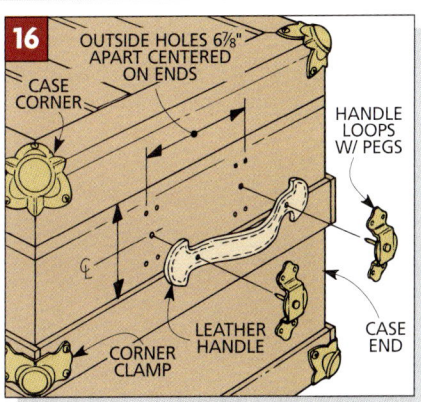

16 OUTSIDE HOLES $6\frac{7}{8}$" APART CENTERED ON ENDS

CASE CORNER

℄

HANDLE LOOPS W/ PEGS

CORNER CLAMP

LEATHER HANDLE

CASE END

DESIGNER'S NOTEBOOK

As it is, the trunk is just a hinged box. You can give the interior a finished look by adding a lining of aromatic cedar to keep moths away. Then build a tray to match the trunk and to hold smaller items.

CONSTRUCTION NOTES:

■ Even after the trunk has been completed, you still have a couple of options for customizing it — a cedar lining and a lift-out tray. It's easy to add one or both of these options.

Trunks and storage chests are often lined with aromatic cedar to keep moths and other insects out of the stored items. Installing a cedar lining is easy. And every time you open the lid, you'll get to enjoy that wonderful fragrance.

In addition, you can build a lift-out tray that sits just below the lid. Its purpose is to organize smaller keepsakes and prevent them from getting lost under a pile of sweaters or blankets.

CEDAR LINING & TRAY

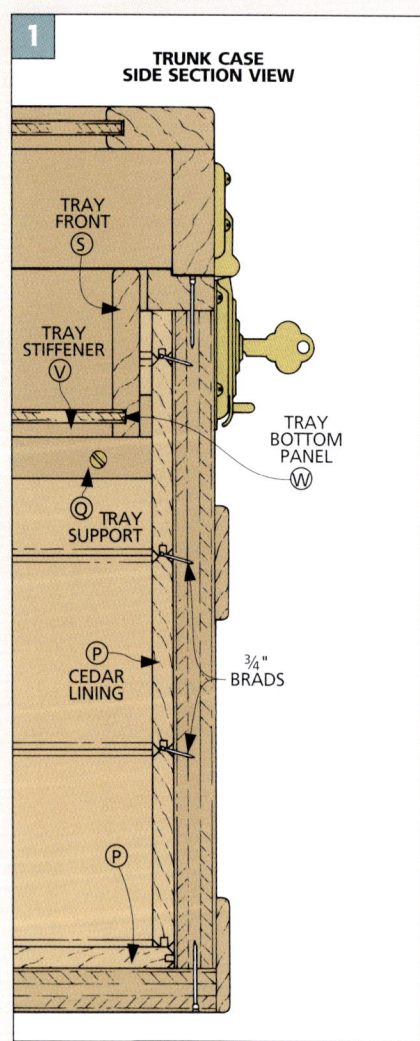

1 TRUNK CASE SIDE SECTION VIEW

TRAY FRONT (S)

TRAY STIFFENER (V)

TRAY BOTTOM PANEL (W)

TRAY SUPPORT (Q)

CEDAR LINING (P)

3/4" BRADS

(P)

(P)

■ To line the trunk with cedar, I bought 3/8"-thick tongue-and-groove aromatic cedar closet lining (P). (You can find this at most home centers.) The package I used contained about 15 square feet of cedar — just right, since the inside of the trunk has a surface area of about 13 square feet. (I didn't line the lid. That would make it too heavy.)

■ The lining is simply nailed to the case using 3/4"-long brads *(Figs. 1 and 2a)*. Do not use glue. The cedar needs to expand and contract with changes in humidity.

■ There are a couple of tricks that can help you with the installation. First, drive the brads at an angle into the tongues of

the lining *(Figs. 1 and 2a)*. Then sink the heads flush or slightly below the surface of the tongue so the groove in the next piece will fit over the tongue.

■ Another trick is the sequence in which each piece is attached. I found the best appearance by nailing lining to the bottom of the trunk first. Second, I worked on the front and back of the case. Then, I finished off by lining the ends *(Fig. 2)*.

Note: You'll likely have to rip a piece to fit at the top of the case *(Fig. 2)*.

■ Also, to get the most out of one package of cedar, use the short cut-off pieces. For the nicest-looking effect, stagger the cut-offs between full-length pieces *(Fig. 2)*.

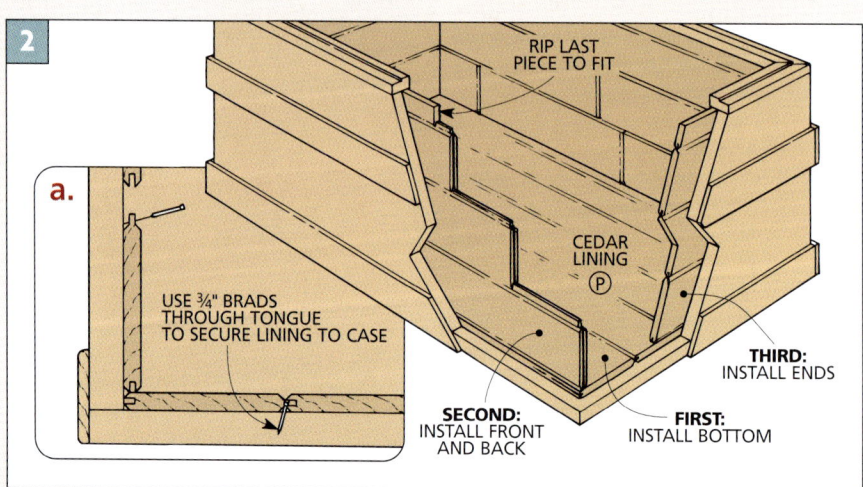

2

RIP LAST PIECE TO FIT

a.

USE 3/4" BRADS THROUGH TONGUE TO SECURE LINING TO CASE

CEDAR LINING (P)

THIRD: INSTALL ENDS

SECOND: INSTALL FRONT AND BACK

FIRST: INSTALL BOTTOM

Note: After time, the cedar fragrance may taper off. A light sanding will restore the aroma to full strength.

■ That's all there is to installing the lining. If you want to add the lift-out tray, begin by ripping two tray supports (Q) to a width of 1" from ³⁄₄"-thick stock.

Note: If you didn't add the cedar lining to the trunk, cut the supports 1³⁄₈" wide.

■ Then cut the supports to length so they fit tightly across the ends *(Fig. 3)*. Now screw each support in place with brass woodscrews *(Figs. 3 and 3a)*.

■ To allow clearance between the lid stay hardware and the tray, and to center the tray left-to-right, I added two tray spacers (R) *(Fig. 3)*. Cut the ³⁄₄"-thick spacers to a width of ¹⁄₂" (or ⁷⁄₈" wide if the trunk is unlined). Then cut them to the same length as the tray supports. Finally, glue the spacers in place *(Fig. 3a)*.

■ Next, work can begin on the lift-out tray. The tray consists of a hardwood frame with splined miter corners, and a hardwood "stiffener" that separates two plywood panels that make up the bottom of the tray *(Fig. 4)*.

■ I began making the tray frame by ripping the tray front (S) and back (S) from ¹⁄₂"-thick stock to a finished width of 3" *(Fig. 4)*. The tray ends (T) are 2" wider (5" wide) to allow for the handles that are cut in them later *(Fig. 5)*.

■ The pieces are cut to length so the tray will lift out of the trunk easily. Miter the tray front and back so they're 1" shorter than the inside length of the trunk rim. (This allows room for the lid stay hardware.) Then miter the tray ends ¹⁄₈" less than the inside width of the trunk rim.

■ Next, cut ³⁄₁₆"-deep kerfs into each mitered end, and make hardwood tray splines (U) to fit the kerfs *(Fig. 4a)*.

■ At this point, you're ready to make the handles. First lay out the profile of the handle on one tray end *(Fig. 5)*.

■ I fastened both tray ends together face to face with carpet tape so I could cut out both profiles at once. Using a jig saw or a band saw, cut around the outside of the handle. Then bore two ¹⁄₂" holes and complete the inside with a jig saw.

■ When both handles have been shaped, round over the top edges and the inside of the cut-outs with a ¹⁄₈" roundover bit.

■ Make the tray stiffener (V) by cutting a piece of ³⁄₄" stock to a width of 2". Then cut it ¹⁄₂" longer than the inside width of the tray frame *(Fig. 6)*. (This allows for a tenon on each end.)

■ The plywood tray panels fit into grooves in the stiffener and the tray frame. Cut these grooves along the bottom edge of each frame piece *(Fig. 4b)* and centered on both sides of the stiffener *(Fig. 6)*.

■ Now cut ¹⁄₄"-long stub tenons on both ends of the tray stiffener *(Fig. 6)*. (Cut these to thickness so they fit in the panel grooves.) Then rout ¹⁄₈" roundovers around the top and bottom edges of the tray stiffener *(Fig. 6)*.

■ With the tenons cut on the stiffener, cut two equal-size tray bottom panels (W) from ¹⁄₄" plywood. Cut the panels to size so they fit in the grooves of the tray.

■ With all the tray parts cut, I glued the tray together with the splines, panels, and stiffener in place. Then clamp the unit square until the glue dries.

MATERIALS LIST

NEW PARTS

P	Cedar Lining	³⁄₈ x 15 sq. ft.
Q	Tray Supports (2)	³⁄₄ x 1 - 18
R	Tray Spacers (2)	¹⁄₂ x ³⁄₄ - 18
S	Tray Front/Back (2)	¹⁄₂ x 3 - 32¹⁄₂
T	Tray Ends (2)	¹⁄₂ x 5 - 17³⁄₈
U	Tray Splines (4)	¹⁄₈ x 3 - ⁵⁄₁₆
V	Tray Stiffener (1)	³⁄₄ x 2 - 16⁷⁄₈
W	Tray Btm. Panels (2)	¹⁄₄ ply - 16⁷⁄₈ x 15¹⁄₂

HARDWARE SUPPLIES

(6) No. 6 x 1³⁄₄" Fh brass woodscrews
(1 box) ³⁄₄" brads

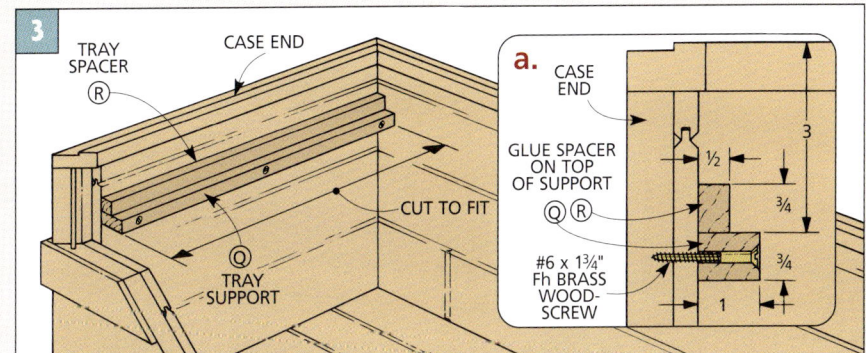

3 TRAY SPACER (R) · CASE END · CUT TO FIT · TRAY SUPPORT (Q)

a. CASE END · GLUE SPACER ON TOP OF SUPPORT (Q) (R) · #6 x 1³⁄₄" Fh BRASS WOODSCREW · 3 · ¹⁄₂ · ³⁄₄ · ³⁄₄ · 1

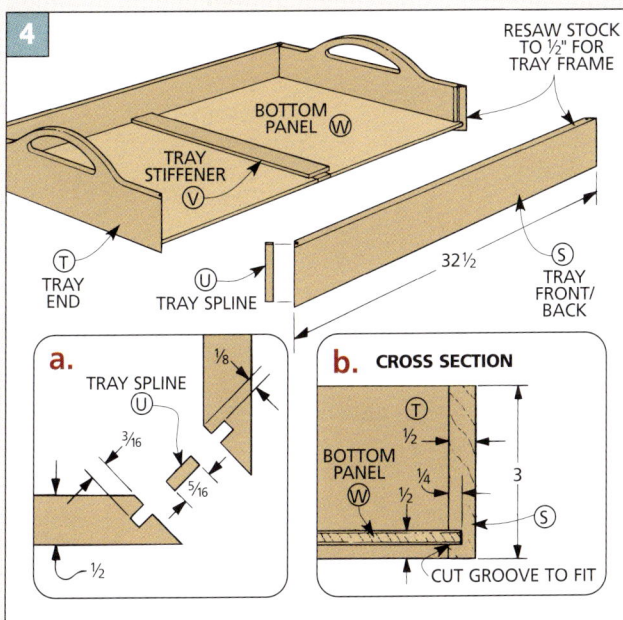

4 RESAW STOCK TO ¹⁄₂" FOR TRAY FRAME · BOTTOM PANEL (W) · TRAY STIFFENER (V) · TRAY END (T) · TRAY SPLINE (U) · 32¹⁄₂ · TRAY FRONT/BACK (S)

a. TRAY SPLINE (U) · ¹⁄₈ · ³⁄₁₆ · ⁵⁄₁₆ · ¹⁄₂

b. CROSS SECTION · (T) · ¹⁄₂ · BOTTOM PANEL (W) · ¹⁄₄ · ¹⁄₂ · 3 · (S) · CUT GROOVE TO FIT

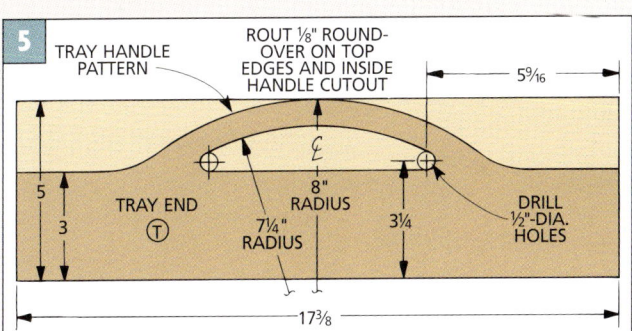

5 TRAY HANDLE PATTERN · ROUT ¹⁄₈" ROUNDOVER ON TOP EDGES AND INSIDE HANDLE CUTOUT · 5⁹⁄₁₆ · TRAY END (T) · 8" RADIUS · 7¹⁄₄" RADIUS · 3¹⁄₄ · DRILL ¹⁄₂"-DIA. HOLES · 5 · 3 · 17³⁄₈

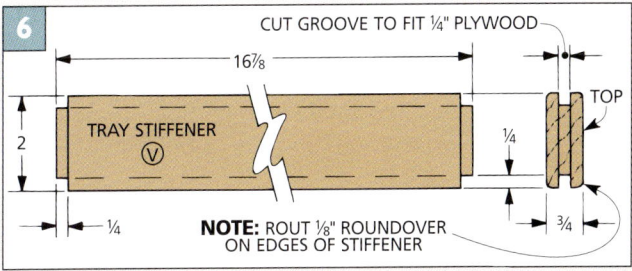

6 CUT GROOVE TO FIT ¹⁄₄" PLYWOOD · 16⁷⁄₈ · TRAY STIFFENER (V) · TOP · 2 · ¹⁄₄ · ³⁄₄ · ¹⁄₄ · **NOTE:** ROUT ¹⁄₈" ROUNDOVER ON EDGES OF STIFFENER

Jewelry Chest

When you need a curved piece of plywood, it may be easier to start from scratch and make your own than to bend a store-bought piece. The simple secret to shop-made plywood may surprise you.

Bending plywood can be tricky. For this Jewelry Chest, I wanted to avoid the extra work that steam or kerf bending requires. So I decided to *make* a piece of curved plywood instead. (It's actually easier than it sounds.)

Another reason for making my own plywood was for appearance. Most hardwood plywood has one good face, and one face covered with a lower grade veneer. But I wanted the lid of this chest to look as good open as when it's closed.

LID. To make my own curved plywood for the lid, I started with a piece of walnut veneer for the inside and outside faces. Then I experimented with a number of different materials for the core. I ended up using the least expensive grade of "wood" I could find — posterboard.

One nice thing about using posterboard is that it bends easily without breaking. And when glued into "plies," posterboard is surprisingly strong.

BACK AND SIDE PANELS. After buying the veneer for the lid, I realized there was going to be enough left over to veneer the side and back panels for the case. But they aren't made the same as the lid.

Since the sides and back aren't curved, I glued the veneer to hardboard instead. And because the inside faces aren't visible, I only veneered the outside face.

SATIN LINER. The lid tray and drawers in my chest are lined (see inset photo). You could leave yours unlined, but it's easy to do in just a few steps. First, I cut a piece of mat board slightly undersize to fit each drawer. Then I covered the mat board with batting and satin — wrapping the satin over the edges.

EXPLODED VIEW

OVERALL DIMENSIONS:
12¼"W x 8¼"D x 13¼"H

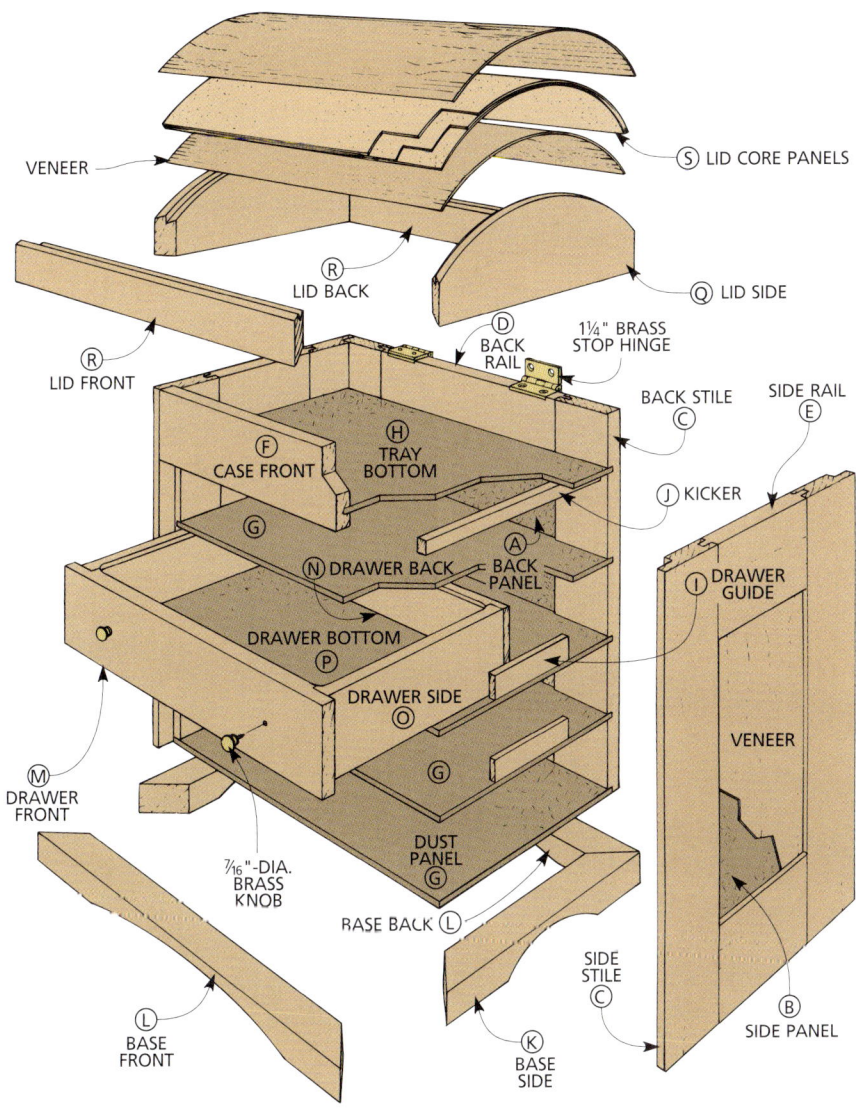

VENEER

Ⓢ LID CORE PANELS

Ⓡ LID BACK

Ⓡ LID FRONT

Ⓠ LID SIDE

Ⓓ BACK RAIL

1¼" BRASS STOP HINGE

BACK STILE Ⓒ

SIDE RAIL Ⓔ

Ⓕ CASE FRONT

Ⓗ TRAY BOTTOM

Ⓙ KICKER

Ⓖ

Ⓝ DRAWER BACK

Ⓐ BACK PANEL

Ⓘ DRAWER GUIDE

DRAWER BOTTOM

Ⓟ

DRAWER SIDE

Ⓞ

VENEER

Ⓜ DRAWER FRONT

⁷⁄₁₆"-DIA. BRASS KNOB

DUST PANEL Ⓖ

Ⓖ

BASE BACK Ⓛ

Ⓛ BASE FRONT

SIDE STILE Ⓒ

Ⓚ BASE SIDE

SIDE STILE Ⓒ

Ⓑ SIDE PANEL

MATERIALS LIST

WOOD

A	Back Panel (1)	⅛ hdbd. - 8 x 6½
B	Side Panels (2)	⅛ hdbd. - 4½ x 6½
C	Back/Side Stiles (6)	½ x 2 - 10
D	Back Rails (2)	½ x 2 - 8
E	Side Rails (4)	½ x 2 - 4½
F	Case Front (1)	½ x 2 - 11½
G	Dust Panels (4)	⅛ hdbd. - 7³⁄₁₆ x 11⅜
H	Tray Bottom (1)	⅛ hdbd. - 7⅜ x 11⅜
I	Drawer Guides (6)	³⁄₁₆ x ½ - 4
J	Kickers (2)	¼ x ⅜ - 7
K	Base Sides (2)	¾ x 1 - 8¼
L	Base Front/Bk. (2)	¾ x 1 - 12¼
M	Drawer Fronts (4)	¾ x 1¹⁵⁄₁₆ - 11⅜
N	Drawer Backs (4)	½ x 1¹³⁄₁₆ - 10⅞
O	Drawer Sides (8)	½ x 1¹³⁄₁₆ - 6⅝
P	Drawer Btms. (4)	⅛ hdbd. - 6⅝ x 10⅜
Q	Lid Sides (2)	½ x 2½ - 8
R	Lid Front/Back (2)	½ x 1¾ rgh. - 12
S	Lid Core Panels (3)	10-ply posterboard - 9¼ x 12½ rough

HARDWARE SUPPLIES

(1) 24" x 24" walnut veneer
(8) ⁷⁄₁₆"-dia. brass knobs
(2) 1¼" brass stop hinges w/ screws
(1) ¹⁄₁₆"-thick mat board (12" x 14")

CUTTING DIAGRAM

½ x 4½ - 60 WALNUT (1.9 Sq. Ft.)

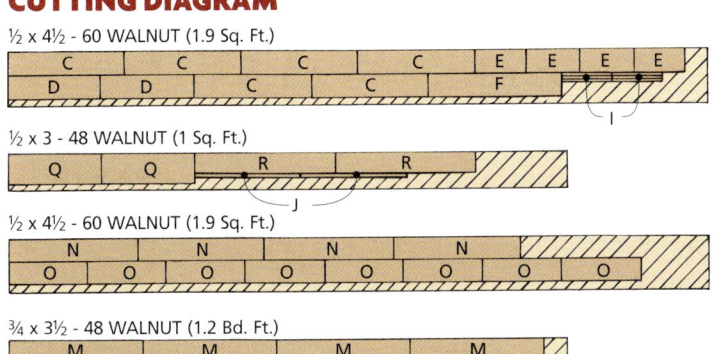

C C C C E E E E
D D C C F
I

½ x 3 - 48 WALNUT (1 Sq. Ft.)

Q Q R R
J

½ x 4½ - 60 WALNUT (1.9 Sq. Ft.)

N N N N
O O O O O O O O

¾ x 3½ - 48 WALNUT (1.2 Bd. Ft.)

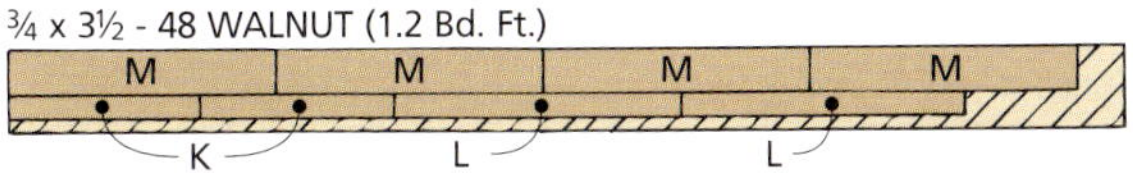

M M M M
K L L

⅛" TEMPERED HARDBOARD - 24 x 48

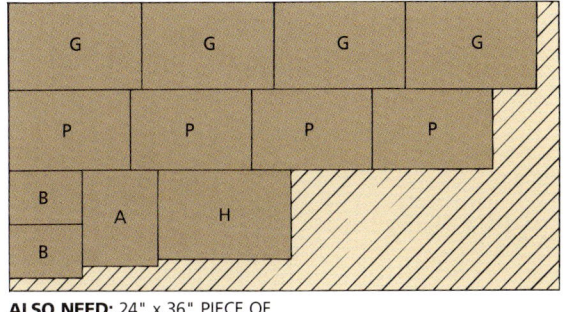

G G G G
P P P P
B
B A H

ALSO NEED: 24" x 36" PIECE OF 10-PLY POSTERBOARD FOR PART S

CASE

The sides and back of this Jewelry Chest are framed panels with stub tenon and groove joinery holding the frame pieces together. When building framed panels like these, you need to know how thick the panels are going to be before cutting the joinery. Since the panels were going to be veneered, I started with them. Then I moved on to the frame.

VENEERED PANELS. Begin by cutting one back panel (A) and two side panels (B) to finished dimensions *(Fig. 1)*. (For these, I used ⅛"-thick hardboard.)

Then to cover each piece of hardboard, cut a piece of veneer slightly larger than the panel. I used flexible veneer (see Sources on page 126).

After the veneer is glued in place, trim it flush with the edges of the panel. (For more on gluing and trimming veneer, see the Shop Tip on the next page.)

FRAMES. With the veneered panels complete, the stiles and rails of the frames can be cut to fit around the panels. The first thing to do is to cut the back and side stiles (C) and the back and side rails (D, E) to finished size *(Fig. 1)*.

Note: I cut all of the hardwood pieces (except the thicker drawer fronts and base pieces) from ½"-thick walnut.

Next, to hold the panels in the frames, cut a ¼"-deep groove centered on the inside edge of each stile and rail *(Fig. 2)*.

After the grooves are cut, stub tenons can be cut on the ends of each rail *(Fig. 2)*. Then all three frame and panel units can be glued up and assembled.

RABBET SIDE FRAMES. Now the side frames are rabbeted to accept the back panel and a ½"-thick case front. These rabbets are cut on the inside edges of each side assembly *(Fig. 3)*.

CASE FRONT. With the back and side assemblies complete, rip a case front (F) to width *(Fig. 3)*. Then cut it to length to match the width of the back frame.

DUST PANELS AND TRAY BOTTOM. The chest is divided with four dust panels (creating four drawer openings) and a tray bottom. The tray bottom and three of the dust panels fit in grooves cut inside the case. The fourth dust panel fits in a rabbeted edge cut along the bottom of the case *(Figs. 3 and 4)*.

I cut the ³⁄₁₆"-deep grooves and rabbets with one pass over the blade.

Note: Cut the grooves and rabbets on the inside faces of the sides and back. Also, cut a groove on the inside face of the case front (F) *(Fig. 3)*.

To find the size of the dust panels (G) and tray bottom (H), I dry-assembled the case and measured the opening, adding the depth of the grooves *(Fig. 4)*. Then I cut the pieces to size.

CASE ASSEMBLY. Next, glue and clamp the case with the dust panels and the tray bottom in place *(Fig. 4)*.

Note: I used a spacer block to help keep the case square. This spacer is cut to the same length as the case front.

DRAWER GUIDES AND KICKERS. To prevent the drawers from catching the side panels as they're pushed in, I installed two drawer guides (I) in the top three drawer compartments *(Fig. 4)*. After the guides are cut to fit between the stiles in the side panels, glue them in place.

To keep the top drawer from "kicking up" when it's pulled out, glue two kickers (J) to the tray bottom *(Fig. 4)*.

Note: Glue the kickers to the tray bottom only, not to the sides. This will allow the sides to expand and contract with changes in humidity.

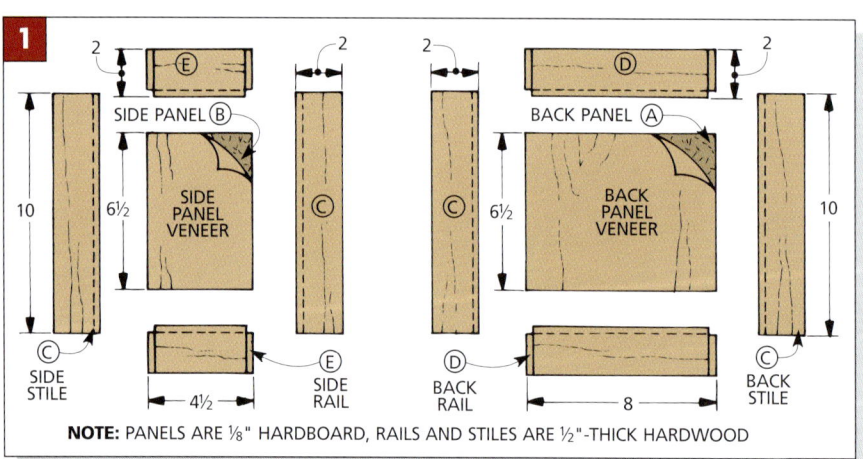

1

SIDE PANEL (B)

SIDE PANEL VENEER

BACK PANEL (A)

BACK PANEL VENEER

10 — 6½ — 10

SIDE STILE (C)

SIDE RAIL (E) — 4½

BACK RAIL (D) — 8

BACK STILE (C)

NOTE: PANELS ARE ⅛" HARDBOARD, RAILS AND STILES ARE ½"-THICK HARDWOOD

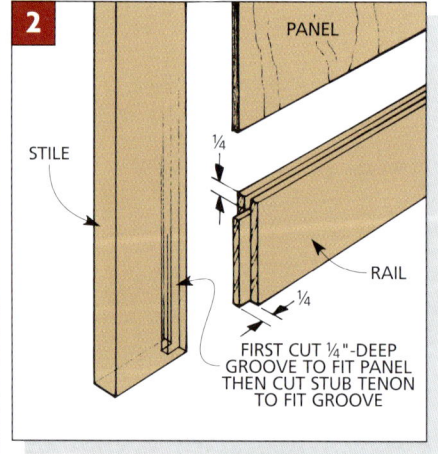

2

PANEL

STILE

RAIL

¼ ¼

FIRST CUT ¼"-DEEP GROOVE TO FIT PANEL THEN CUT STUB TENON TO FIT GROOVE

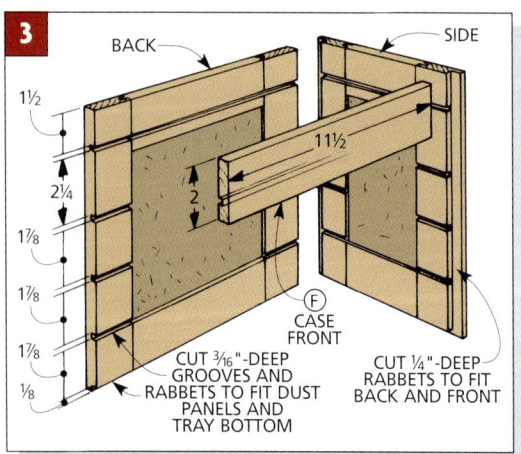

3

BACK

SIDE

1½
2¼
1⅞
1⅞
1⅞
⅛

11½

2

CASE FRONT (F)

CUT ³⁄₁₆"-DEEP GROOVES AND RABBETS TO FIT DUST PANELS AND TRAY BOTTOM

CUT ¼"-DEEP RABBETS TO FIT BACK AND FRONT

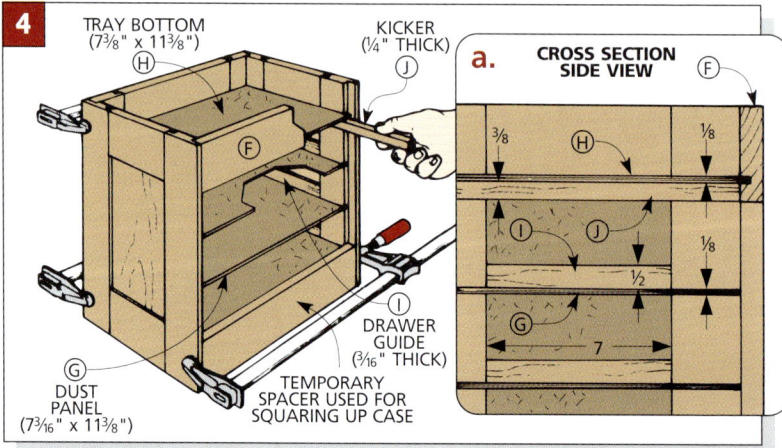

4

TRAY BOTTOM (7⅜" x 11⅜") (H)

KICKER (¼" THICK) (J)

F

DUST PANEL (7³⁄₁₆" x 11⅜") (G)

TEMPORARY SPACER USED FOR SQUARING UP CASE

a. CROSS SECTION SIDE VIEW (F)

⅜ (H) ⅛

(I) (J) ⅛

½

DRAWER GUIDE (³⁄₁₆" THICK) (I)

(G) 7

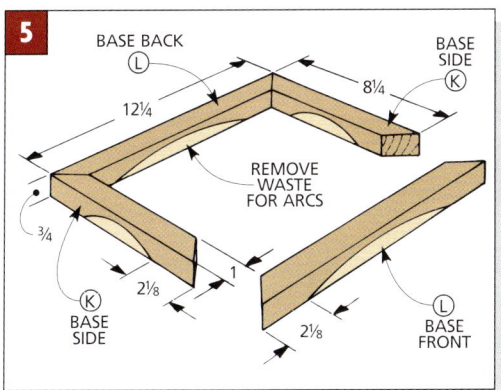

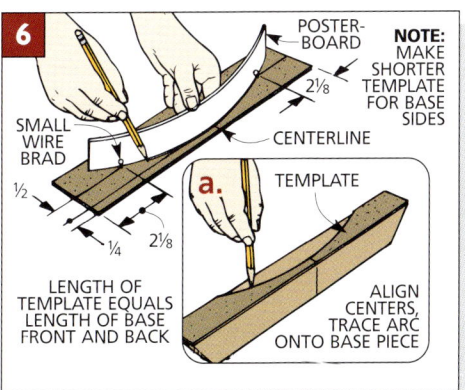

BASE

With the case complete, the next step is to add a ³⁄₄"-thick base under the chest. This consists of four pieces mitered to length to form a small lip below the chest.

BASE PIECES. To begin on the base, first rip enough stock to finished width (1") for all the base pieces *(Fig. 5)*.

Next, measure the width and depth of the case (near the bottom dust panel). Then add ¹⁄₄" to each measurement to determine the length of the base pieces. Now miter the base sides (K), front (L), and back (L) to finished length *(Fig. 5)*.

ARCS. At this point, the base could be glued to the bottom of the case. But to add a decorative touch and allow some finger space for picking up the chest, I cut an arc on each base piece.

To lay out the arcs, I made two templates out of hardboard. First, cut one piece of hardboard to the same length as the base front and back *(Fig. 6)*. Then cut another piece of hardboard to equal the length of the base sides.

Now a shallow arc can be laid out on each template. To do this, first draw a centered line along the length of the template *(Fig. 6)*. Then nail two small wire brads on this line and into a piece of scrap to help you hold the template.

Next, draw a centerline perpendicular to the first line, between the brads, and make a mark ¹⁄₄" from the top edge. Bend a flexible straightedge (such as a piece of posterboard) from the two brads to this mark and draw the arc. Then cut out the templates and trace the arcs onto the base pieces *(Figs. 5 and 6a)*.

ASSEMBLY. After the arcs have been cut and sanded smooth, you can glue and clamp the base pieces together. Then glue the base to the bottom of the case and the bottom dust panel *(Fig. 7)*.

SHOP TIP . . . . *Applying Veneer*

When veneering a project, I typically use flexible veneer. (Other types are available, but they don't work as well.)

Flexible veneer has a paper backing that keeps the thin hardwood veneer from cracking as it's rolled onto a project. (Don't try to remove this paper.)

Before applying veneer, clean up the substrate. For the best bond, the substrate must be smooth and free of voids. If there are any voids or gaps in the surface, use a wood filler to level them out.

After applying a filler, sand the surface smooth. Then, remove any dust by wiping the surface with denatured alcohol.

Now, cover both the veneer and the substrate with two coats of non-flammable, solvent-based contact cement.

After the cement dries (in about 15 minutes), the veneer is ready to be applied to the substrate.

But a word of caution. As soon as the two cemented surfaces touch, they're stuck for good.

So, to avoid premature sticking, first cover the dried substrate with a sheet of waxed paper. Then position the veneer.

When the veneer is down, slowly pull out the waxed paper *(Fig. 1)*. As you're removing the paper, flatten the veneer with a roller. This improves the glue bond and squeezes out air bubbles.

After the waxed paper is removed, I roll out the veneer again, starting in the center and rolling towards the edges.

After rolling out the bubbles, the veneer can be trimmed to the edges of the workpiece with a razor knife *(Fig. 2)*.

Note: Before trimming the edges, inspect the grain direction on the veneer. Then cut with the grain to avoid tearout.

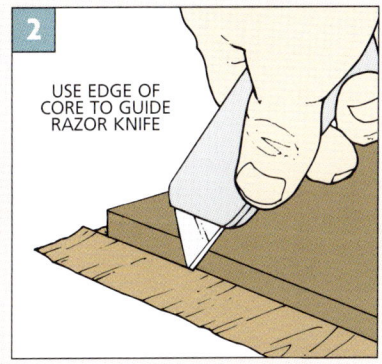

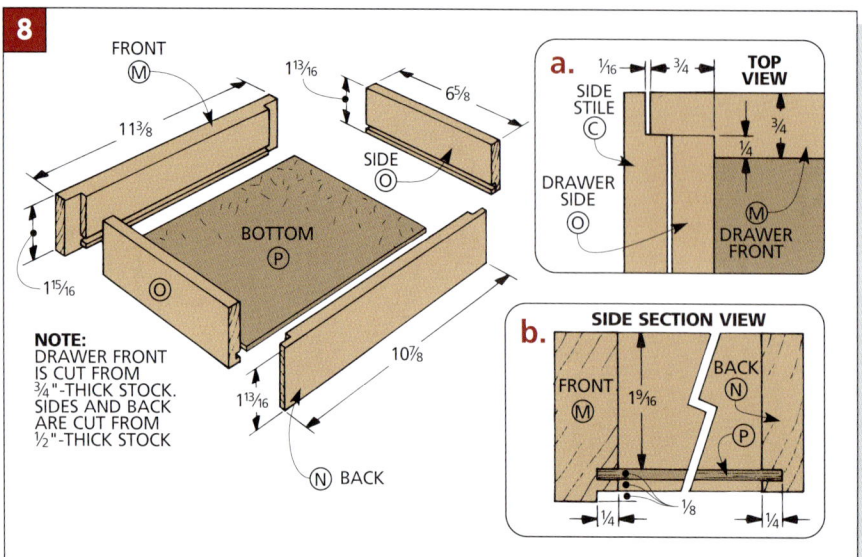

8

FRONT Ⓜ

1¹³⁄₁₆

11³⁄₈

SIDE Ⓞ

6⁵⁄₈

BOTTOM Ⓟ

Ⓞ

1¹⁵⁄₁₆

10⁷⁄₈

1¹³⁄₁₆

Ⓝ BACK

NOTE:
DRAWER FRONT
IS CUT FROM
³⁄₄"-THICK STOCK.
SIDES AND BACK
ARE CUT FROM
½"-THICK STOCK

a.

1⁄16 ³⁄₄

TOP VIEW

SIDE STILE Ⓒ

³⁄₄

¼

DRAWER SIDE Ⓞ

DRAWER FRONT Ⓜ

b. **SIDE SECTION VIEW**

FRONT Ⓜ

1⁹⁄₁₆

BACK Ⓝ

Ⓟ

¼   ⅛   ¼

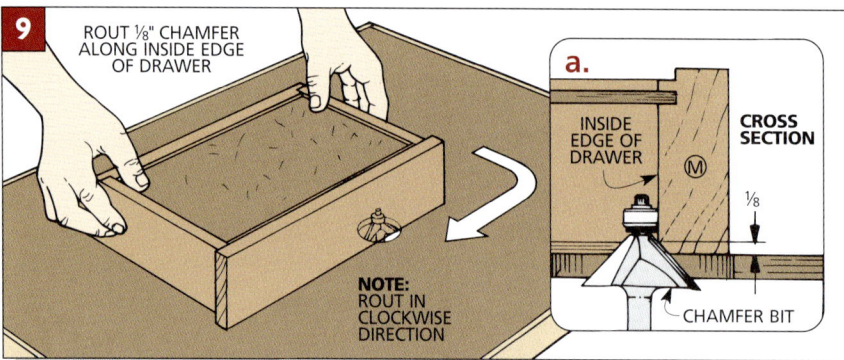

9

ROUT ⅛" CHAMFER
ALONG INSIDE EDGE
OF DRAWER

NOTE:
ROUT IN
CLOCKWISE
DIRECTION

a.

INSIDE EDGE OF DRAWER

Ⓜ

CROSS SECTION

⅛

CHAMFER BIT

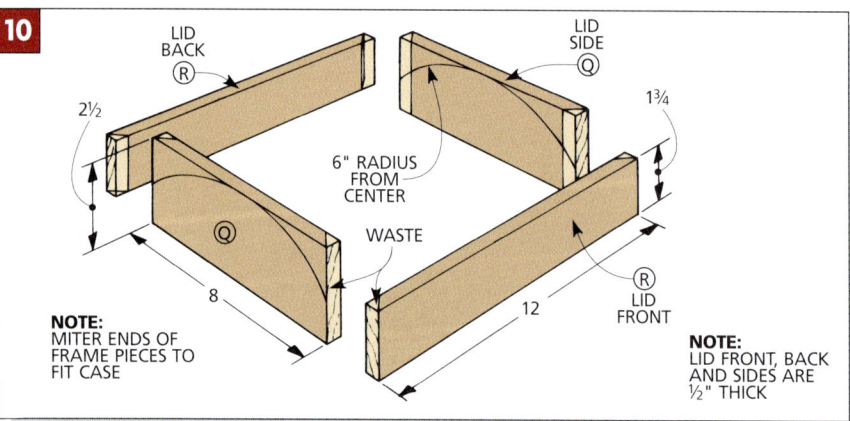

10

LID BACK Ⓡ

LID SIDE Ⓠ

2½

6" RADIUS FROM CENTER

Ⓠ

1¾

WASTE

8

Ⓡ LID FRONT

12

NOTE:
MITER ENDS OF
FRAME PIECES TO
FIT CASE

NOTE:
LID FRONT, BACK
AND SIDES ARE
½" THICK

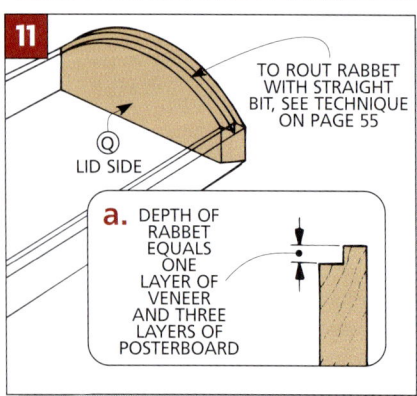

11

TO ROUT RABBET
WITH STRAIGHT
BIT, SEE TECHNIQUE
ON PAGE 55

Ⓠ
LID SIDE

a. DEPTH OF
RABBET
EQUALS
ONE
LAYER OF
VENEER
AND THREE
LAYERS OF
POSTERBOARD

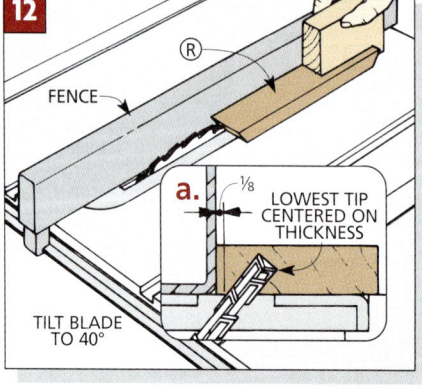

12

Ⓡ

FENCE

a.

⅛

LOWEST TIP
CENTERED ON
THICKNESS

TILT BLADE
TO 40°

DRAWERS

The drawers in this chest look like simple rabbeted drawers. And they are. But because the drawer fronts act as drawer stops, I began with them.

FRONTS. Begin by cutting the drawer fronts (M) to size. They're cut ⅛" shorter than the opening in the front of the case (rabbet-to-rabbet) *(Fig. 8).* This allows a ¹⁄₁₆" gap on each side *(Fig. 8a).*

To keep the drawer fronts flush with the front of the case, rabbets are cut along the sides and bottom edge of each drawer front *(Figs. 8a and 8b).* (The bottom rabbet also hides the dust panel.)

Note: When cutting the rabbets, it's best to sneak up on the depth, testing the fit until each front is exactly flush with the case *(Fig. 8a).*

BACKS AND SIDES. Now, cut the drawer backs (N) and sides (O) to size. Then cut ¼"-deep rabbets in each back (N) to accept the sides.

To hold the drawer bottom (P) in place, cut a groove on the inside face of each drawer piece *(Fig. 8).*

Note: Since the drawer front isn't flush at the bottom, I referenced from the top edges when I laid out the grooves.

BOTTOMS. I used ⅛" hardboard for the bottoms. Once they're cut to size, glue and clamp the drawers together. Then chamfer the top inside edges to soften the sharp corners *(Fig. 9).*

CURVED LID

After installing the drawers, work can begin on the lid. The lid is basically a mitered frame with curved sides.

LID FRAME. To begin on the lid frame, first rip the lid sides (Q) to finished width *(Fig. 10).* Then rip the lid front (R) and back (R) to rough width.

Next, miter the pieces to length to fit on top of the case *(Fig. 10).*

CURVED SIDES. Now the curved edge on the lid sides can be cut *(Fig. 10).* First, lay out the arc on one blank. Then, to ensure that both pieces end up the same, tape the blanks together with carpet tape.

Once the arcs are cut on the band saw, sand the curved edges smooth. Then complete each lid side by routing a ³⁄₁₆" rabbet for the curved top on the inside edge *(Fig. 11).* (The Technique box on the next page shows how I did this.)

RABBETING FRONT AND BACK. The next step is cutting rabbets in the lid front (R) and back (R) to align with the rabbets

Routing a rabbet on a curved piece isn't difficult. A router table and a rabbeting bit with a pilot bearing will do the job just fine.

But what about an odd-sized rabbet like the side pieces on the chest lid? To get around this, I used a 1/4" straight bit on the router table and a zero-clearance fence with the router bit partially "buried" in the fence *(Fig. 1)*.

ADJUST BIT. After mounting the straight bit in the router table, raise it to the desired height (width). For the side of the chest I set it to 1/4" *(Fig. 1a)*.

CUT NOTCH. I used a piece of plywood scrap with a notch cut in the center for a zero-clearance fence *(Figs. 1 and 1a)*.

To cut the notch, clamp the scrap piece to the router table fence. Then turn on the router and push the fence into the bit

until 1/16" of the bit is buried in the fence *(Fig. 1a)*. (This leaves 3/16" exposed to cut the rabbet to the desired depth.)

Next, clamp the fence to the table. Then draw a line on the fence directly over the center of the bit (This is a reference mark used for routing the rabbet.)

ROUT RABBET. Now rout the rabbet (routing right-to-left), keeping the piece in contact with the fence *(Fig. 2)*.

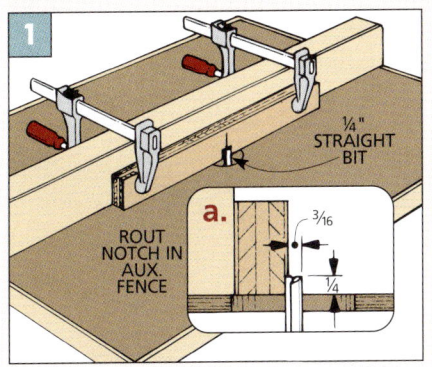

1

1/4"
STRAIGHT
BIT

a.
ROUT
NOTCH IN
AUX.
FENCE
3/16
1/4

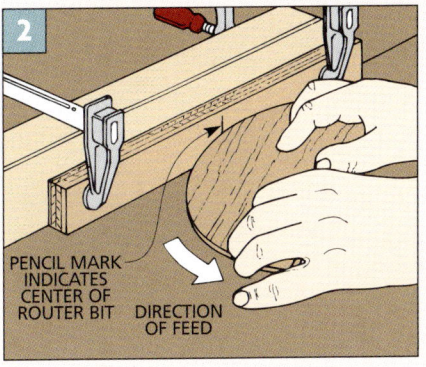

2

PENCIL MARK
INDICATES
CENTER OF
ROUTER BIT
DIRECTION
OF FEED

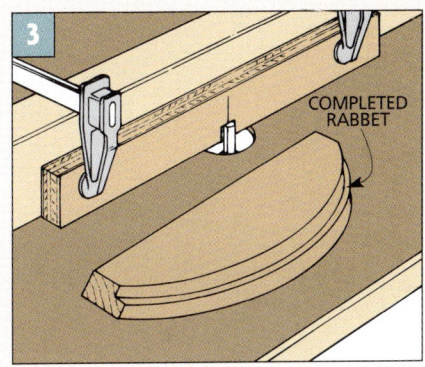

3

COMPLETED
RABBET

in the sides. This is the trickiest part of the project. But it's simple to do when you break it down into three steps.

First, set up the table saw with a 1/4"-wide stacked dado blade tilted to 40°. Next, raise the blade so the lowest tip of the blade is centered on the thickness of the workpiece *(Fig. 12)*. Then position the rip fence 1/8" away from the blade.

Now check the setup by cutting a groove on a piece of scrap. If necessary, readjust the fence or blade. After you're satisfied with the setup, cut a groove across the *inside* face of both the lid front (R) and back (R) *(Fig. 12)*. This completes the first step.

MARK RABBET. Now the top edge of the rabbet can be marked and cut. But to do this, you'll first have to figure out where to position the rip fence. It's not that hard to do, but it does take a little extra thought. Here's how I did it.

First, hold the end of the lid front and a lid side piece together *(Fig. 13)*. Line up the bottom edge of the groove you just cut in the lid front with the bottom edge of the rabbet in the lid side. Then just transfer the location of the curved edge to the end of the lid front *(Fig. 13a)*.

Next, using the mark on the lid front, position the rip fence *(Fig. 14a)*. Now trim off the notched edge of both the lid front and back to create the 3/16"-deep rabbet. This completes the second step.

FINISHED WIDTH. Again, using one of the side pieces, mark the finished width of the front and back pieces. Only this time, make your mark on the bottom edge *(Fig. 13a)*. After you've marked the pieces, set up the table saw to trim them to finished width *(Fig. 15)*.

GLUING FRAME. With the lid front and back pieces complete, all that's left on the curved lid frame is to glue and clamp it together. I like to use a band clamp for a mitered frame. And to keep the frame square, set the lid on the case and use four corner blocks *(Fig. 16)*.

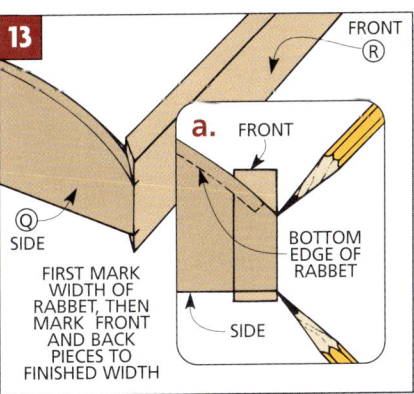

13

FRONT
R

a. FRONT

Q
SIDE

BOTTOM
EDGE OF
RABBET

SIDE

FIRST MARK
WIDTH OF
RABBET, THEN
MARK FRONT
AND BACK
PIECES TO
FINISHED WIDTH

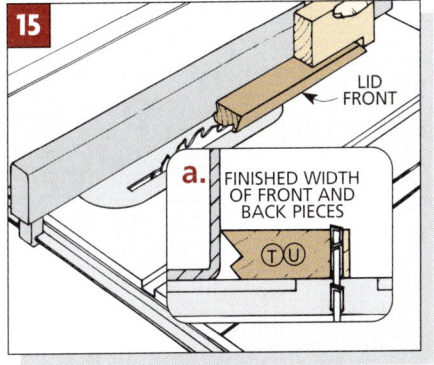

15

LID
FRONT

a. FINISHED WIDTH
OF FRONT AND
BACK PIECES

T U

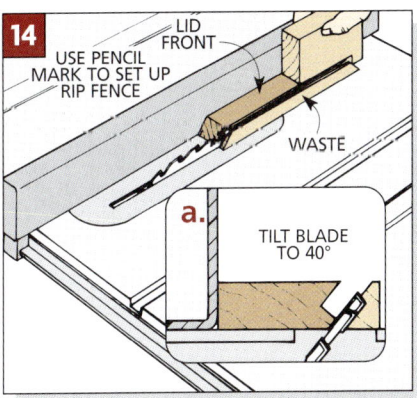

14

LID
FRONT

USE PENCIL
MARK TO SET UP
RIP FENCE

WASTE

a.
TILT BLADE
TO 40°

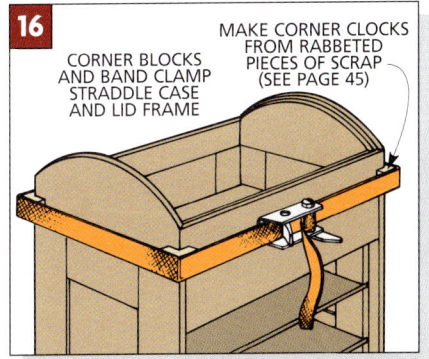

16

MAKE CORNER CLOCKS
FROM RABBETED
PIECES OF SCRAP
(SEE PAGE 45)

CORNER BLOCKS
AND BAND CLAMP
STRADDLE CASE
AND LID FRAME

Making the curved lid of the Jewelry Chest may seem difficult. But it's not. The secret is a posterboard core formed on this bending jig.

The jig consists of a frame made of two curved ends, two support bars, and a clamping form made of one layer of $1/16$"-thick mat board (see drawing). (Mat board is available at art supply stores.)

After cutting the ends to shape and screwing the support bars between them, the form is simply bent over the frame and stapled down.

Then, to prevent glue from sticking to the mat board, I covered it with two coats of polyurethane finish.

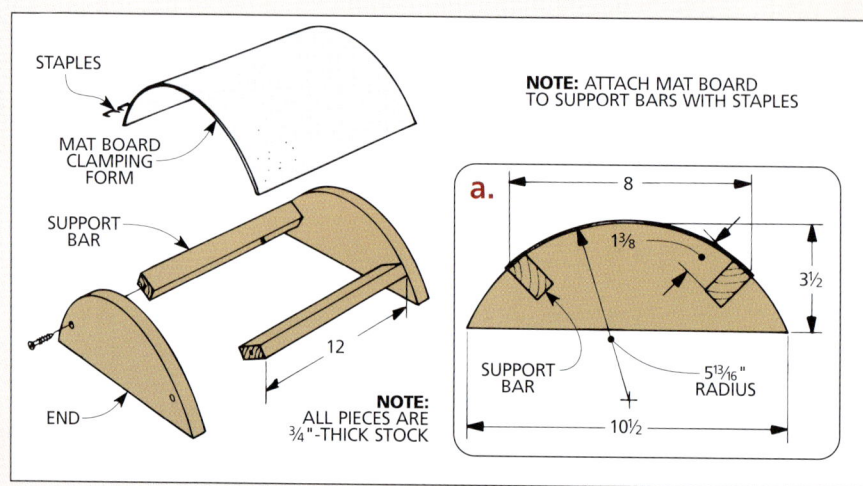

STAPLES

MAT BOARD CLAMPING FORM

SUPPORT BAR

END

NOTE: ALL PIECES ARE $3/4$"-THICK STOCK

NOTE: ATTACH MAT BOARD TO SUPPORT BARS WITH STAPLES

a.

8

$1^3/8$

$3^1/2$

SUPPORT BAR

$5^{13}/16$ " RADIUS

$10^1/2$

17 FIRST: GLUE UP THREE LAYERS OF POSTERBOARD

CENTER CORE ON FORM

SECOND: WHILE GLUE IS STILL WET, PLACE ON FORM

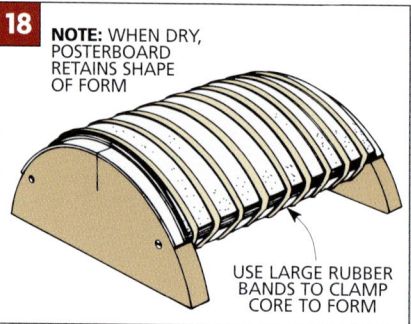

18 NOTE: WHEN DRY, POSTERBOARD RETAINS SHAPE OF FORM

USE LARGE RUBBER BANDS TO CLAMP CORE TO FORM

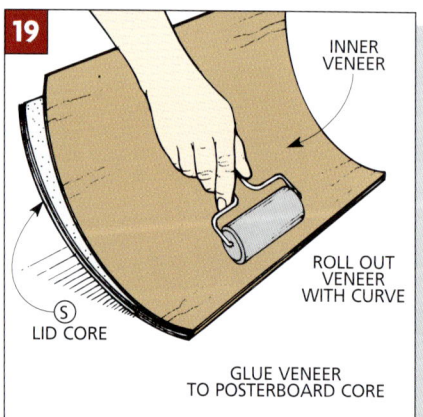

19 INNER VENEER

ROLL OUT VENEER WITH CURVE

(S) LID CORE

GLUE VENEER TO POSTERBOARD CORE

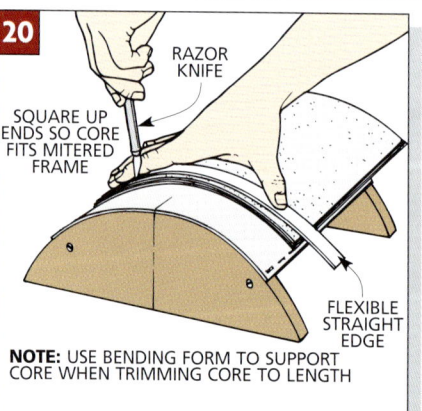

20 RAZOR KNIFE

SQUARE UP ENDS SO CORE FITS MITERED FRAME

FLEXIBLE STRAIGHT EDGE

NOTE: USE BENDING FORM TO SUPPORT CORE WHEN TRIMMING CORE TO LENGTH

21

SQUARE UP SIDES SO CORE FITS MITERED FRAME

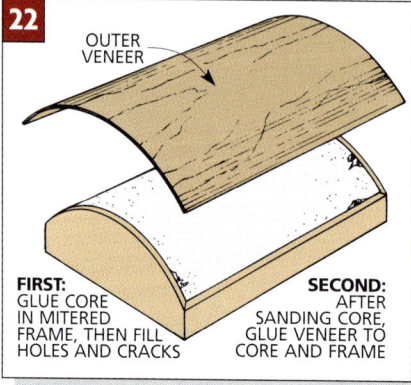

22 OUTER VENEER

FIRST: GLUE CORE IN MITERED FRAME, THEN FILL HOLES AND CRACKS

SECOND: AFTER SANDING CORE, GLUE VENEER TO CORE AND FRAME

CURVED TOP

To make the curved top for the Jewelry Chest, begin by cutting three pieces of 10-ply posterboard for the lid core (S). Cut these slightly larger than the rabbeted opening in the lid frame.

I used posterboard because it bends easily and I didn't have to worry about breaking it. Another great feature of posterboard is that it's amazingly strong after it's been glued into "plies."

To make the curve in the lid panel, you'll need a bending jig (see box above). Once the jig is built, glue the three lid core pieces together. You want to cover each surface with enough yellow glue so it bends easily *(Fig. 17)*. Then clamp the core to the jig *(Fig. 18)*. When the core dries, it will hold the shape of the form.

VENEER. Now, glue an oversize piece of veneer to the bottom of the core *(Fig. 19)*. (The outside face will be veneered after the core is attached to the lid side pieces.)

INSTALL CORE. Next, cut the core to fit the rabbets in the lid frame. To cut the core, I used a flexible straightedge (a scrap piece of posterboard) and the bending jig *(Figs. 20 and 21)*. Then glue the core in the frame *(Fig. 22)*.

After the core is glued in place, you may find that the core and the curved lid sides (Q) aren't flush. If this is the case, you'll need to lightly sand the curved edges or the core until they are flush. Also, fill any holes or blemishes with wood filler. (Be sure to fill any gaps between the core and the sides as well.) Once the filler dries completely, sand the surface smooth again.

VENEER LID. Now you can veneer the lid. First cut a piece of veneer slightly oversize to cover the core and the frame. Then glue it in place *(Fig. 22)*. As with the panels, use a razor knife to trim the veneer flush with the lid sides *(Fig. 23)*.

FINISH AND HARDWARE. Finally, I wiped on three coats of a combination oil/wax finish. Then to complete the chest, install the knobs and hinges *(Fig. 24)*. The hinges are mortised into both the back frame and the lid. ■

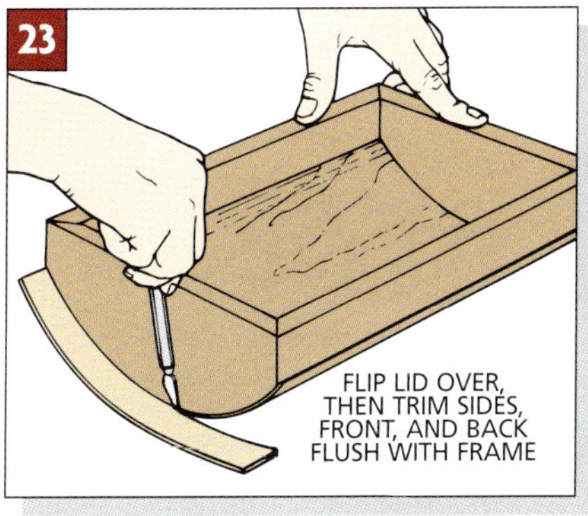

23
FLIP LID OVER, THEN TRIM SIDES, FRONT, AND BACK FLUSH WITH FRAME

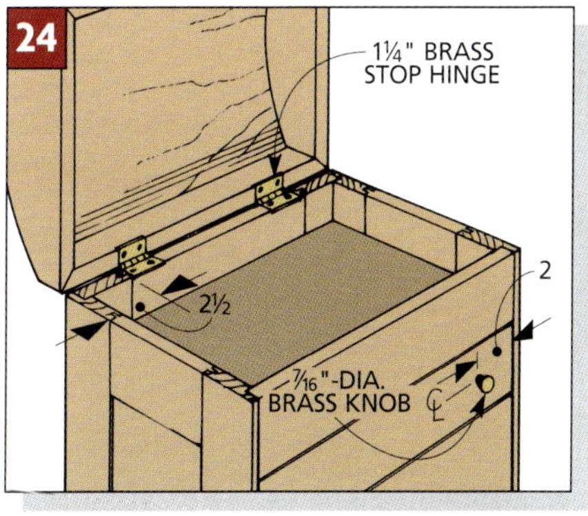

24
1¼" BRASS STOP HINGE
2½
2
7⁄16"-DIA. BRASS KNOB

DESIGNER'S NOTEBOOK

A flat top simplifies construction while moldings add visual appeal to this version of the chest.

CONSTRUCTION NOTES:

■ After gluing up the base, a bottom molding is built to go between the base and the case. The moldings (T, U) are just ½"-thick stock with a ¼" roundover on each face *(Fig. 1a)*. Then the pieces are mitered to fit above the base with a ¼" overhang *(Figs. 1 and 1a)*.

■ Next, the base and moldings can be glued to each other, then to the case.

■ The lid panel (V) is two pieces of ¼" hardboard glued face to face *(Fig. 2a)*. The outside faces of the panel are covered with veneer.

MATERIALS LIST

CHANGED PARTS		
Q	Lid Sides (2)	½ x 1¾ - 8
NEW PARTS		
T	Btm. Side Moldings (2)	½ x 1¼ - 8¾
U	Btm. Fr./Bk. Moldings (2)	½ x 1¼ - 12¾
V	Lid Panels (2)	¼ hdbd. - 7½ x 11½
W	Lid Molding	½ x ⅜ - 48 rgh.

Note: Don't need part S

■ The front and back lid pieces (R) are the same as before. The lid sides (Q) are ripped to the same width as the front and back *(Fig. 2a)*.

■ Next, cut a groove in the inside face of each lid piece so the lid panel will end up flush with the top of the lid *(Fig. 2a)*.

■ Miter the lid pieces to length *(Fig. 2)*. Then dry-assemble them and measure for the final size of the lid panel.

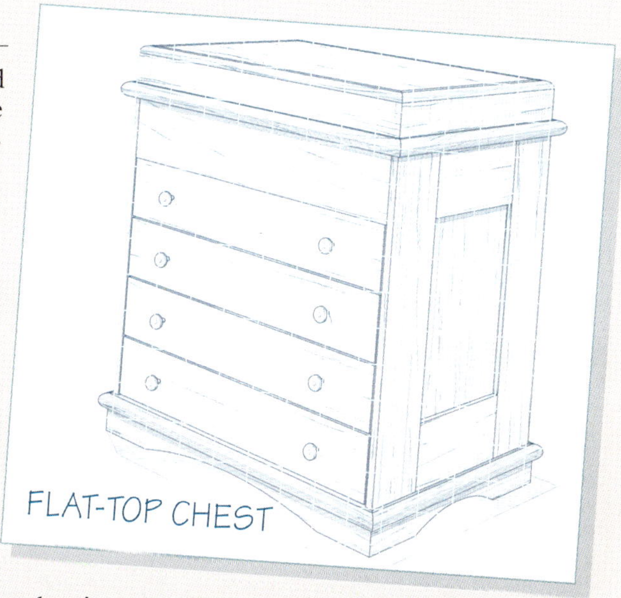

FLAT-TOP CHEST

■ After cutting the lid panel to size, cut a ¼" rabbet around it to leave a tongue to fit the grooves in the lid pieces.

■ Once the lid is assembled, the top edge can be rounded over *(Fig. 2a)*.

■ To make the half-round molding (W) around the bottom of the lid, I routed ¼" roundovers on both edges of a ½"-thick blank, then ripped the strips of ⅜"-wide molding from each edge *(Fig. 2a)*.

■ The molding is then mitered to fit around the lid and glued, flush with the bottom edge of the lid *(Fig. 2a)*.

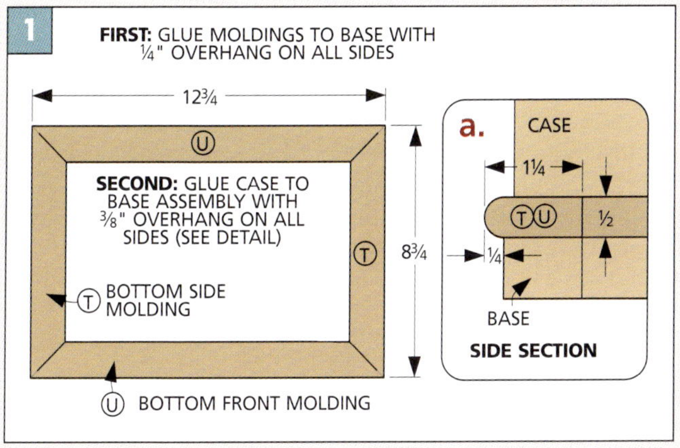

1
FIRST: GLUE MOLDINGS TO BASE WITH ¼" OVERHANG ON ALL SIDES
12¾
U
SECOND: GLUE CASE TO BASE ASSEMBLY WITH ⅜" OVERHANG ON ALL SIDES (SEE DETAIL)
T
8¾
T BOTTOM SIDE MOLDING
U BOTTOM FRONT MOLDING
a.
CASE
1¼
T U
½
¼
BASE
SIDE SECTION

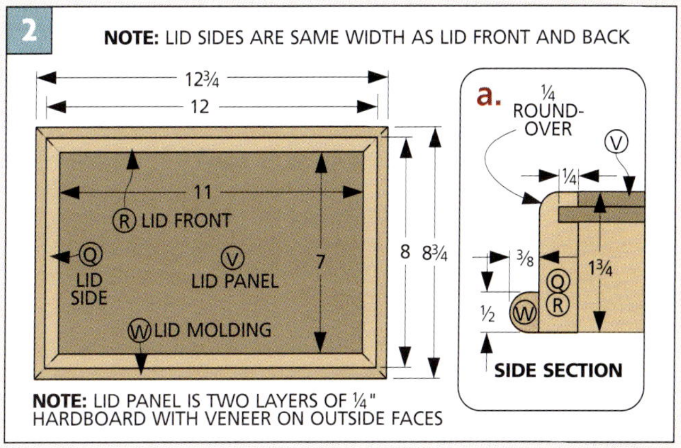

2
NOTE: LID SIDES ARE SAME WIDTH AS LID FRONT AND BACK
12¾
12
11
R LID FRONT
Q LID SIDE
V LID PANEL
7
W LID MOLDING
8
8¾
NOTE: LID PANEL IS TWO LAYERS OF ¼" HARDBOARD WITH VENEER ON OUTSIDE FACES
a.
¼ ROUND-OVER
V
¼
⅜
1¾
½
Q
R
W
SIDE SECTION

Bedside Chest

This chest may have heirloom-quality details, but the joinery is straightforward. And we'll show you step-by-step how to build the bracket feet. Or go with an optional mitered apron for the base.

There are quite a few eye-catching details on this formal chest — the frame and panel top, the pull-out tray, the "raised" drawers with their ogee profiles and brass hardware, and the bracket feet. But if you look closely, you'll see that most of these details are pretty straightforward to build.

PLYWOOD PANELS. Take the top of the case, as a quick example. The panel is plywood, so there is no solid wood panel to glue up. (In fact, all the panels on this chest are plywood.) And the solid wood frame wraps around the top panel with tongue and groove joinery, which is cut entirely on the table saw.

PULL-OUT TRAY. Tucked inside the top of the cabinet is a pull-out tray (see inset photo). It's covered with plastic laminate so it's the perfect place to set a mug or glass. And any spills wipe right off.

DRAWERS. The drawers are joined with machine-cut half-blind dovetails. Each drawer slides in and out on an under-drawer guide that straddles a runner. This system ensures the drawers pull out straight. The one drawback is that the drawers are unsupported at their outside edges, so they want to tip side-to-side. I found a simple way to prevent this using small plastic bumpers that fit above and below the drawers.

BRACKET FEET. When it comes right down to it, the only detail that's a bit out of the ordinary are the bracket feet. But don't worry. There's no carving involved. The curved profiles are roughed out with the table saw and band saw. The final shaping is done with a block plane and sandpaper. So it's really not all that difficult. And we've made it even easier with the step-by-step article that begins on page 122. (If you're still not convinced, there's also an optional base featured on page 67 or you could opt for manufactured feet as shown on page 65.)

EXPLODED VIEW

OVERALL DIMENSIONS:
33 1/2W x 20 3/8D x 29 3/4H

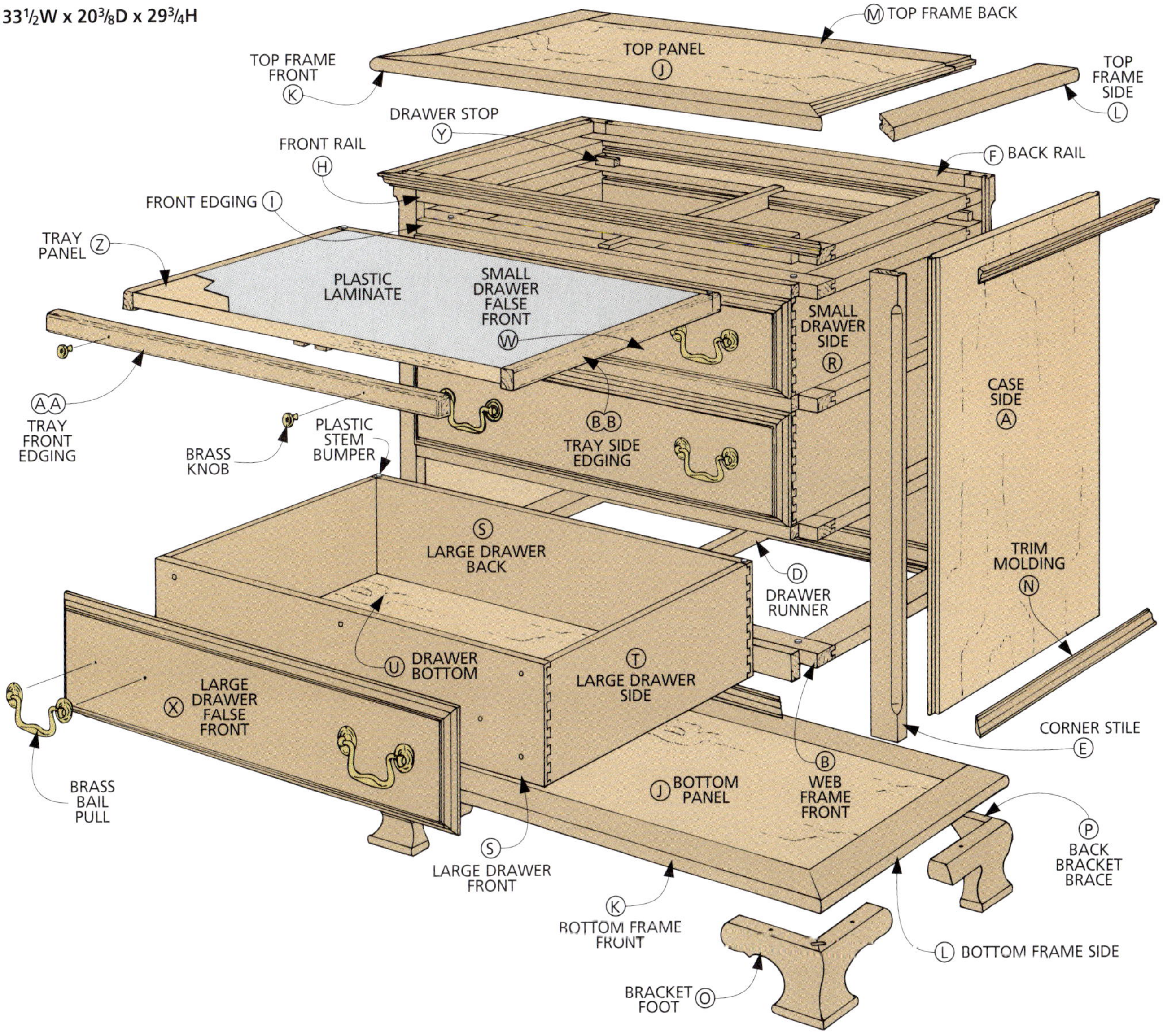

MATERIALS LIST

WOOD

- **A** Case Sides (2) — 3/4 ply - 18 1/4 x 24 1/4
- **B** Web Frm. Fr./Bk. (10) — 3/4 x 1 3/4 - 29 3/4
- **C** Web Frame Sides (10) — 3/4 x 1 3/4 - 14 3/4
- **D** Drawer Runners (4) — 3/4 x 1 - 17 1/2
- **E** Corner Stiles (4) — 3/4 x 1 3/4 - 24 1/4
- **F** Back Rails (2) — 3/4 x 1 3/4 - 28
- **G** Back Panel (1) — 1/4 ply - 28 x 21 1/2
- **H** Front Rails (2) — 3/4 x 1 1/2 - 27 1/4
- **I** Front Edging (3) — 3/4 x 3/4 - 27 1/4
- **J** Top/Btm. Panels (2) — 3/4 ply - 15 1/4 x 28
- **K** Top/Btm. Frame Fr. (2) — 3/4 x 2 3/4 - 32 3/4
- **L** Top/Btm. Frm. Sides (4) — 3/4 x 2 3/4 - 20
- **M** Top/Btm. Frame Bk. (2) — 3/4 x 2 3/4 - 28
- **N** Trim Molding — 3/4 x 11/16 - 164
- **O** Bracket Foot Blanks (3) — 1 1/2 x 4 1/2 - 16
- **P** Back Brkt. Braces (2) — 3/4 x 3 - 3
- **Q** Small Drwr. Fr./Bk. (2) — 1/2 x 5 1/4 - 27 1/8
- **R** Small Drwr. Sides (2) — 1/2 x 5 1/4 - 17 1/4
- **S** Large Drwr. Fr./Bk. (4) — 1/2 x 6 1/8 - 27 1/8
- **T** Large Drwr. Sides (4) — 1/2 x 6 1/8 - 17 1/4
- **U** Drawer Bottoms (3) — 1/4 ply - 17 1/8 x 26 5/8
- **V** Drawer Guides (3) — 5/16 x 1 1/2 - 16 5/8
- **W** Small Drwr. False Fr. (1) — 3/4 x 5 1/4 - 27 1/8
- **X** Large Drwr. False Fr. (2) — 3/4 x 6 1/8 - 27 1/8
- **Y** Drawer Stops (8) — 1/2 x 1/2 - 1 1/2
- **Z** Tray Panel (1) — 3/4 ply - 16 3/8 x 25 5/8
- **AA** Tray Fr./Bk. Edg. (2) — 3/4 x 1 - 27 1/8
- **BB** Tray Side Edging (2) — 3/4 x 1 - 16 3/8
- **CC** Tray Guides (2) — 7/32 x 3/4 - 16 3/8

HARDWARE SUPPLIES

- (10) No. 8 x 1 1/4" Fh woodscrews
- (18) No. 8 x 3/4" Rh woodscrews
- (30) No. 8 washers
- (22) 1"-long wire brads
- (16) Plastic stem bumpers
- (6) 3 1/2" brass bail pulls w/ screws
- (2) 5/8"-dia. brass knobs
- (1 pc.) Plastic laminate 17" x 26" rough

CUTTING DIAGRAM

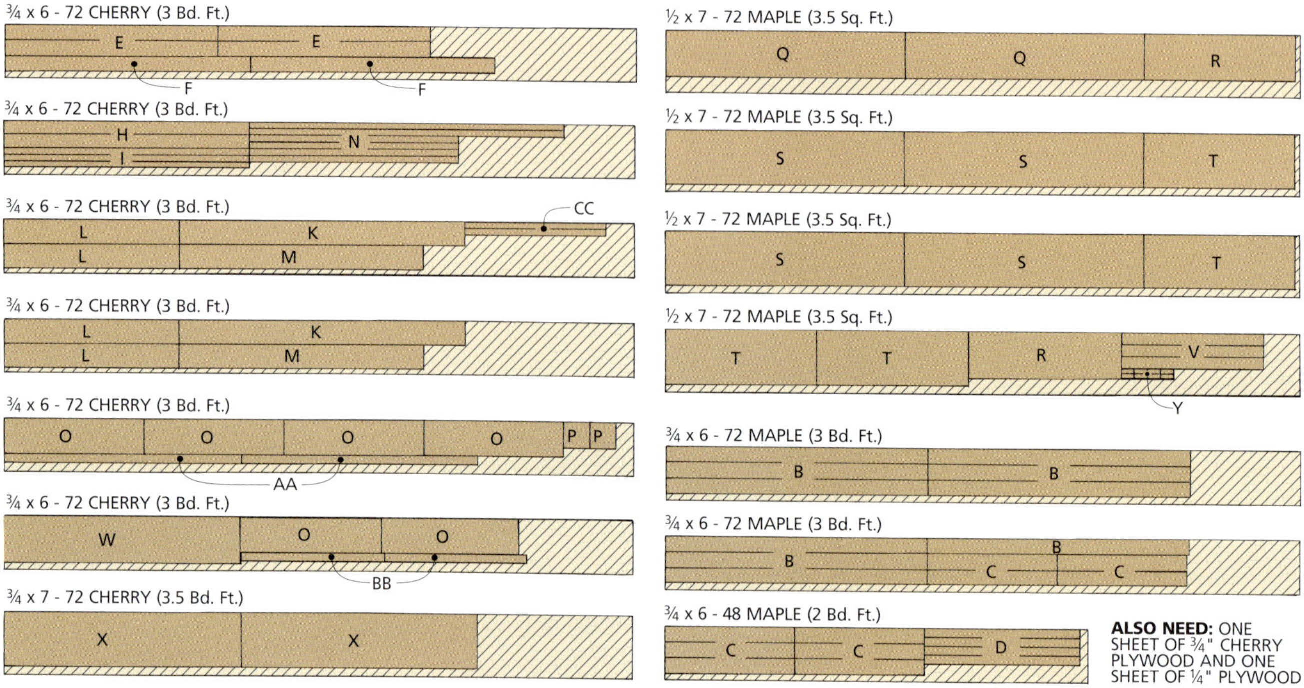

¾ x 6 - 72 CHERRY (3 Bd. Ft.)
E
E
F
F
¾ x 6 - 72 CHERRY (3 Bd. Ft.)
H
I
N
¾ x 6 - 72 CHERRY (3 Bd. Ft.)
CC
L
K
L
M
¾ x 6 - 72 CHERRY (3 Bd. Ft.)
L
K
L
M
¾ x 6 - 72 CHERRY (3 Bd. Ft.)
O
O
O
O
P
P
AA
¾ x 6 - 72 CHERRY (3 Bd. Ft.)
W
O
O
BB
¾ x 7 - 72 CHERRY (3.5 Bd. Ft.)
X
X
½ x 7 - 72 MAPLE (3.5 Sq. Ft.)
Q
Q
R
½ x 7 - 72 MAPLE (3.5 Sq. Ft.)
S
S
T
½ x 7 - 72 MAPLE (3.5 Sq. Ft.)
S
S
T
½ x 7 - 72 MAPLE (3.5 Sq. Ft.)
T
T
R
V
Y
¾ x 6 - 72 MAPLE (3 Bd. Ft.)
B
B
¾ x 6 - 72 MAPLE (3 Bd. Ft.)
B
B
C
C
¾ x 6 - 48 MAPLE (2 Bd. Ft.)
C
C
D

ALSO NEED: ONE SHEET OF ¾" CHERRY PLYWOOD AND ONE SHEET OF ¼" PLYWOOD

CASE SIDES

The chest case has two side panels sandwiching five simple web frames *(Fig. 1)*.

CASE SIDES. I started by cutting the case sides (A) to size from ¾" cherry plywood *(Fig. 1)*. This way, when it's time to join the web frames to the panels later, you can simply glue them in place without having to worry about wood movement.

To make it easy to position the web frames, I cut five dadoes across the inside of each panel *(Figs. 1a and 2)*. These dadoes should match the thickness of your stock for the web frames (¾").

With the dadoes cut, the next step is to cut tongues on the front and back edges of the panels *(Fig. 3)*. And since

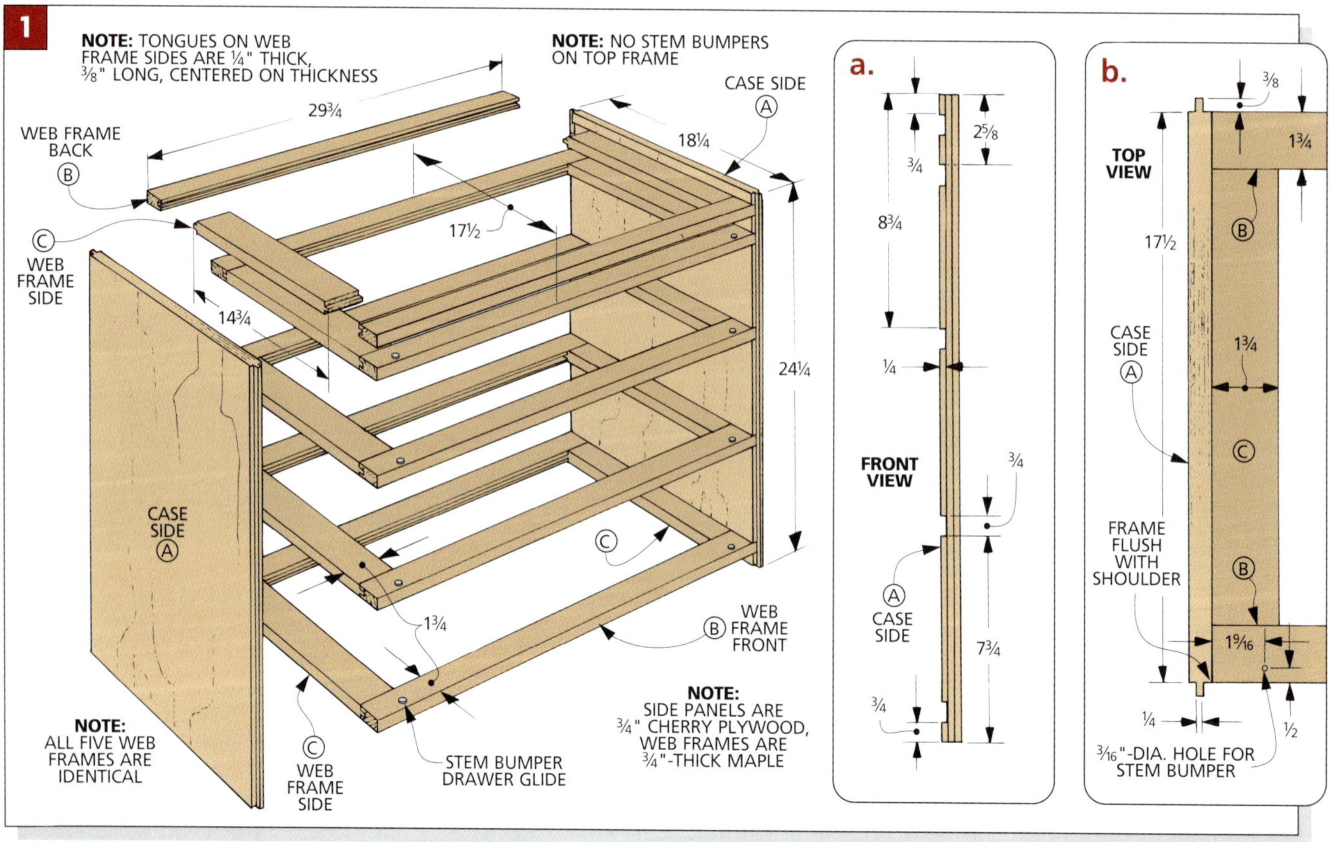

1
NOTE: TONGUES ON WEB FRAME SIDES ARE ¼" THICK, ⅜" LONG, CENTERED ON THICKNESS
NOTE: NO STEM BUMPERS ON TOP FRAME
29¾
WEB FRAME BACK
B
C
WEB FRAME SIDE
CASE SIDE
A
18¼
17½
14¾
24¼
CASE SIDE
A
C
1¾
WEB FRAME FRONT
B
NOTE: ALL FIVE WEB FRAMES ARE IDENTICAL
C
WEB FRAME SIDE
STEM BUMPER DRAWER GLIDE
NOTE: SIDE PANELS ARE ¾" CHERRY PLYWOOD, WEB FRAMES ARE ¾"-THICK MAPLE
a.
2⅝
¾
8¾
¼
FRONT VIEW
¾
A
CASE SIDE
7¾
¾
b.
⅜
1¾
TOP VIEW
B
17½
CASE SIDE
A
1¾
C
FRAME FLUSH WITH SHOULDER
B
1⁹⁄₁₆
¼
½
³⁄₁₆"-DIA. HOLE FOR STEM BUMPER

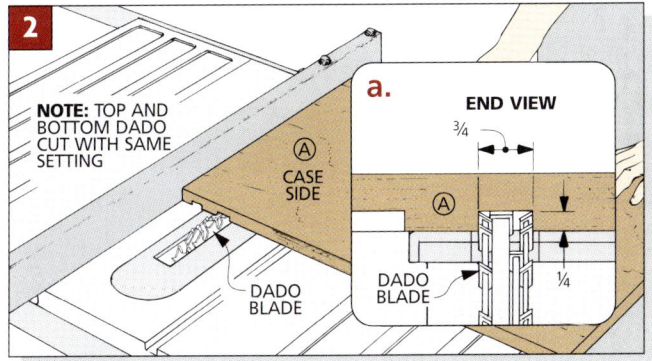

2 NOTE: TOP AND BOTTOM DADO CUT WITH SAME SETTING

a. END VIEW

CASE SIDE (A)

DADO BLADE

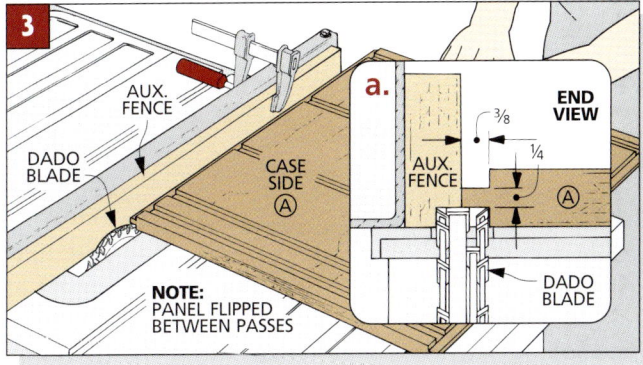

3 AUX. FENCE

DADO BLADE

CASE SIDE (A)

NOTE: PANEL FLIPPED BETWEEN PASSES

a. END VIEW

AUX. FENCE

DADO BLADE

each tongue is formed in two passes (flipping the panel between passes), it's best to sneak up on the final height of the blade until the tongue is $1/4$" thick.

WEB FRAMES

Now the panels can be set aside while you make the five web frames that fit between them *(Fig. 1)*. These are open frames joined with stub tenons and grooves. And since the front edges will be covered with edging, the frames can be built with a less expensive wood, like maple or poplar.

Before cutting the frame pieces to size, make sure the stock will fit snug in the dadoes you cut in the side panels. Then rip the web frame fronts (B), backs (B), and sides (C) to width ($1^3/4$"). The fronts and backs are cut to finished length ($29^3/4$"), but when cutting the sides, you'll want to make sure the assembled frame matches the shoulder-to-shoulder dimension of the side panels ($17^1/2$") *(Fig. 1b)*.

With the web frame pieces cut to size, the only thing left to do is cut the stub tenon and groove joints that hold them together. (Refer to the Technique box on page 115.) The grooves are cut on the front and back pieces. The sides receive the stub tenons *(Fig. 1)*.

ASSEMBLY. Now the web frames are ready to be glued together. This isn't especially difficult, but you'll want to make sure the side pieces are flush with the ends of the front and back pieces. This way, you can be sure to get a good glue joint when gluing the frames between the two sides. Plus, it's a good idea to double-check that the frames are flat when you clamp them together.

Before assembling the case, you'll need to add plastic stem bumpers to the front corners of the web frames *(Fig. 1b* and the photo on page 66). These bumpers will support the sides of the drawers later, and they need to be added now because after the case is assembled,

there won't be enough room to get a drill between the frames.

Now the web frames can be glued between the sides. I concentrated on the front of the assembly, making sure the frames were flush with the shoulder of the tongues on the side panels *(Fig. 1b)*.

DRAWER RUNNERS. Next, you can begin work on the four drawer runners (D) *(Fig. 4)*. (The top frame doesn't need a runner.) These $3/4$"-thick pieces are ripped 1" wide and cut to match the depth of the web frames ($17^1/2$").

There are two things to do with these runners before they can be glued in place. First, you'll need to cut a notch on each end to match the width of the frame pieces ($1^3/4$"). What you want to end up

with is a runner that stands $1/4$" proud of the top of the web frame *(Fig. 4b)*.

I did this with multiple passes over a dado blade, supporting the runners with a long auxiliary fence and sneaking up on the length of the notch by using the rip fence as a stop *(Fig. 4c)*.

With the notches cut, the leading edge of each runner needs to be chamfered so the drawer will slide over it *(Fig. 4a)*. Then it can be glued into the case.

The position here is critical. You want the runner parallel to the side panels and perfectly centered. To do this, just cut scrap spacers to fit on either side of the runner. Sneak up on the size of these spacers so the runner fits snug between them. Then glue the runner in place.

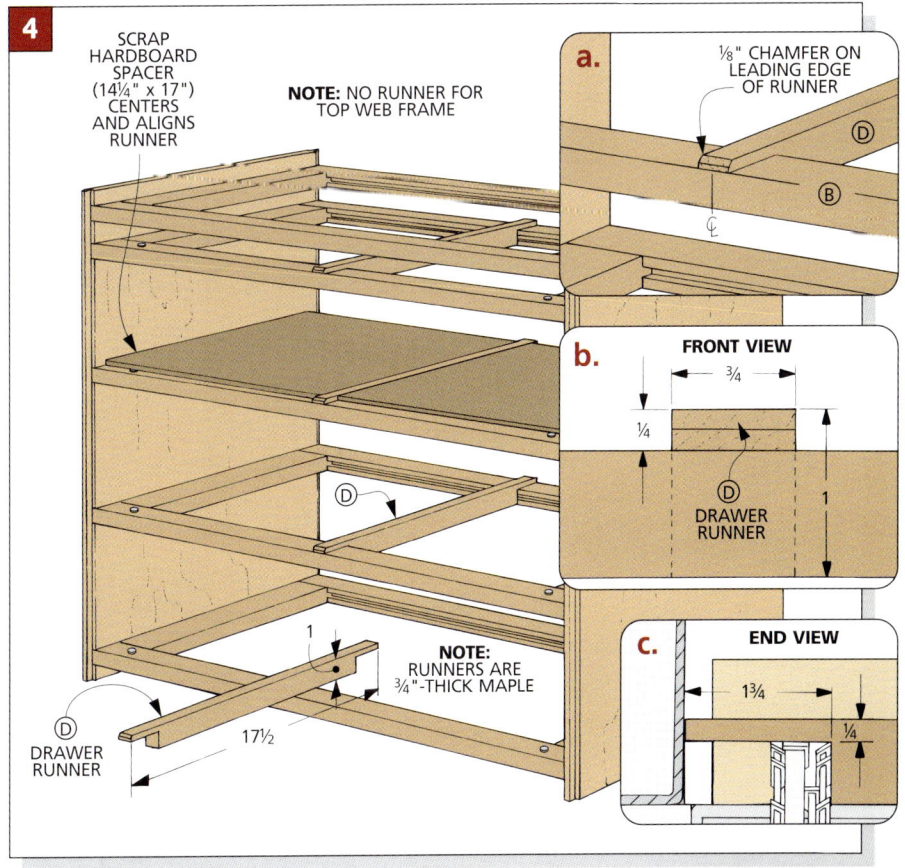

4 SCRAP HARDBOARD SPACER ($14^1/4$" x 17") CENTERS AND ALIGNS RUNNER

NOTE: NO RUNNER FOR TOP WEB FRAME

a. $1/8$" CHAMFER ON LEADING EDGE OF RUNNER

(D)

(B)

(D) DRAWER RUNNER

NOTE: RUNNERS ARE $3/4$"-THICK MAPLE

(D) DRAWER RUNNER

$17^1/2$

b. FRONT VIEW

$3/4$

$1/4$

(D) DRAWER RUNNER

1

c. END VIEW

$1^3/4$

$1/4$

BACK PANEL

With the case of the Bedside Chest assembled, it's time to dress up the front and back edges. The back is a simple frame and panel assembly with a piece of 1/4" plywood for the panel *(Fig. 5)*. Then the front of the case gets individual pieces of edging to cover the plywood and the dadoes for the web frames.

CORNER STILES. Both the front and back have identical corner stiles (E) for covering the edges of the plywood case sides *(Fig. 5)*. Each is a 3/4"-thick piece of cherry that's ripped 1 3/4" wide and cut to length to match the height of the case.

All four stiles need to have a groove cut on their inside face *(Fig. 5c)*. These grooves are cut to fit over the tongues on the side panels. The tongues on my panels were right at 1/4" thick, so I was able to use a 1/4" dado blade in the table saw. Regardless of how you cut them, the grooves should be cut to fit over the tongues and positioned so the edge of each corner stile ends up flush with the face of the side panel *(Figs. 5a and 5b)*.

When these grooves have been cut, you can glue the two front stiles to the side panels and the web frames. But there's still some more work to be done with the back stiles.

BACK FRAME AND PANEL. The back corner stiles become part of the frame and panel assembly that's glued to the back of the case. The trick with this assembly is that, once it's glued up, it has to fit over the tongues already cut on the back of the side panels — which means you have to be careful to get the overall size of the panel right.

The dimension to focus on is the length of the back rails (F) *(Fig. 5)*. And the best way to do this is to measure right from the case. First, dry-clamp the stiles to the case. Then measure between them to find the shoulder-to-shoulder length of the back rails. Now add 3/4" to this dimension to account for both stub tenons, and you have the final length of the back rails. (Mine ended up 28" long.)

After the rails have been cut to length, you can cut the grooves on the inside edge of all four pieces *(Fig. 5d)*. This groove is different from the one you cut on the face of the stile earlier. It's centered on the edge and is sized to hold a 1/4" plywood panel. Because the plywood was under 1/4" thick, I used a combination blade to cut these grooves so I could sneak up their width.

Next, you can cut the stub tenons on the ends of the back rails *(Fig. 5e)*. Then the frame can be dry-assembled, and the back panel (G) can be cut to size.

After the back is assembled, you can glue it to the case. The grooves in the assembly should fit onto the tongues on the case and align it automatically. But if

5

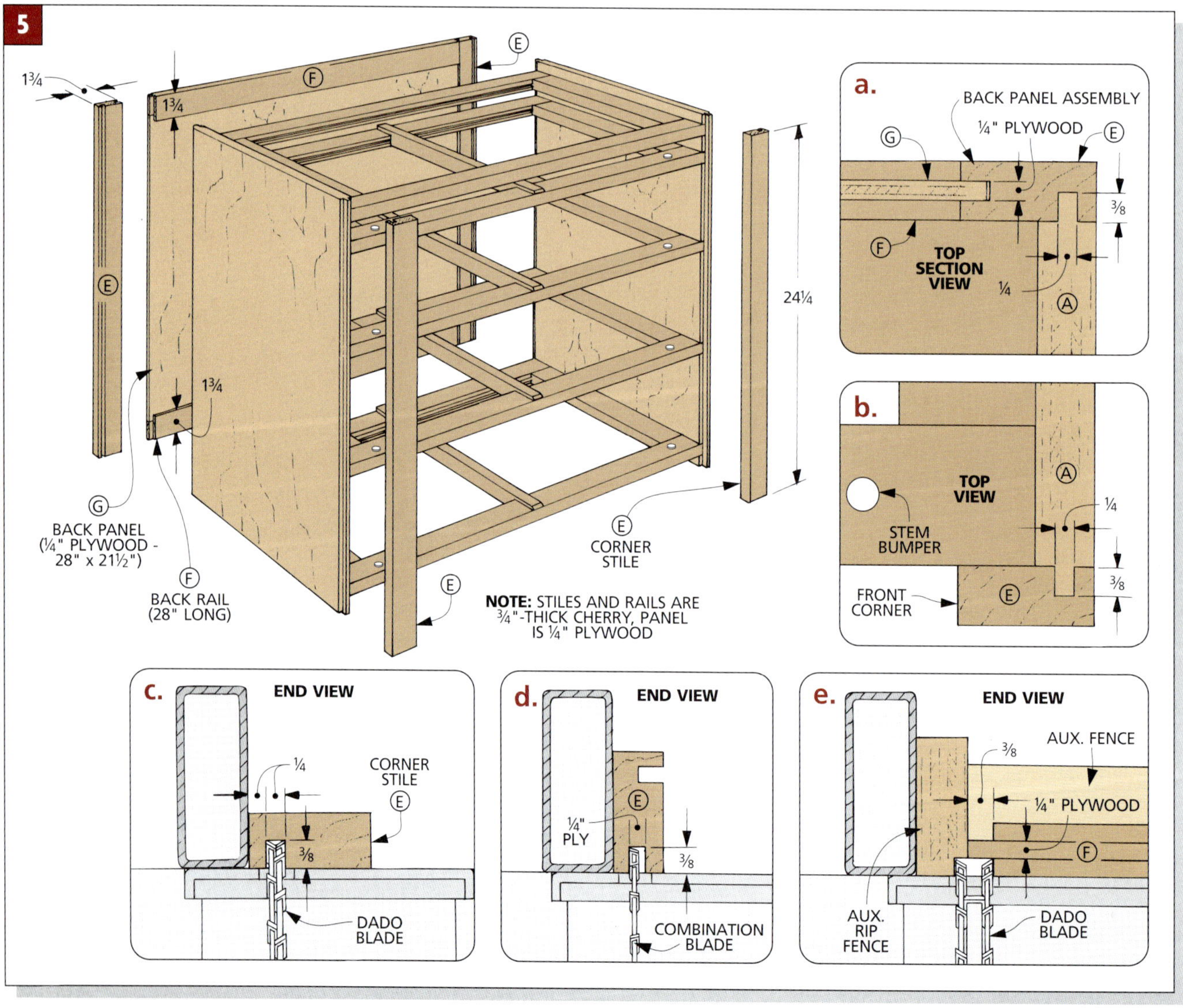

1¾
F
E
1¾
E
24¼
1¾
G
BACK PANEL (¼" PLYWOOD - 28" x 21½")
F
BACK RAIL (28" LONG)
E
CORNER STILE
E
NOTE: STILES AND RAILS ARE ¾"-THICK CHERRY, PANEL IS ¼" PLYWOOD
a.
BACK PANEL ASSEMBLY
¼" PLYWOOD
G
E
⅜
F
TOP SECTION VIEW
¼
A
b.
TOP VIEW
A
¼
STEM BUMPER
⅜
FRONT CORNER
E
c.
END VIEW
¼
CORNER STILE
E
⅜
DADO BLADE
d.
END VIEW
E
¼" PLY
⅜
COMBINATION BLADE
e.
END VIEW
AUX. FENCE
⅜
¼" PLYWOOD
F
AUX. RIP FENCE
DADO BLADE

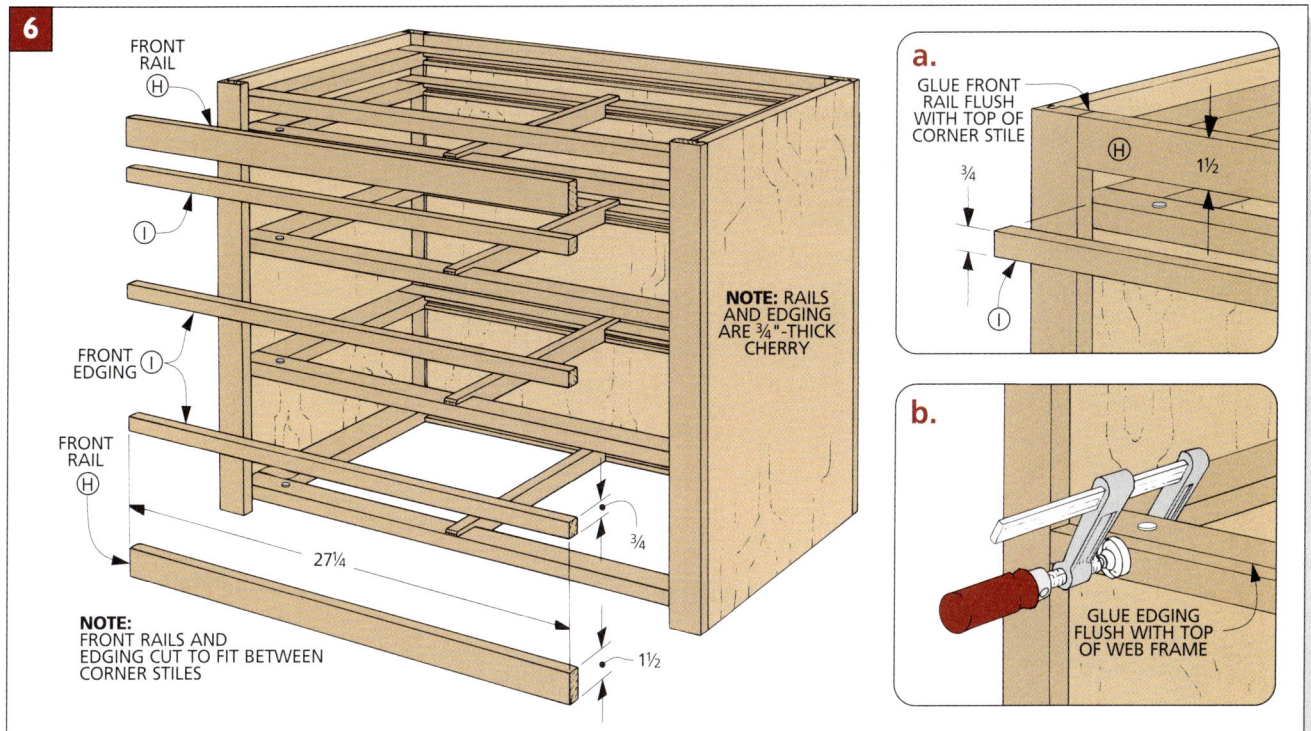

6

FRONT RAIL (H)

FRONT EDGING (I)

FRONT RAIL (H)

FRONT EDGING (I)

27¼

¾

1½

NOTE:
FRONT RAILS AND EDGING CUT TO FIT BETWEEN CORNER STILES

NOTE: RAILS AND EDGING ARE ¾"-THICK CHERRY

a.

GLUE FRONT RAIL FLUSH WITH TOP OF CORNER STILE

(H)

¾

1½

(I)

b.

GLUE EDGING FLUSH WITH TOP OF WEB FRAME

---

your tongues and grooves don't line up perfectly, you may need to widen the grooves enough to get the back to fit well. Then you'll have to plane the edges of the stiles flush with the case sides.

FRONT EDGING

With the back in place, you can turn your attention to the front of the case. Here, the edges of the maple web frames need to be covered with cherry to match the rest of the case *(Fig. 6)*.

FRONT RAILS AND EDGING. To do this, I cut two front rails (H) to cover the openings at the top and bottom and three front edging pieces (I) for the web frames in the middle. All these pieces are ¾" thick and are cut to length to fit between the front corner stiles (27¼").

The bottom rail and edging are glued flush with the tops of the web frames *(Fig. 6b)*. But the top rail is glued flush with the top of the corner stile *(Fig. 6a)*.

CHAMFER CORNERS. The last thing to do at this point is rout a stopped chamfer on each of the corner stiles *(Fig. 7)*. These chamfers stop and start even with the upper and lower front rails. Nothing fancy is needed to do this. I just marked the start and stop points on the stiles and then routed between them. And when you're done, you'll want to do a little cleanup at the ends of the chamfers (see the Shop Tip box at right).

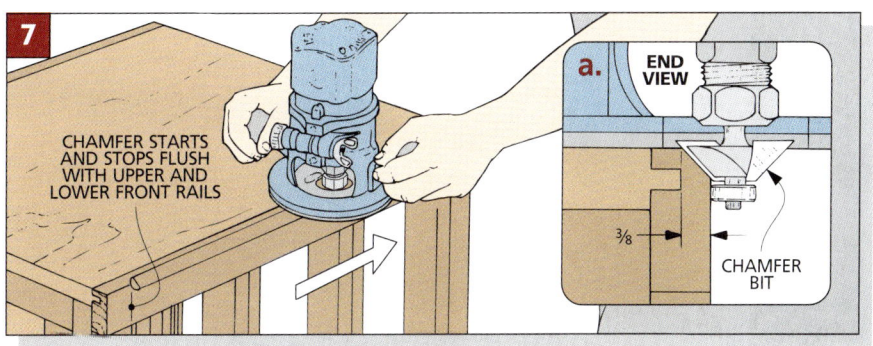

7

CHAMFER STARTS AND STOPS FLUSH WITH UPPER AND LOWER FRONT RAILS

a. END VIEW

⅜

CHAMFER BIT

SHOP TIP ..... *Chamfer Cleanup*

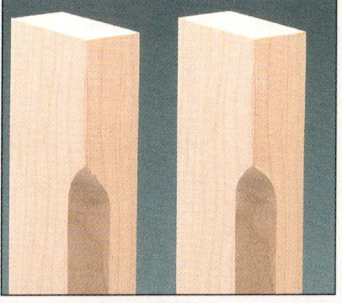

The stopped chamfers on the chest add a decorative touch. There's just one thing to be aware of. When you rout a stopped chamfer, its end isn't symmetrical (left work-piece in left photo above). One side ends up a little "flat." But here's a quick solution.

Simply wrap a piece of adhesive sandpaper around a dowel and carefully sand the end (right photo). It won't take much work to get both sides looking the same (right workpiece in first photo).

TOP & BOTTOM

The case is now ready for its top and bottom assemblies *(Fig. 8)*. To keep things as easy as possible, these frame and panels are identical — or nearly identical. The only difference is that the top assembly has a small shadow line routed around the panel *(Fig. 8b)*.

PANEL. The first thing to do is cut the top and bottom panels (J) to size *(Fig. 8)*. Both panels are ³⁄₄" cherry plywood. To hold and align the frame pieces, each panel has a tongue cut around all four edges *(Fig. 8a)*. I cut these on the table saw just like the tongues on the side panels (refer to *Fig. 3* on page 61).

At this point, the bottom panel can be set aside while you cut on the top panel the tiny shadow line I mentioned earlier. This shadow line is simply a ¹⁄₁₆"-wide rabbet. And to get a really clean cut, I decided to use a straight bit at the router table *(Fig. 9)*. But be sure to start with the ends of the panel first so if there's any chipout, it'll be cleaned up when you cut the rabbet on the long edges.

FRAME. With the panels complete, it's time to work on their frames. The first thing to do is plane some stock to match the thickness of the plywood exactly. (Remember that ³⁄₄" plywood is usually a hair less than ³⁄₄" thick.) Then you can rip the top and bottom frame fronts (K), sides (L), and backs (M) 2³⁄₄" wide *(Fig. 8b)*. I also cut the pieces extra long to allow working room when the pieces are cut to fit around the panel.

The next thing is to cut the grooves that will fit over the tongues on the panels *(Fig. 8a)*. This way, as you miter and cut the frame pieces to finished length, you can check your progress by fitting the pieces around the actual panels.

Once the grooves were cut, I started cutting the pieces to length. I began with the sides, mitering one end but still leaving the pieces extra long. Then with the sides dry-clamped to the panel, you can miter both ends of the front pieces, sneaking up on their final length.

Now before you cut the sides to length, the frame backs (M) need a little work. They're cut to the same overall length as the panel (28"). Then tongues are cut on the ends to fit the grooves in the frame side pieces.

Finally, add the frame backs to the dry assembly. Then cut the frame sides to length so they end up flush with the back of the frame *(Fig. 8b)*.

ASSEMBLY. After the frame sides have been cut to length, the top and bottom frames can be assembled. Then you can rout a bullnose profile around the front and side edges *(Fig. 10)*.

To attach the panels to the case, you can't use screws, so I simply glued them flush with the back and centered side-to-side. Then I made some trim molding (N) *(Fig. 11)*, and glued and nailed it to the case with small wire brads *(Fig. 8a)*.

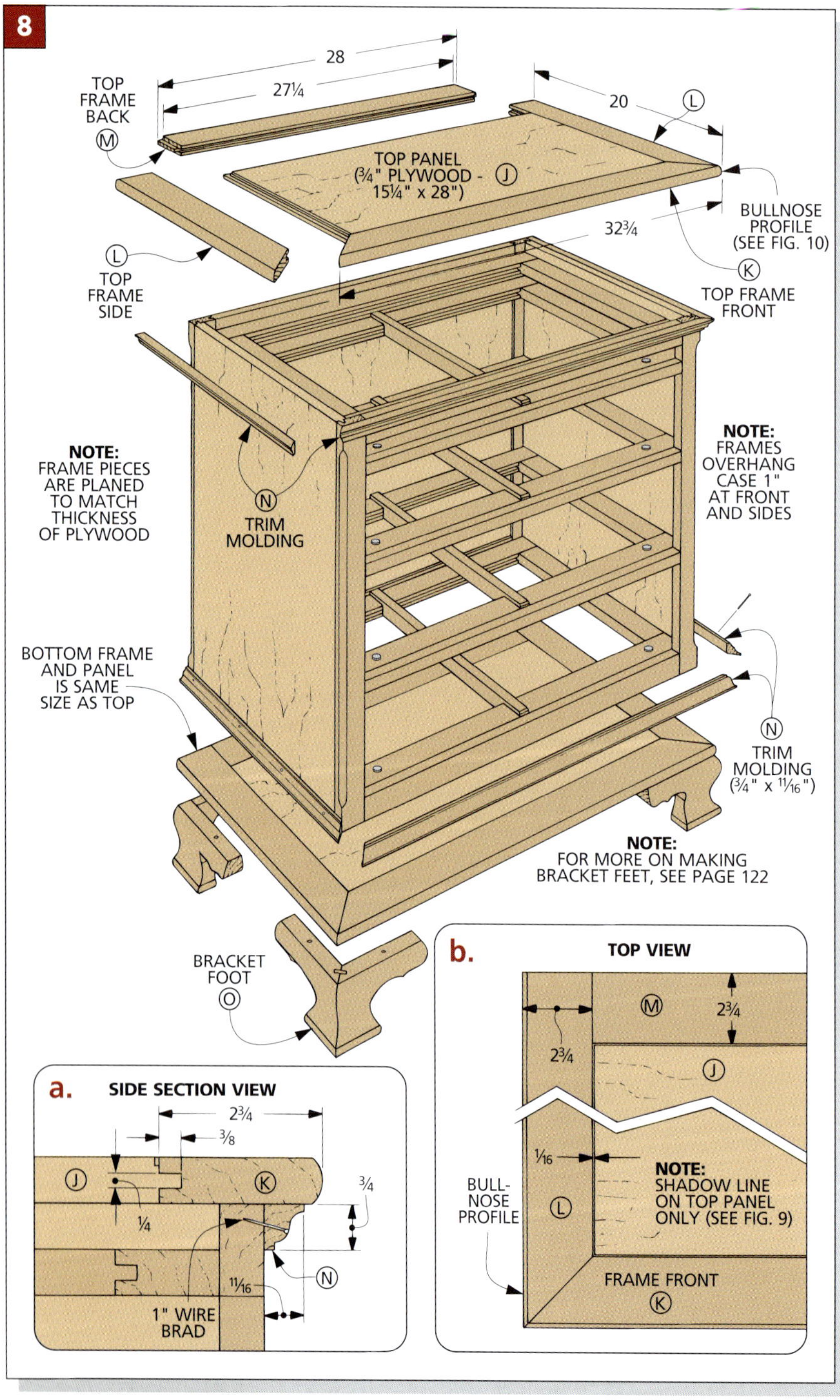

8

TOP FRAME BACK
(M)

28
27¼

TOP PANEL
(¾" PLYWOOD -
15¼" x 28")
(J)

20
(L)

BULLNOSE PROFILE
(SEE FIG. 10)

32¾

(K)
TOP FRAME FRONT

(L)
TOP FRAME SIDE

NOTE: FRAMES OVERHANG CASE 1" AT FRONT AND SIDES

NOTE: FRAME PIECES ARE PLANED TO MATCH THICKNESS OF PLYWOOD

(N)
TRIM MOLDING

BOTTOM FRAME AND PANEL IS SAME SIZE AS TOP

(N)
TRIM MOLDING
(¾" x ¹¹⁄₁₆")

BRACKET FOOT
(O)

NOTE: FOR MORE ON MAKING BRACKET FEET, SEE PAGE 122

a. SIDE SECTION VIEW

2¾
³⁄₈
(J)
(K)
¾
¼
(N)
¹¹⁄₁₆
1" WIRE BRAD

b. TOP VIEW

(M)
2¾
2¾
(J)
¹⁄₁₆
BULL-NOSE PROFILE
(L)
NOTE: SHADOW LINE ON TOP PANEL ONLY (SEE FIG. 9)
FRAME FRONT
(K)

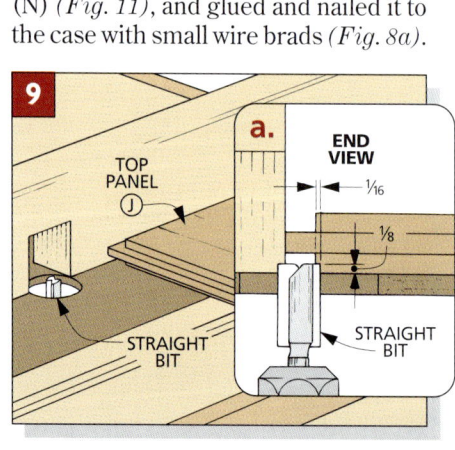

9

a.
TOP PANEL
(J)
END VIEW
¹⁄₁₆
⅛
STRAIGHT BIT
STRAIGHT BIT

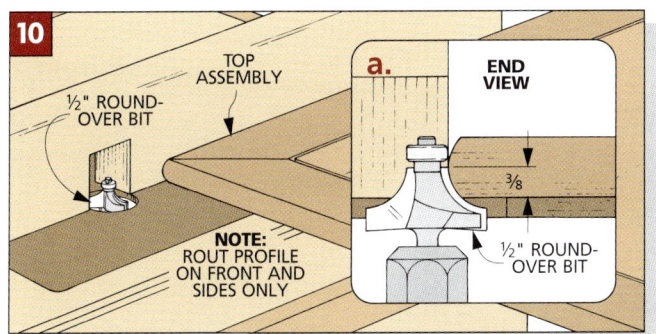

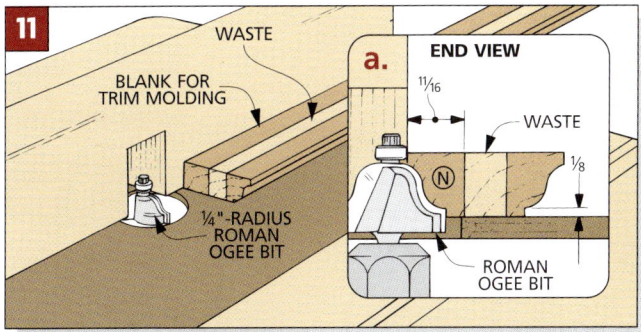

BRACKET FEET

With the cabinet completed, you're ready to make the bracket feet (O). These are fashioned from $1\frac{1}{2}$"-thick, glued-up blanks. And even though there are quite a few steps to follow, these feet are easier to build than you might think.

You'll start with three extra-long blanks that will yield two feet each. By cutting two feet from one long blank, the grain will "wrap around" each corner at the front of the cabinet. And you only need six feet because a triangular brace (instead of a foot) is attached to the feet at the back of the cabinet *(Fig. 12)*.

To make the bracket feet, the first step is to cut a cove on each blank at the table saw. After mitering the pieces to length, the profiles are laid out *(Fig. 13)* and cut to shape on the band saw. Then some handwork with a plane and sandpaper brings the feet to final shape. A step-by-step article to walk you through the process starts on page 122.

Note: As an alternative to making feet, ready-made bracket feet can be purchased (see photo below). A list of sources can be found on page 126. And if you'd like to consider another option

Instead of making bracket feet, you can purchase them. They are available in a variety of wood species to match your project. (See page 126 for sources.)

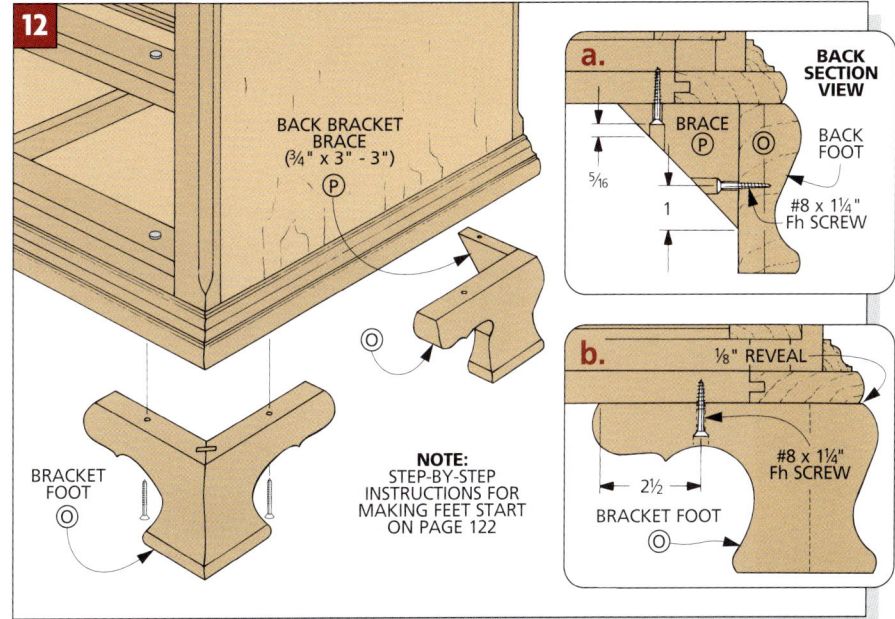

entirely, take a look at the base described in the Designer's Notebook on page 67.

No matter whether you make them or buy them, mounting the bracket feet couldn't be any easier. All you have to do is drill some counterbored pilot holes at the drill press and then screw the feet to the bottom of the case *(Fig. 12b)*. And as I mentioned before, the back bracket braces (P) are screwed and glued to the back feet before the assembly is secured to the cabinet *(Fig. 12a)*.

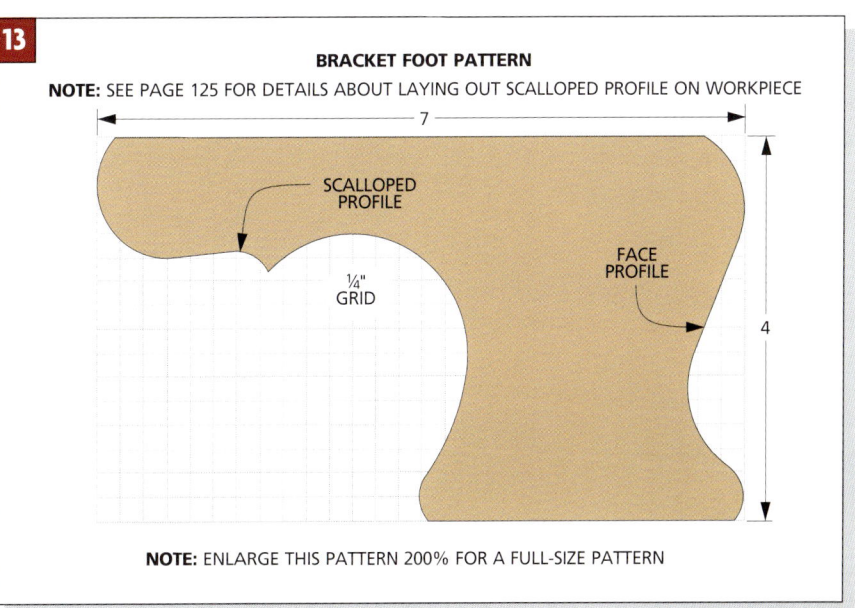

14

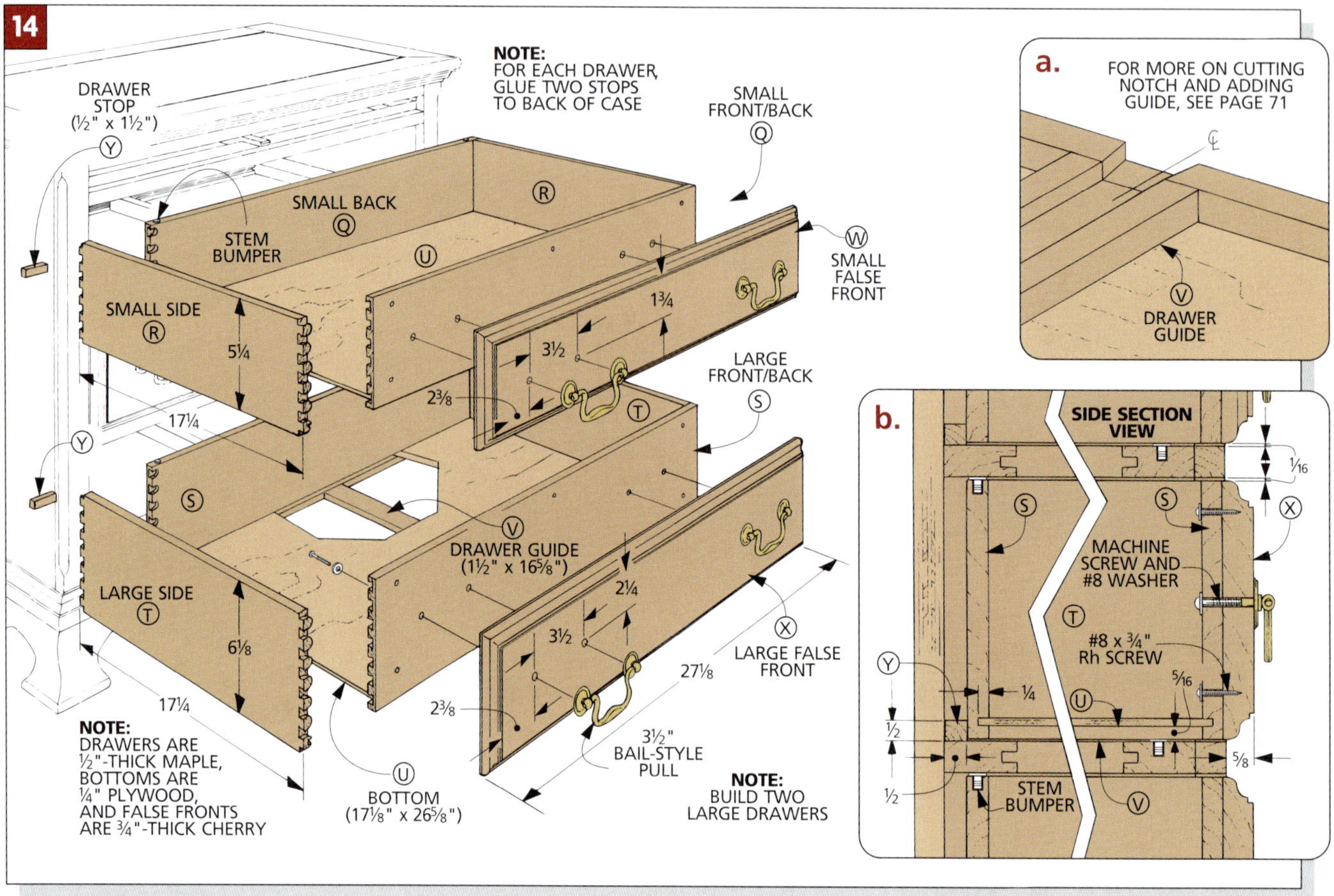

NOTE: FOR EACH DRAWER, GLUE TWO STOPS TO BACK OF CASE
DRAWER STOP (½" x 1½")
Y
SMALL FRONT/BACK
Q
SMALL BACK
Q
R
STEM BUMPER
U
W
SMALL FALSE FRONT
SMALL SIDE
R
5¼
1¾
3½
2⅜
17¼
LARGE FRONT/BACK
S
T
Y
S
V
DRAWER GUIDE (1½" x 16⅝")
LARGE SIDE
T
2¼
6⅛
3½
27⅛
X
LARGE FALSE FRONT
17¼
2⅜
NOTE: DRAWERS ARE ½"-THICK MAPLE, BOTTOMS ARE ¼" PLYWOOD, AND FALSE FRONTS ARE ¾"-THICK CHERRY
U
BOTTOM (17⅛" x 26⅝")
3½" BAIL-STYLE PULL
NOTE: BUILD TWO LARGE DRAWERS
a.
FOR MORE ON CUTTING NOTCH AND ADDING GUIDE, SEE PAGE 71
V
DRAWER GUIDE
b.
SIDE SECTION VIEW
1⁄16
S
S
X
MACHINE SCREW AND #8 WASHER
T
#8 x ¾" Rh SCREW
Y
5⁄16
¼
U
½
⅝
½
STEM BUMPER
V

DRAWERS

With the case completed, I moved on to building the three drawers. One is a bit shorter than the other two, but other than this, the drawers are identical.

CUT TO SIZE. The first thing to do is cut the drawer pieces to size from ½"-thick maple *(Fig. 14)*. Even though false fronts will be added later, I still sized the small front (Q), back (Q), and sides (R) and the large fronts (S), backs (S), and sides (T)

A dovetail jig makes quick work of building drawers that have a classic look when they are pulled open. For tips on getting the best results with a dovetail jig, see the Joinery article on page 70.

to fit the drawer openings, allowing for a $\frac{1}{16}$" gap along each edge. (This also allows the dovetails to work out evenly.)

DOVETAILS. The drawers are joined with routed, half-blind dovetails (see photo below left). If you have one of these jigs, then you already know the routine, but on page 70 there are a few tips that will make this process even easier.

Next I cut the grooves for the ¼" plywood drawer bottoms. Usually, I'll center these grooves on the bottom pin (and socket), but this time, there will be guides underneath these panels, so I wanted the bottoms (U) exactly $\frac{5}{16}$" from the bottoms of the drawers *(Fig. 14b)*. Before gluing the drawers together, it's a good idea to drill six shank holes in each front for attaching the false fronts later.

GUIDES. Now it's time to work on the drawer guides (V) *(Fig. 14a)*. Actually, the first thing you're going to want to do is cut a notch on the back bottom edge of each drawer to allow it to slip over the drawer runner. Then a guide is added to the bottom of each drawer so the drawer slides smoothly in and out of the cabinet. (Cutting the notches, and making and adding the guides are covered in detail in the Technique article on page 71.)

FALSE FRONTS. Now that the guides are in place, you'll be able to slide the drawers into the case and position the small false front (W) and large false fronts (X). These ¾"-thick cherry pieces are simple enough. They are cut to match the size of the drawer fronts. Then an ogee fillet profile is routed around the edges of each one *(Fig. 15)*.

Positioning the false fronts on the drawers and screwing them in place can take three hands. So I use carpet tape to

Since the runners and guides are centered under the drawers, plastic stem bumpers on the web frames and upper rear corners of the drawers prevent wood parts from wearing on each other.

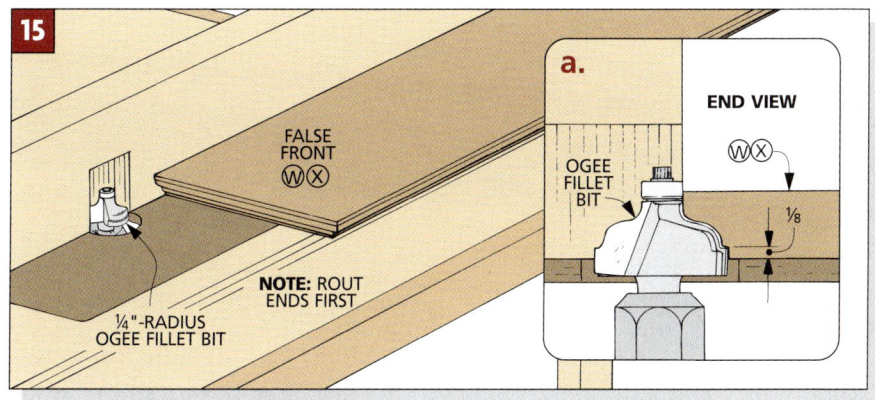

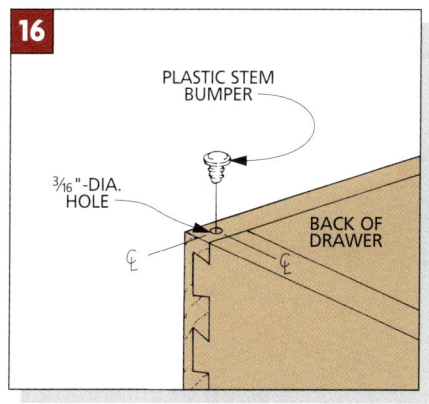

hold the false fronts in position temporarily. Then I can mark the positions of the pilot holes through the shank holes drilled earlier. After that, the false fronts are just screwed in place.

STOPS AND HARDWARE. Finally, I added two stops (Y) to each drawer opening, sizing them so the drawers stood $5/8''$ proud of the case *(Fig. 14b)*. Then the drawer pulls can be attached to

the front, and a stem bumper like the ones added to the web frames (see photo on opposite page) can be added to each upper back corner to keep the drawer from sagging *(Fig. 16)*.

DESIGNER'S NOTEBOOK

Bracket feet are distinctive, but for a base that's easier to make, this apron is an attractive alternative.

CONSTRUCTION NOTES:

■ This profiled base is simply a mitered apron sized so the completed frame will stand $5/16''$ proud at the front and sides. (This way, the base aligns with the trim molding above the bottom panel.)

MATERIALS LIST

NEW PARTS
DD	Front Apron (1)	$3/4 \times 4 - 32 1/8$
EE	Side Aprons (2)	$3/4 \times 4 - 19 11/16$
FF	Splines (2)	$1/4$ hdbd. - $1 1/2 \times 4$
GG	Cleat (1)	$3/4 \times 3/4 - 72$ rough

Note: Do not need part O.

HARDWARE SUPPLIES
(10) No. 8 x 1 1/2" Fh woodscrews
(11) No. 8 x 1 1/4" Fh woodscrews

■ After mitering the front apron (DD) to length, miter one end of each side apron (EE) and crosscut the side aprons to length.
■ Now the curved profiles can be laid out and cut (detail 'a' below). The real trick here is creating the straight line between the profiles. To do this, I used carpet tape to attach a straight piece of scrap along the layout line between the curves. The scrap guides the bearing of a flush trim bit in a router to trim the edge straight.
■ The rounded inside corners of the profile are chopped square with a chisel.
■ The apron pieces are glued together with splines (FF) to reinforce the miters.

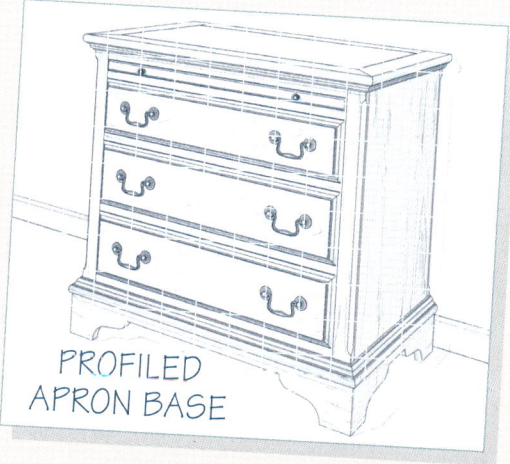

PROFILED APRON BASE

■ To attach the base to the chest, cleats (GG) are added to the front and sides, with back bracket braces (P) at the rear (refer to *Fig. 12* on page 65).

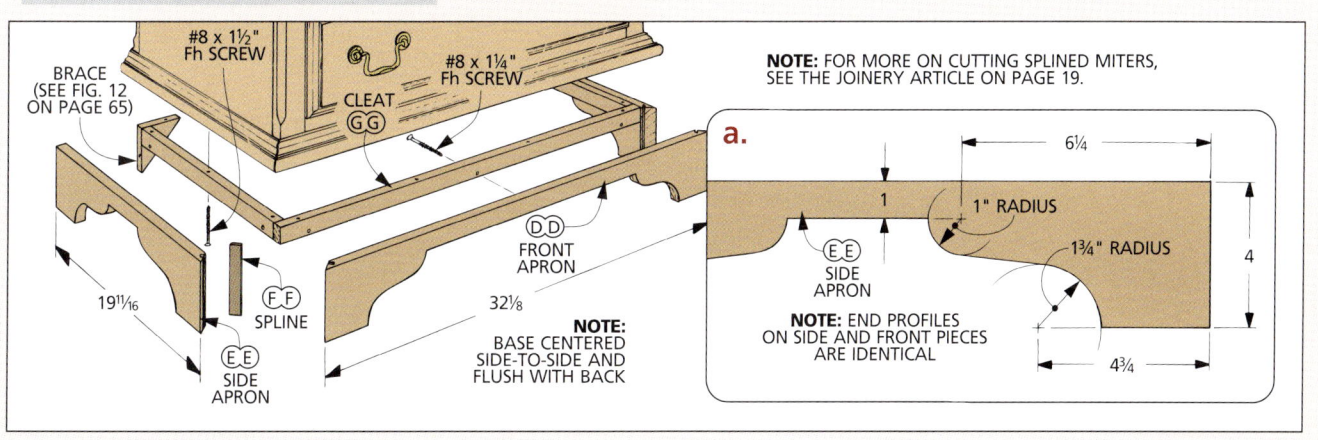

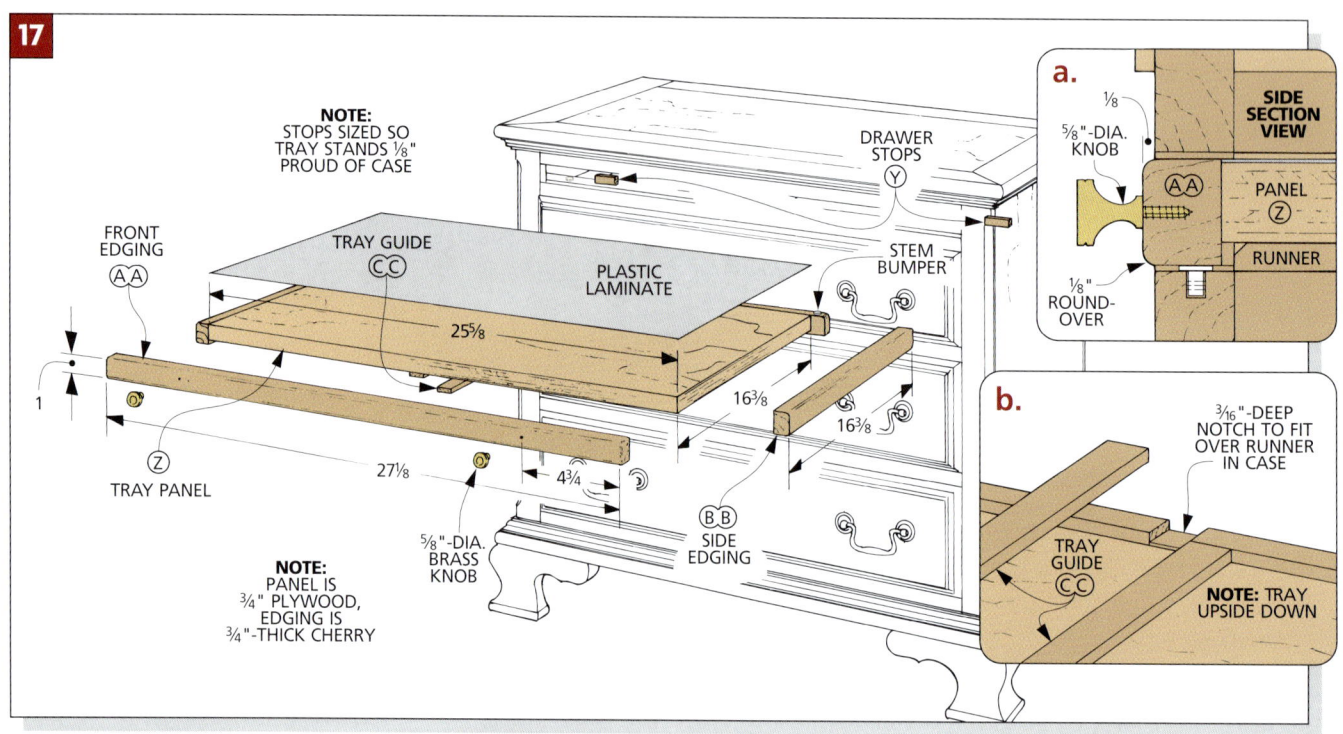

17

NOTE:
STOPS SIZED SO
TRAY STANDS ⅛"
PROUD OF CASE

DRAWER
STOPS
Ⓨ

FRONT
EDGING
Ⓐ Ⓐ

TRAY GUIDE
Ⓒ Ⓒ

PLASTIC
LAMINATE

STEM
BUMPER

25⅝

16⅜

16⅜

1

Ⓩ

TRAY PANEL

27⅛

4¾

Ⓑ Ⓑ
SIDE
EDGING

⅝"-DIA.
BRASS
KNOB

NOTE:
PANEL IS
¾" PLYWOOD,
EDGING IS
¾"-THICK CHERRY

a.

SIDE
SECTION
VIEW

⅛

⅝"-DIA.
KNOB

Ⓐ Ⓐ

PANEL
Ⓩ

RUNNER

⅛"
ROUND-
OVER

b.

³⁄₁₆"-DEEP
NOTCH TO FIT
OVER RUNNER
IN CASE

TRAY
GUIDE
Ⓒ Ⓒ

NOTE: TRAY
UPSIDE DOWN

PULL-OUT TRAY

All that remains to complete the chest is to build the pull-out tray above the drawers. This piece is really quite simple. It's a ¾" plywood panel with a piece of plastic laminate that provides a durable, easy-to-clean surface. Strips of solid wood wrap around the panel to hide the plywood edges *(Fig. 17)*.

TRAY PANEL. The first thing to do is cut the tray panel (Z) to size *(Fig. 17)*. I ripped the plywood 16⅜" wide. But I should probably point out that the tray won't stand proud as far as the drawers (⅝"). It only sticks out ⅛" past the front face of the case. (A couple of "drawer" stops added later will take care of this.) As for the length of the panel, it's the same as the width of a drawer (27⅛"), minus 1½" to allow for a strip of edging on each side. (My panel ended up 25⅝" long.)

LAMINATE. Next I added the plastic laminate to the panel. Not only does it protect the wood from hot cups of coffee and tea, it's also easy to clean.

When applying plastic laminate, I like to use contact cement. The important thing is to make sure you've got enough cement on both surfaces to get a good, consistent bond. Then applying and trimming the laminate is no big deal (see the Technique box on the opposite page).

EDGING. After the laminate has been trimmed flush, it's time to add the edging to cover the plywood. The height (width) of these ¾"-thick solid wood pieces will determine the gaps above and below the tray *(Fig. 17a)*. So I ripped the front (AA), back (AA), and side edging (BB) pieces 1" wide (tall) and left them extra long.

Since the front and back pieces will overlap the sides, I glued the sides to the panel first and trimmed them flush. Then

the same can be done for the front and back edging pieces.

The last thing I did before working on the tray guides was to rout a ⅛" roundover around all the edges and sand the corners *(Fig. 17a)*.

TRAY GUIDES. Now it's time to get the tray to slide in and out of the case smoothly. The system used here is similar to that used on the drawers. The tray is supported and guided from below by a drawer runner at the center and plastic stem bumpers on the web frame.

What's different here is how the tray rests on the runner. Instead of actually resting on the guides like the drawers do, the tray rests on the notch you'll cut in the back. This means, you want to be pretty careful that the notch in the back of the tray ends up ³⁄₁₆" deep *(Fig. 17b)*.

To guide the tray, I made a pair of thin tray guides (CC) to fit under the panel and

When a little bit of extra surface is needed, a pull-out tray is right at your fingertips. It's a plywood panel covered with plastic laminate and trimmed with solid wood.

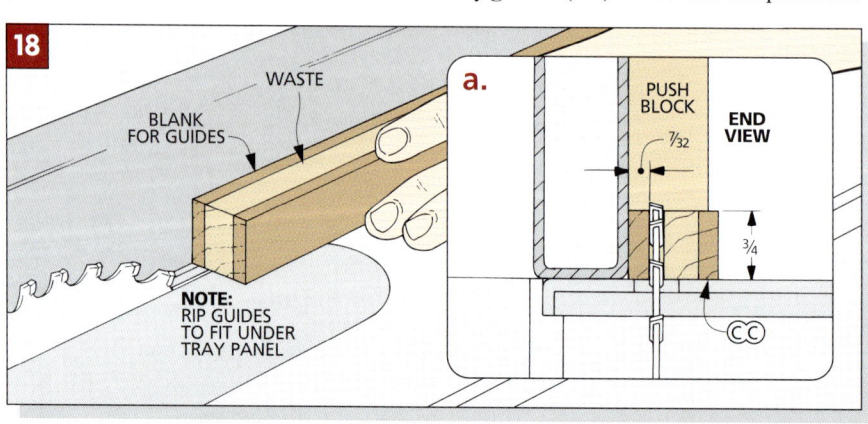

18

WASTE

BLANK
FOR GUIDES

a.

PUSH
BLOCK

END
VIEW

⁷⁄₃₂

¾

NOTE:
RIP GUIDES
TO FIT UNDER
TRAY PANEL

Ⓒ Ⓒ

ride against either side of the runner *(Fig. 19a)*. Mine ended up only $^7/_{32}$" tall. I found that rather than plane these thin pieces down, it's a lot easier to rip them to this width from the edge of a piece of $^3/_4$"-thick stock *(Fig. 18)*.

To mount the guides, the first thing I did was find the center of the tray and draw a line $^3/_8$" over from the center. (This is half the width of the centered runner inside the case.) To double check the accuracy of this line, I removed the drawers and slid the tray into its opening, centering it from side-to-side. Now you should be able to look up under the tray and see if your layout line aligns with the edge of the drawer runner. If it does, you can remove the tray and glue one of the guides flush with this line.

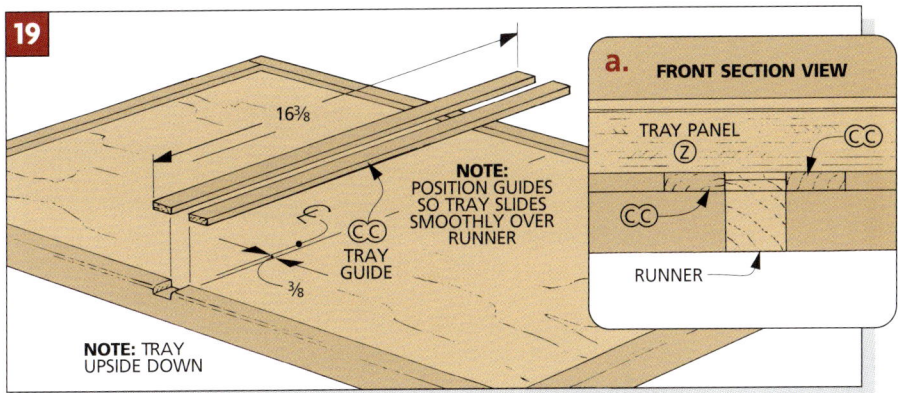

19

16⅜

NOTE: POSITION GUIDES SO TRAY SLIDES SMOOTHLY OVER RUNNER

TRAY GUIDE

⅜

NOTE: TRAY UPSIDE DOWN

a. FRONT SECTION VIEW

TRAY PANEL (Z)

CC

CC

RUNNER

To position the other guide, I used a scrap the same thickness as the drawer runner to act as a spacer. When the glue dried, I tested the fit and sanded the guides as needed for smooth operation.

HARDWARE AND FINISH. Finally, before adding the knobs to the tray *(Fig. 17)* and the bail pulls to the drawers (refer to *Fig. 14* on page 66), I wiped on several coats of tung-oil varnish. ∎

TECHNIQUE . . . . *Applying Laminate*

The pull-out tray is probably my favorite feature on the Bedside Chest. It provides a convenient place to set a drink or snack. And because the tray is covered with plastic laminate, you don't need to worry about a spill ruining the finish.

Applying the laminate is easy with the help of contact cement, a roller, and a flush trim bit to even up the edges.

CONTACT CEMENT

I prefer to use contact cement to attach laminates and veneers to a substrate. That's because once the mating surfaces touch, that's where they will stay. There's no sliding around on wet, slippery glue. And the instant bond allows you to roll out the surface to remove air bubbles.

The first thing to do is to apply a coat of contact cement to both the laminate and the plywood tray panel. You'll know the pieces are ready to be joined when you touch the cement and it feels tacky, but doesn't stick to your finger.

SPACERS. When you're ready to fasten the laminate to the tray, you don't want the pieces to touch until the laminate is properly positioned over the plywood. To allow you to do this, set some dowels on the panel to serve as spacers *(Step 1)*.

ROLL OUT

Once the laminate is in position, start from one end and remove one dowel at a time. Use a roller to press the laminate down as you go *(Step 1)*.

After the dowels have been removed, use the roller to work from the center of the panel out to the edges. This will help remove any trapped air bubbles. And don't be afraid to really bear down on the roller. The more pressure you apply, the better the bond will be.

TRIM FLUSH

After the laminate is in place, it needs to be trimmed to match the plywood panel. This is easy to do with a flush trim bit in a hand-held router. Just set the bit so the bearing rides on the plywood and rout around the panel *(Step 2)*. Then lightly sand the edges. But be careful not to round the edges. You want a tight joint between the panel and the edging.

1 Before positioning the laminate, lay a series of dowels on the plywood. Then when the laminate is properly placed, remove the dowels one at a time and roll the laminate down.

2 The laminate is oversize to allow some "play" room when you position it over the plywood. Trim the laminate even with the plywood with a flush trim bit in a hand-held router.

When building drawers, you just can't beat a half-blind dovetail jig. With a couple of test pieces and a few minutes, you can be set up to rout dovetails quickly and accurately. And whether you only have one drawer to build or an entire chest full, you know they'll be as strong as possible — and they'll look good too.

Most half-blind dovetail jigs work about the same way, but the instruction manual won't tell you everything. So I've come up with a few tips that'll help you rout dovetails cleanly and accurately.

Note: A couple of additional Shop Tips can be found on pages 77 and 91.

LABEL CAREFULLY

The more drawers you're building, the more you need to be organized to avoid mistakes. So I've gotten into the habit of labeling my drawer pieces carefully. I

identify each piece, number each corner sequentially around the drawer, and mark the bottom edges *(Fig. 1)*.

But I've also found that labeling the jig makes the process even more foolproof. The left side is numbered to remind me that the first and third corners are routed against this stop. The second and fourth corners are routed on the right side. And in each case, the bottom edges are positioned against the stops.

SCORING PASS

When routing dovetails, chipout on the side pieces is pretty common, but there's an easy way to avoid this problem (or at least minimize it). I like to make a light, skim pass from *right to left* before I begin following the template *(Fig. 2)*. This skim cut establishes the shoulder, removing the material that would tend to chip out.

DOUBLE CHECK

After turning off the router, your first impulse is to take the pieces out of the jig and test their fit. But resist that urge and visually double-check the sockets first to make sure they're all the same depth. This way, if you've inadvertently stopped one short, you can finish routing the pieces without any trouble.

SANDPAPER SOLUTION

Every once in a while, I would finish routing dovetails only to find that one of the workpieces had shifted slightly in the jig. And of course it was ruined.

Then I came up with a simple solution. I put adhesive-backed sandpaper on the base of the jig and the clamping bars *(Fig. 1)*. This gives these parts a firm grip on the workpieces.

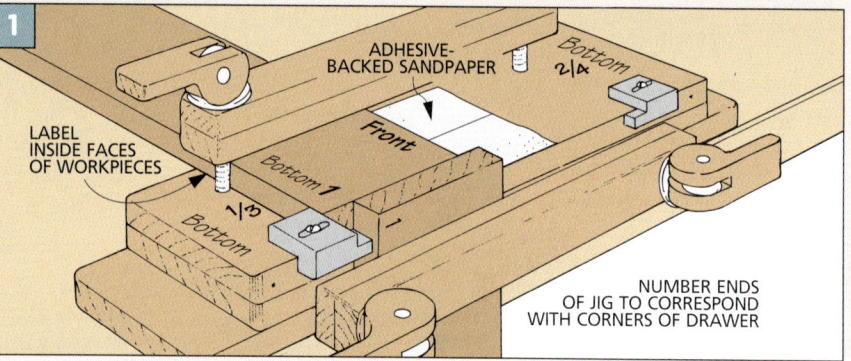

LABEL INSIDE FACES OF WORKPIECES

ADHESIVE-BACKED SANDPAPER

Bottom 2|4

Front

Bottom 1

1|3

Bottom

NUMBER ENDS OF JIG TO CORRESPOND WITH CORNERS OF DRAWER

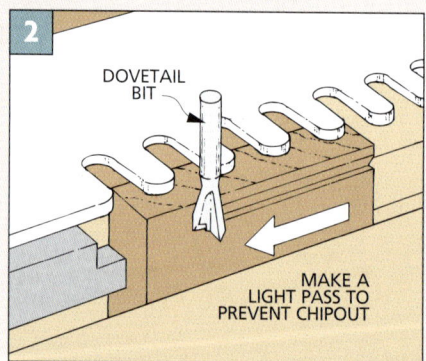

DOVETAIL BIT

MAKE A LIGHT PASS TO PREVENT CHIPOUT

TROUBLESHOOTING GUIDE

Setting up to rout machine-cut dovetails is always a trial and error effort. Before you start routing your actual workpieces, you'll need to cut a few test joints on some scrap to "fine-tune" the joint. Depending on how the test joints fit, you may need to adjust the jig, the bit, or the workpieces to get a perfect fit. Shown below are a few common problems (and solutions).

Too Loose. *If the pins are too loose in the sockets, increase the depth of the bit.*
Too Tight. *If the joint is too tight, decrease the depth of the bit.*

Too Deep. *If the pins go too deep into the sockets, move the template forward.*
Too Shallow. *If they're not deep enough, move the template back.*

Offset. *If the workpieces don't align at the top or bottom, that usually means they may not have been positioned tight against the stops of the jig.*

TECHNIQUE .. *Center Drawer Guides*

Under the drawers of the Bedside Chest, there are thin, U-shaped guides that fit over the runners inside the case (see main photo at right). Positioning these guides correctly is important because it affects how well the drawers will slide.

NOTCH. Before you begin working on the guides, you'll need to cut a notch in the back of the drawer (inset photo). This allows the drawer to fit over the drawer runner. After laying out the notch centered on the back bottom edge of the drawer, I made two vertical cuts $3/16$" down with a hand saw *(Fig. 1)*. Then I removed the waste with a chisel, scoring the bottom edge of the notch and paring down to remove the waste *(Fig. 1a)*. Later, after the guides are installed, you'll "flare" the notch to make it easier to slide the drawer into the case (see inset photo).

GUIDES. At this point, you can start on the drawer guides (V). I planed some stock to $5/16$" thick so the pieces would fit under the drawer bottoms *(Fig. 2)*.

Each guide needs a centered groove so it'll slide over the runners in the case.

I cut this groove in two passes over a dado blade, flipping the guide between passes *(Fig. 2)*. Sneak up on the final width of the groove, testing the fit on the runners in the case after each pass.

There's one thing to watch. With a thin piece like this, you'll find that the dado blade will tend to lift it up off the saw. So to keep it pressed down safely, I made a long push block out of a 2x4 and nailed a cleat to the end to "hook" the workpiece *(Fig. 2)*.

MOUNT GUIDES. The last step is to mount each guide to its drawer. To do this accurately, first draw a centerline on the drawer bottom and then measure $3/4$" in both directions to lay out the edges of the $1\frac{1}{2}$"-wide guide *(Fig. 3)*. (I drew the lines using a square and then checked that they were centered by measuring out from each line to the drawer side.)

Finally, to double-check the layout lines, I centered the drawer in its opening and looked underneath to make sure the lines were centered over the drawer runner. Then I glued the guide to the drawer bottom, using a few heavy objects to apply some pressure.

Note: If you find that the notch in the back of the drawer and the groove in the runner don't align perfectly, just use a chisel to widen the transition from the notch to the runner.

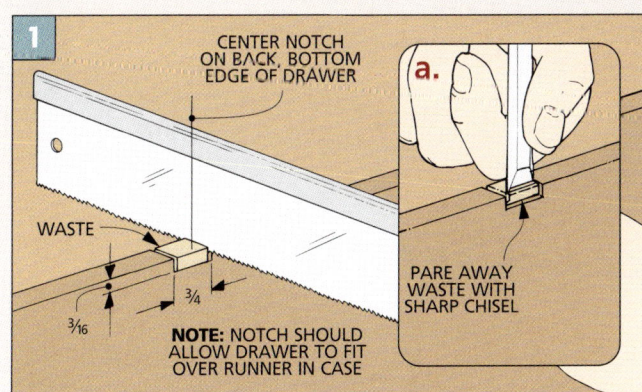

1
CENTER NOTCH ON BACK, BOTTOM EDGE OF DRAWER
a.
WASTE
$3/4$
$3/16$
PARE AWAY WASTE WITH SHARP CHISEL
NOTE: NOTCH SHOULD ALLOW DRAWER TO FIT OVER RUNNER IN CASE

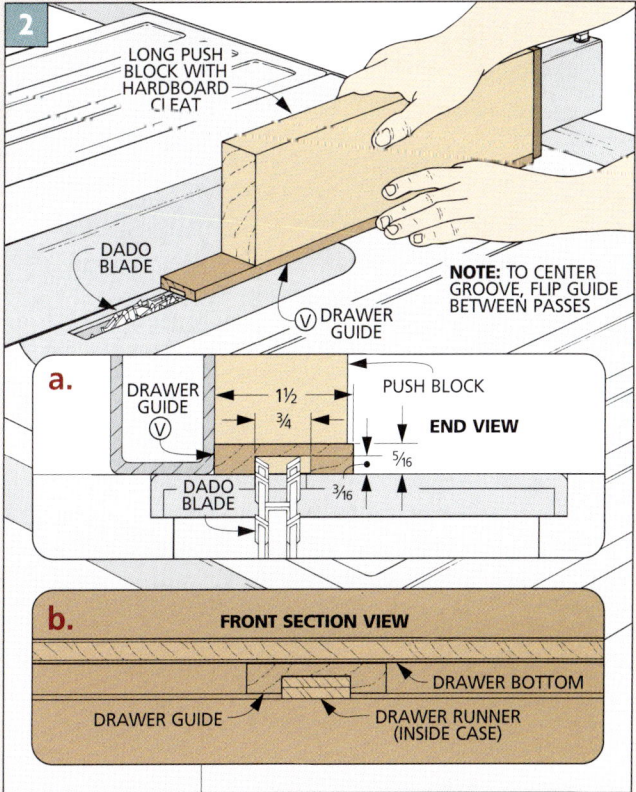

2
LONG PUSH BLOCK WITH HARDBOARD CLEAT
DADO BLADE
(V) DRAWER GUIDE
NOTE: TO CENTER GROOVE, FLIP GUIDE BETWEEN PASSES
a.
DRAWER GUIDE (V)
$1\frac{1}{2}$
$3/4$
PUSH BLOCK
END VIEW
$5/16$
DADO BLADE
$3/16$
b.
FRONT SECTION VIEW
DRAWER BOTTOM
DRAWER GUIDE
DRAWER RUNNER (INSIDE CASE)

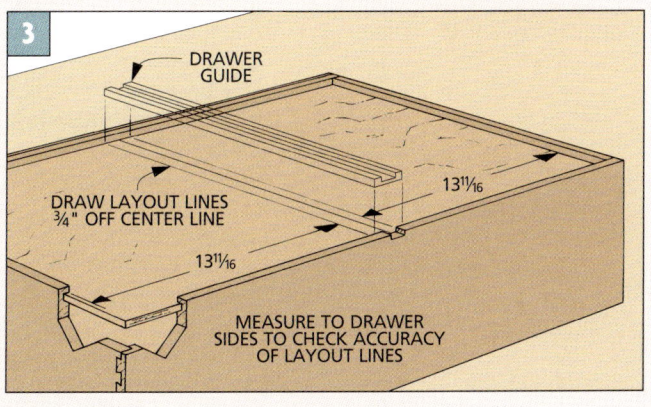

3
DRAWER GUIDE
DRAW LAYOUT LINES $3/4$" OFF CENTER LINE
$13^{11}/_{16}$
$13^{11}/_{16}$
MEASURE TO DRAWER SIDES TO CHECK ACCURACY OF LAYOUT LINES

Blanket Chest

Filled with classic details, this handsome chest boasts a large storage compartment and three drawers. An alternate design uses frame and panel construction with tongues and grooves instead of dovetails.

Have you noticed how dovetails are always hidden away on the corners of a drawer? To me, it's never seemed quite right that one of the strongest and most attractive joints you can make is typically kept from view.

This hasn't always been the case. In the 18th and 19th centuries, country furniture often featured dovetail joinery because of its strength and durability. Craftsmen used the exposed dovetails as both an integral part of their design, and as a display of their skill.

So it was only natural to use this "country" furniture style when I designed this Blanket Chest with exposed dovetail

joints. And as you can see, the dovetails aren't all spaced the same. There are two narrow tails at the top and bottom of the chest with three wider tails in the middle. (And while I was at it, I used through dovetails for the drawers as well.)

DOVETAIL JIG. But I have to admit, I did use a bit of modern technology to cut this traditional joint — a through dovetail jig. It allowed me to cut variably-spaced dovetails and still get a perfect fit at each corner. More about using this type of jig can be found on page 76.

Unlike many modern blanket chests, with their overabundance of frills, this chest is straightforward and functional.

The drawers are a good example. Unlike the false fronts that are tacked onto the cases of the modern versions, all three drawers in this chest actually work.

WOODS. In keeping with the traditional theme, I decided to build the chest out of cherry, a wood that was very abundant, and often used during this period.

Although moths aren't the problem today they were at one time, I added cedar to the bottom of the main box. Every time I open the lid, the aromatic scent fills the room. But after time, the scent of the cedar will taper off. To bring it back, just give the cedar a light sanding to expose some fresh wood.

EXPLODED VIEW

OVERALL DIMENSIONS:
43⁷⁄₈W x 18⁵⁄₈D x 22¹⁄₄H

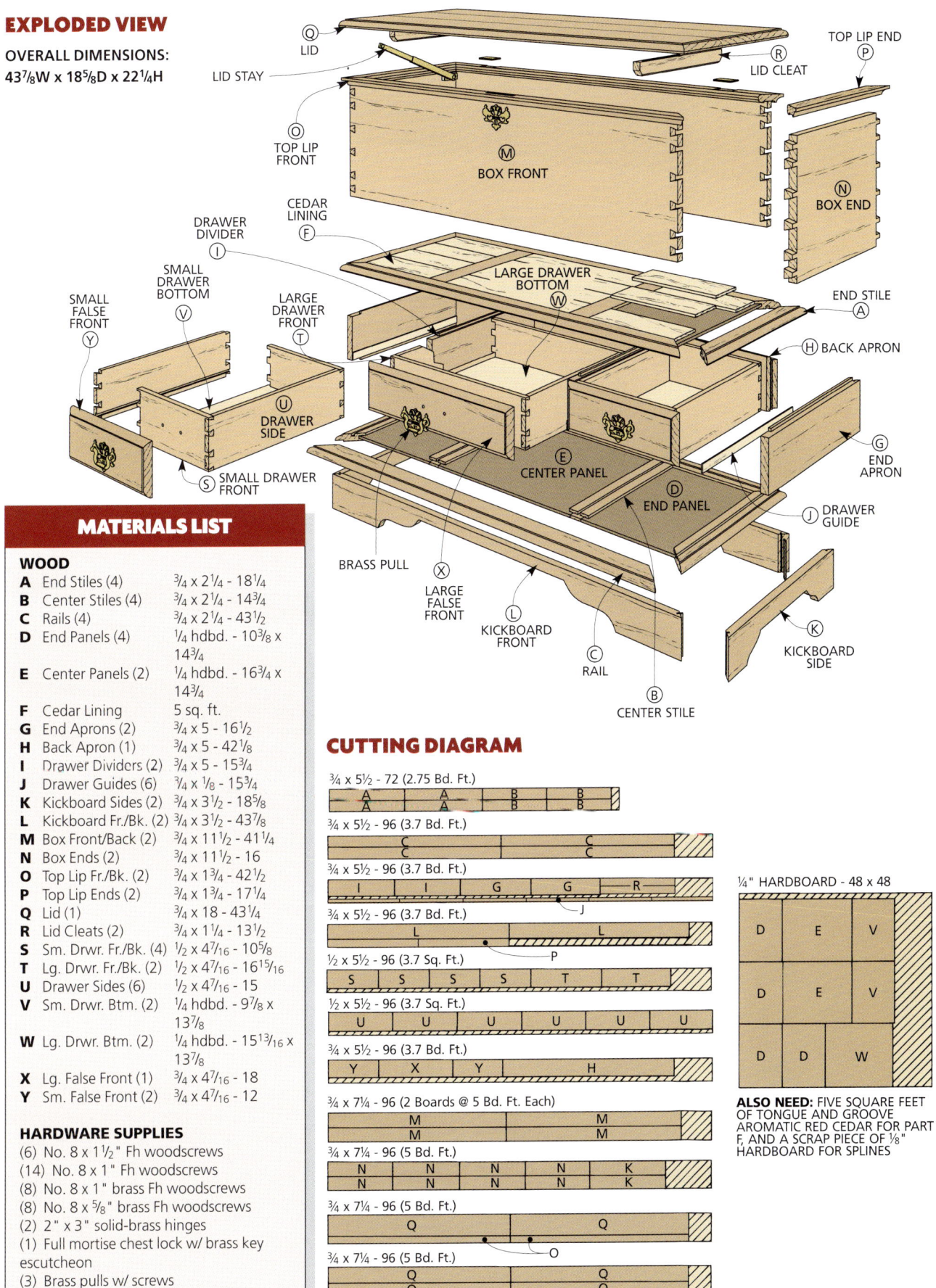

Labels on exploded view: LID (Q), LID STAY, TOP LIP END (P), LID CLEAT (R), TOP LIP FRONT (O), BOX FRONT (M), BOX END (N), CEDAR LINING (F), DRAWER DIVIDER (I), SMALL DRAWER BOTTOM, LARGE DRAWER BOTTOM (W), END STILE (A), SMALL FALSE FRONT (Y), LARGE DRAWER FRONT (T), BACK APRON (H), DRAWER SIDE (U), END APRON (G), SMALL DRAWER FRONT (S), CENTER PANEL (E), END PANEL (D), DRAWER GUIDE (J), BRASS PULL, LARGE FALSE FRONT (X), KICKBOARD FRONT (L), RAIL (C), CENTER STILE (B), KICKBOARD SIDE (K)

MATERIALS LIST

WOOD

A	End Stiles (4)	³⁄₄ x 2¹⁄₄ - 18¹⁄₄
B	Center Stiles (4)	³⁄₄ x 2¹⁄₄ - 14³⁄₄
C	Rails (4)	³⁄₄ x 2¹⁄₄ - 43¹⁄₂
D	End Panels (4)	¹⁄₄ hdbd. - 10³⁄₈ x 14³⁄₄
E	Center Panels (2)	¹⁄₄ hdbd. - 16³⁄₄ x 14³⁄₄
F	Cedar Lining	5 sq. ft.
G	End Aprons (2)	³⁄₄ x 5 - 16¹⁄₂
H	Back Apron (1)	³⁄₄ x 5 - 42¹⁄₈
I	Drawer Dividers (2)	³⁄₄ x 5 - 15³⁄₄
J	Drawer Guides (6)	³⁄₄ x ¹⁄₈ - 15³⁄₄
K	Kickboard Sides (2)	³⁄₄ x 3¹⁄₂ - 18⁵⁄₈
L	Kickboard Fr./Bk. (2)	³⁄₄ x 3¹⁄₂ - 43⁷⁄₈
M	Box Front/Back (2)	³⁄₄ x 11¹⁄₂ - 41¹⁄₄
N	Box Ends (2)	³⁄₄ x 11¹⁄₂ - 16
O	Top Lip Fr./Bk. (2)	³⁄₄ x 1³⁄₄ - 42¹⁄₂
P	Top Lip Ends (2)	³⁄₄ x 1³⁄₄ - 17¹⁄₄
Q	Lid (1)	³⁄₄ x 18 - 43¹⁄₄
R	Lid Cleats (2)	³⁄₄ x 1¹⁄₄ - 13¹⁄₂
S	Sm. Drwr. Fr./Bk. (4)	¹⁄₂ x 4⁷⁄₁₆ - 10⁵⁄₈
T	Lg. Drwr. Fr./Bk. (2)	¹⁄₂ x 4⁷⁄₁₆ - 16¹⁵⁄₁₆
U	Drawer Sides (6)	¹⁄₂ x 4⁷⁄₁₆ - 15
V	Sm. Drwr. Btm. (2)	¹⁄₄ hdbd. - 9⁷⁄₈ x 13⁷⁄₈
W	Lg. Drwr. Btm. (2)	¹⁄₄ hdbd. - 15¹³⁄₁₆ x 13⁷⁄₈
X	Lg. False Front (1)	³⁄₄ x 4⁷⁄₁₆ - 18
Y	Sm. False Front (2)	³⁄₄ x 4⁷⁄₁₆ - 12

HARDWARE SUPPLIES

(6) No. 8 x 1¹⁄₂" Fh woodscrews
(14) No. 8 x 1" Fh woodscrews
(8) No. 8 x 1" brass Fh woodscrews
(8) No. 8 x ⁵⁄₈" brass Fh woodscrews
(2) 2" x 3" solid-brass hinges
(1) Full mortise chest lock w/ brass key escutcheon
(3) Brass pulls w/ screws
(1) Lid stay w/ screws

CUTTING DIAGRAM

³⁄₄ x 5¹⁄₂ - 72 (2.75 Bd. Ft.)
A A B B
A A B B

³⁄₄ x 5¹⁄₂ - 96 (3.7 Bd. Ft.)
C C
C C

³⁄₄ x 5¹⁄₂ - 96 (3.7 Bd. Ft.)
I I G G R
J

³⁄₄ x 5¹⁄₂ - 96 (3.7 Bd. Ft.)
L L
P

¹⁄₂ x 5¹⁄₂ - 96 (3.7 Sq. Ft.)
S S S S T T

¹⁄₂ x 5¹⁄₂ - 96 (3.7 Sq. Ft.)
U U U U U U

³⁄₄ x 5¹⁄₂ - 96 (3.7 Bd. Ft.)
Y X Y H

³⁄₄ x 7¹⁄₄ - 96 (2 Boards @ 5 Bd. Ft. Each)
M M
M M

³⁄₄ x 7¹⁄₄ - 96 (5 Bd. Ft.)
N N N K
N N N K

³⁄₄ x 7¹⁄₄ - 96 (5 Bd. Ft.)
Q Q
O

³⁄₄ x 7¹⁄₄ - 96 (5 Bd. Ft.)
Q Q
Q Q

¹⁄₄" HARDBOARD - 48 x 48
D E V
D E V
D D W

ALSO NEED: FIVE SQUARE FEET OF TONGUE AND GROOVE AROMATIC RED CEDAR FOR PART F, AND A SCRAP PIECE OF ¹⁄₈" HARDBOARD FOR SPLINES

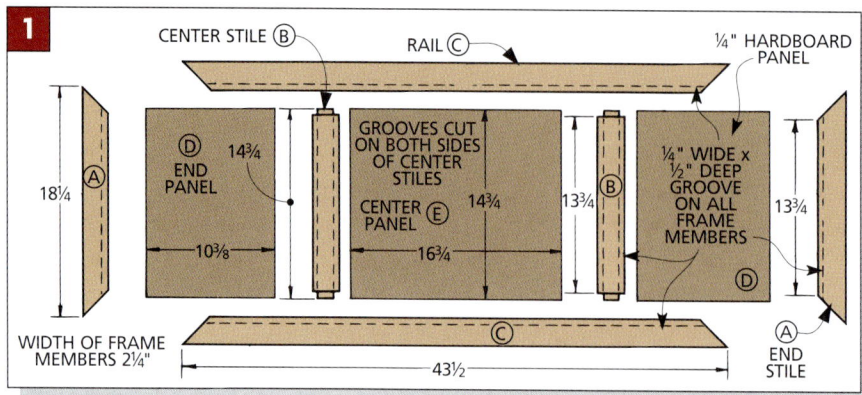

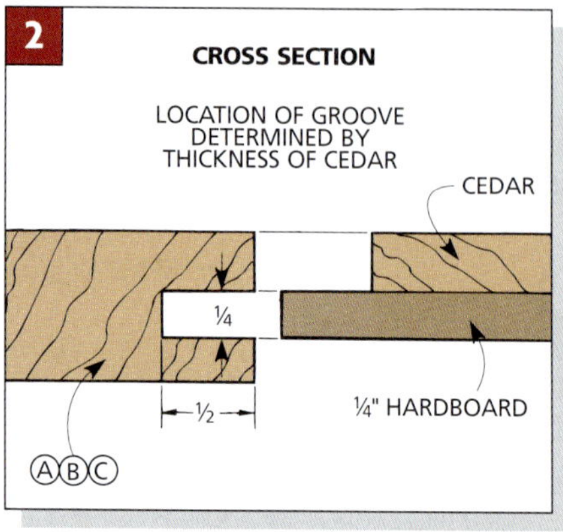

1

CENTER STILE (B) RAIL (C) ¼" HARDBOARD PANEL

GROOVES CUT ON BOTH SIDES OF CENTER STILES

(D) 14¾ END PANEL

CENTER PANEL (E) 14¾

13¾ (B)

¼" WIDE x ½" DEEP GROOVE ON ALL FRAME MEMBERS

13¾ (D)

18¼ (A)

10⅜ 16¾

(A) END STILE

WIDTH OF FRAME MEMBERS 2¼"

(C)

43½

2

CROSS SECTION

LOCATION OF GROOVE DETERMINED BY THICKNESS OF CEDAR

CEDAR

¼

½

¼" HARDBOARD

(A)(B)(C)

FRAMES

The Blanket Chest consists of three stacked sections: the main box, the drawer carcase below it, and the kickboard assembly. I started by building the two frames that form the top and bottom of the drawer carcase.

Both of these frames are identical, and use a typical web frame construction with hardboard panels (*Fig. 1*).

The first step is to cut the end stiles (A), center stiles (B), and rails (C) for the upper and lower frames 1" to 2" longer than their finished lengths, and to a final width of 2¼" (*Fig. 1*).

PANEL GROOVES. Next, grooves are cut on the inside edge of each frame member to hold the hardboard panels. The grooves are positioned so that when the cedar closet lining is added later, it will be flush with the frame (*Fig. 2*). Also,

note that the grooves are cut on the inside edges of the rails and end stiles, and on *both* edges of the center stiles (*Fig. 3*).

Note: Even though the cedar is only added to the upper frame, go ahead and use the same setup to cut the grooves in all twelve frame pieces.

APRON AND DIVIDER GROOVES. The next step is to cut ¼" x ¼" grooves for mounting the aprons and drawer dividers between the frames (*Fig. 3*). For the aprons, cut a groove in each back rail and end stile ¹³/₁₆" from the inside edge (*Fig. 3*). Then, to accept the drawer dividers, cut a groove in each center stile, centered on its width.

MITERING THE FRAME. Once the grooves are cut, the rails and end stiles are mitered to final length. The rails are mitered 43½" long, and the end stiles are mitered to 18¼" (*Fig. 1*).

CUT TENONS. To join the center stiles between the front and back rails, stub tenons are cut on both ends to fit the grooves in the rails. The final shoulder-to-shoulder length of the center stiles should equal the heel-to-heel length of the end stiles (13¾") (*Fig. 1*).

PANELS. Once the tenons are cut, the end panels (D) are cut to size (*Fig. 1*), and dry-assembled with the frame to find the dimensions for the center panel (E). Just measure the center opening (include the depth of the grooves), and cut a ¼" hardboard panel to fit (*Fig. 1*).

ASSEMBLY. Now dry-assemble both frames to make sure that everything fits and the assembly is square. Once everything checks out, glue both frames together with the panels.

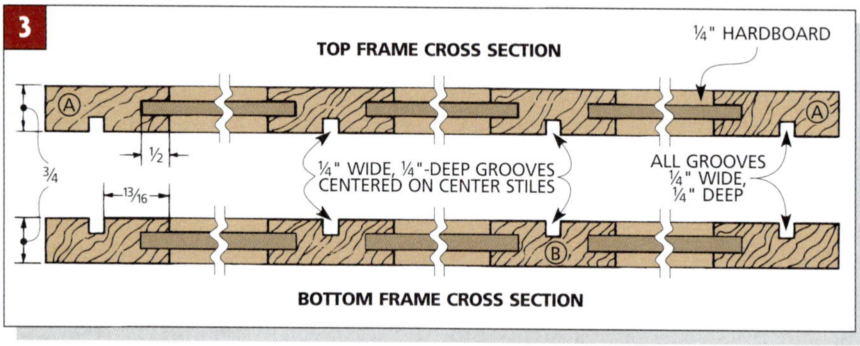

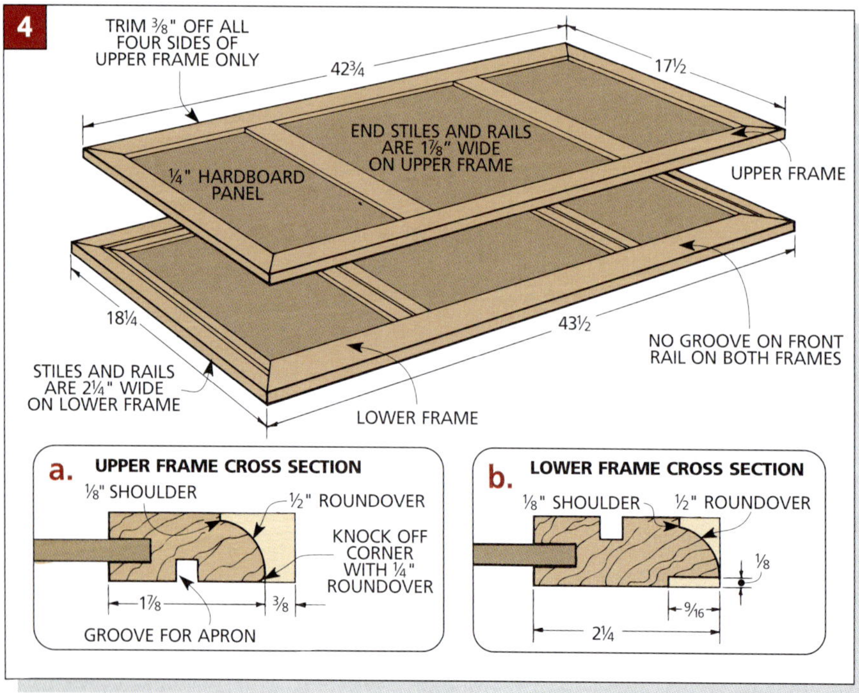

MOLDING

When the frames are dry, ⅜" is trimmed off all four sides of the upper frame only (*Fig. 4*). By cutting an equal amount off all four sides, the grooves for the apron

3

TOP FRAME CROSS SECTION

¼" HARDBOARD

(A) ½ ¾ ¹³/₁₆

¼" WIDE, ¼"-DEEP GROOVES CENTERED ON CENTER STILES

ALL GROOVES ¼" WIDE, ¼" DEEP

(A)

(B)

BOTTOM FRAME CROSS SECTION

4

TRIM ⅜" OFF ALL FOUR SIDES OF UPPER FRAME ONLY

42¾ 17½

END STILES AND RAILS ARE 1⅞" WIDE ON UPPER FRAME

¼" HARDBOARD PANEL

UPPER FRAME

18¼

STILES AND RAILS ARE 2¼" WIDE ON LOWER FRAME

43½

NO GROOVE ON FRONT RAIL ON BOTH FRAMES

LOWER FRAME

a. **UPPER FRAME CROSS SECTION**

⅛" SHOULDER ½" ROUNDOVER

KNOCK OFF CORNER WITH ¼" ROUNDOVER

1⅞ ⅜

GROOVE FOR APRON

b. **LOWER FRAME CROSS SECTION**

⅛" SHOULDER ½" ROUNDOVER

⅛

⁹/₁₆

2¼

and drawer dividers in the upper frame remain perfectly aligned with the grooves in the lower frame. (In this case, the upper frame will be $\frac{3}{4}$" smaller in both dimensions than the lower frame.)

RABBET THE LOWER FRAME. Next, a rabbet is cut on the bottom edge of the lower frame so the kickboard can be joined to it later *(Fig. 4b)*.

ROUT THE EDGE. The top outside edges on both frames receive a roundover with a shoulder. (I did this on the router table.) And watch which face you're routing. On the upper frame, this profile is cut on the face *without* the groove *(Fig. 4a)*. On the lower frame, it's made on the face *with* the groove *(Fig. 4b)*.

After the roundover was cut on both frames, I softened the bottom edges of the upper frame *(Fig. 4a)*. This edge can be removed with a sander, or with a $\frac{1}{4}$" roundover bit set for a very shallow cut.

CEDAR LINING. Next, the red cedar lining (F) is attached to the top of the panels in the upper frame only. First, trim off the tongues and grooves on the edges of the cedar. Then rip five equal-width pieces to fit the panel opening, leaving a small gap between each piece for expansion. Next, cut the cedar to length to fit snug in the panel opening *(Fig. 5)*.

To allow for expansion, just use a spot of glue in the center of each cedar strip. Then I used a set of clamping cauls to reach across the panel *(Fig. 5a)*.

DRAWER CARCASE

Aprons and drawer dividers connect the just-completed web frames *(Fig. 7)*. I started by cutting two end aprons (G), a back apron (H), and two drawer dividers (I) 5" wide, and to rough lengths *(Fig. 6)*.

TONGUES. The next step is to form tongues on the aprons and dividers to fit into the $\frac{1}{4}$"-wide grooves in the frames. On the drawer dividers, rabbets are cut along both edges to form tongues centered on the thickness of the divider *(Fig. 6a)*. The shoulder-to-shoulder width between the rabbets should be $4\frac{1}{2}$".

On the aprons, the tongues are offset so they are flush with the inside face of the apron *(Fig. 6a)*. Be sure that the shoulder-to-shoulder width is exactly the same as on the drawer dividers ($4\frac{1}{2}$").

MITERS. The back corners of the aprons are joined with a splined miter joint *(Fig. 7a)*. The splines reinforce the miters and help keep everything lined up during assembly.

To start, miter the back corner of each end apron so the tongue is on the inside face *(Fig. 7a)*. Then the front edge of each end apron is trimmed so the overall length is $16\frac{1}{2}$" *(Fig. 7)*. Also, notch the tongues on the aprons to fit the grooves in the frames *(Fig. 8)*.

Next, position both end aprons in the upper frame, and miter both ends of the back apron to fit between them.

SPLINED MITER. While the saw is still set at 45°, lower the blade and cut a $\frac{1}{4}$"-deep kerf on the face of each miter for the $\frac{1}{8}$"-thick hardboard splines *(Fig. 7a)*. (See the Joinery box on page 19 for more details about cutting this joint.)

TRIM DIVIDERS. The next thing to do is to trim the drawer dividers to length so the dividers and aprons are flush across the front of the frames.

DRAWER GUIDES. Finally, I ripped drawer guides (J) from the edge of a blank and glued them to the end aprons and drawer dividers *(Figs. 7 and 8)*. Although all the parts for the drawer carcase are finished at this point, it's not assembled until after the kickboard assembly and the main box are built.

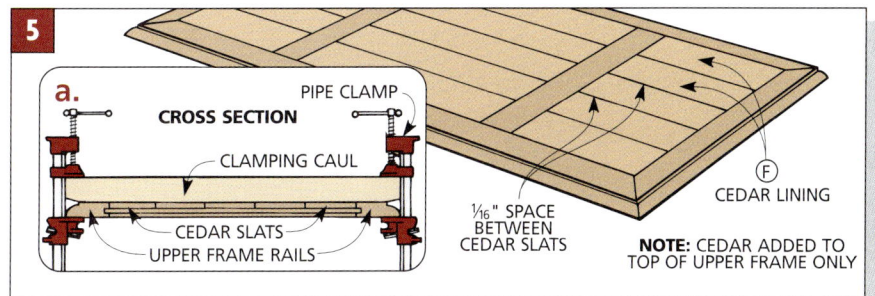

5

a. CROSS SECTION

PIPE CLAMP

CLAMPING CAUL

CEDAR SLATS

UPPER FRAME RAILS

$\frac{1}{16}$" SPACE BETWEEN CEDAR SLATS

CEDAR LINING (F)

NOTE: CEDAR ADDED TO TOP OF UPPER FRAME ONLY

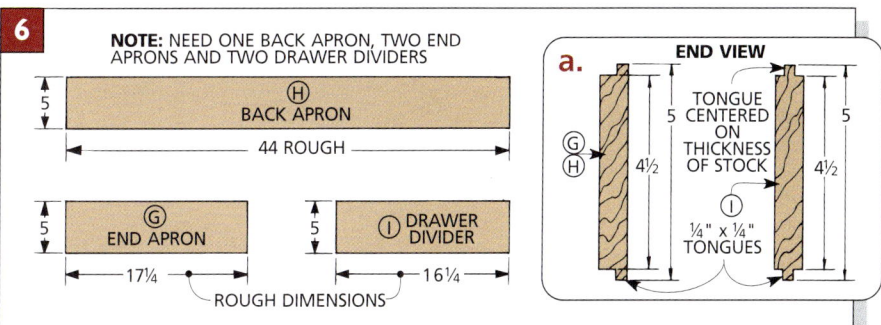

6

NOTE: NEED ONE BACK APRON, TWO END APRONS AND TWO DRAWER DIVIDERS

5

BACK APRON (H)

44 ROUGH

5

END APRON (G)

17¼

5

DRAWER DIVIDER (I)

16¼

ROUGH DIMENSIONS

a. END VIEW

(G) (H)

5

4½

TONGUE CENTERED ON THICKNESS OF STOCK

5

4½

(I)

¼" x ¼" TONGUES

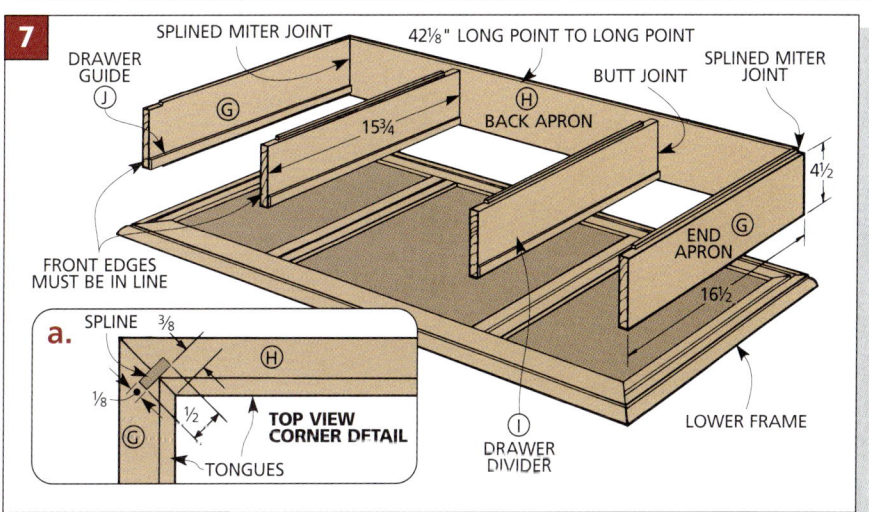

7

SPLINED MITER JOINT

42⅛" LONG POINT TO LONG POINT

SPLINED MITER JOINT

DRAWER GUIDE (J)

(G)

15¾

BUTT JOINT

BACK APRON (H)

4½

FRONT EDGES MUST BE IN LINE

END APRON (G)

16½

a. SPLINE ⅜

⅛

½

(H)

TOP VIEW CORNER DETAIL

(G)

TONGUES

(I) DRAWER DIVIDER

LOWER FRAME

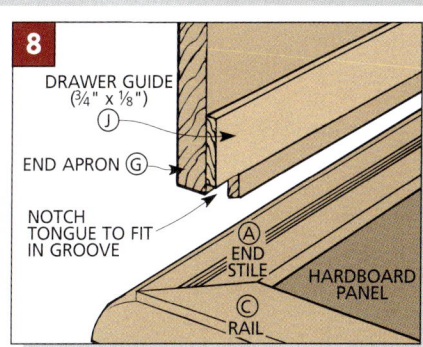

8

DRAWER GUIDE (¾" x ⅛") (J)

END APRON (G)

NOTCH TONGUE TO FIT IN GROOVE

(A) END STILE

HARDBOARD PANEL

(C) RAIL

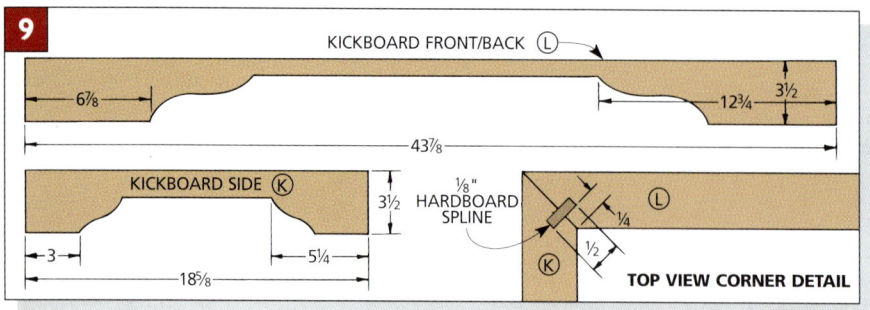

9

KICKBOARD FRONT/BACK (L)

6⅞ 12¾ 3½

43⅞

KICKBOARD SIDE (K)

3½ ⅛" HARDBOARD SPLINE (L) ¼

3 5¼ ½

18⅝ (K) **TOP VIEW CORNER DETAIL**

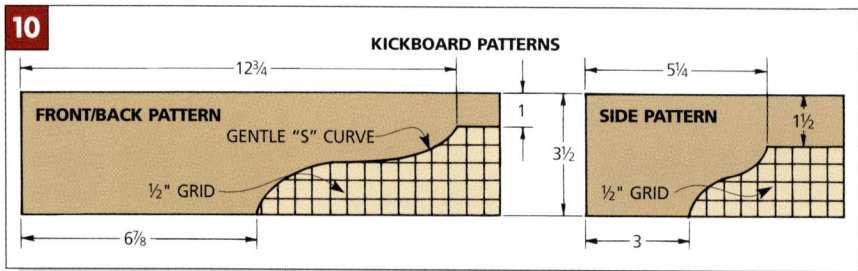

10

KICKBOARD PATTERNS

12¾ 5¼

FRONT/BACK PATTERN 1 **SIDE PATTERN** 1½

GENTLE "S" CURVE 3½

½" GRID ½" GRID

6⅞ 3

KICKBOARD ASSEMBLY

The kickboard assembly consists of a molded frame that's joined with miters and splines *(Fig. 9)*.

The kickboard sides (K), front (L), and back (L) are cut 3½" wide and mitered to fit the rabbet on the bottom face of the lower frame *(Figs. 9 and 13a)*. Then the patterns *(Fig. 10)* are traced onto the pieces, cut out using a band saw, and the kickboard is glued together.

MAIN BOX

With both the drawer carcase and kickboard assembly finished, it's finally time to cut some dovetails.

GLUING UP. The first step is to glue up enough stock to produce two solid-wood

SHOP INFO .... *Machine-Cut Through Dovetails*

Commercial dovetail jigs have taken much of the mystery out of cutting through dovetails. While cutting them by hand isn't difficult, doing it well does take practice. That may be why dovetail jigs are so popular. After the jig is set up (which takes practice also), great-looking, tight joints are pretty much a cinch.

TEMPLATE AND ROUTER. Through-dovetail jigs all operate essentially the same way. The workpiece is clamped vertically in the jig. (With smaller jigs, the jig is clamped to the vertical workpiece.) Next, a template is positioned over the end of the workpiece. A router rides on, and is guided by, the template as the router cuts the tails or pins.

The jigs use stops, spacers, or a combination of both to position the pieces in the jig and position the template over the workpiece. This ensures that the joint will fit together properly and that the edges of the workpieces will be flush.

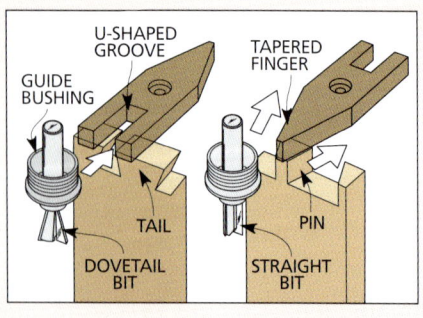

U-SHAPED GROOVE

GUIDE BUSHING

TAPERED FINGER

TAIL PIN

DOVETAIL BIT STRAIGHT BIT

GROOVES AND FINGERS. The way the template works varies slightly from jig to jig, but generally, it consists of two parts: a series of U-shaped grooves and a set of corresponding, tapered fingers.

The U-shaped grooves are used with a dovetail bit when routing the tails. The grooves are sized to accept a guide bushing in the base of the router (or in some cases, a bearing above the cutter of the bit). As the router slides into each groove, the dovetail bit cuts a socket for a pin. What's left are tails (see drawing).

To cut the pins in the mating piece, you switch to a straight bit in the router. For this step, the tapered fingers of the template are positioned over the workpiece. Then the material between the fingers is routed away, leaving a pin under each finger (see drawing).

To get a perfect fit, you'll need to test-fit the joint and probably do some "fine-tuning." Just as with hand-cut dovetails, you adjust the fit by trimming the pins. If they're too loose, you'll have to cut a new piece. If the fit is too tight, the pins can be trimmed slightly in the jig. That's why it's best to cut the pins last. If you need to make adjustments, the template is still positioned for the pins.

THINGS TO CONSIDER. Jigs that cut variably-spaced dovetails like those on the Blanket Chest aren't cheap. (Prices can be more than some power tools.) But with that price tag, you will get some versatility. Most of these jigs will also cut half-blind dovetails. Some also can be used to cut box joints and sliding dovetails. Still, you'll need to balance the cost with how often you'll use the jig.

Accessories can add to the cost. For example, some of the jigs require that you use only specific bits that the manufacturer sells. Typically, the basic bits are included with the jig. If you want to cut different types of joints, or work in a different thickness of stock, you may need to purchase more bits.

And unlike hand-cut dovetails, you can't vary the angle on the sides of the tails. That's determined by the bit you're required to use. However, machine-cut tails will be perfectly symmetrical — something that's tough to do by hand.

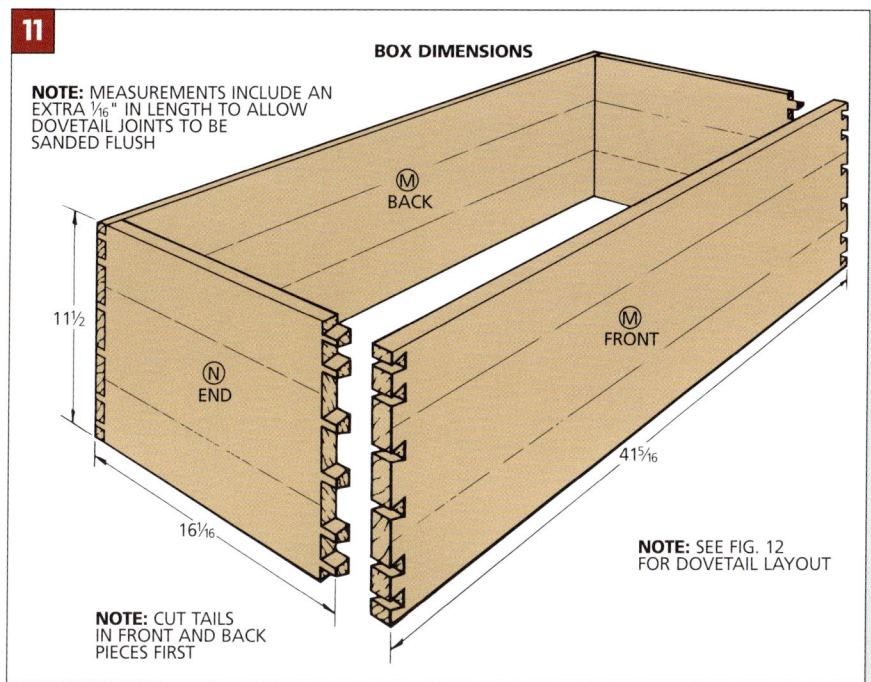

11

BOX DIMENSIONS

NOTE: MEASUREMENTS INCLUDE AN EXTRA 1/16" IN LENGTH TO ALLOW DOVETAIL JOINTS TO BE SANDED FLUSH

Ⓜ BACK

Ⓜ FRONT

11½

Ⓝ END

16¹⁄₁₆

41⁵⁄₁₆

NOTE: SEE FIG. 12 FOR DOVETAIL LAYOUT

NOTE: CUT TAILS IN FRONT AND BACK PIECES FIRST

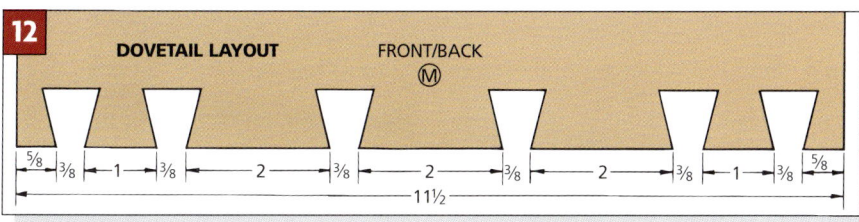

12

DOVETAIL LAYOUT **FRONT/BACK**
Ⓜ

5/8 — 3/8 — 1 — 3/8 — 2 — 3/8 — 2 — 3/8 — 2 — 3/8 — 1 — 3/8 — 5/8

11½

SHOP TIP

Dovetail Clamping Block

The key to clamping a dovetail joint is to apply pressure directly over the tails. To help with this, I cut a series of notches in a piece of scrap to create "fingers" that line up with the tails of the joint.

ALIGN CLAMPING FINGERS OVER TAILS

DOVETAIL CLAMPING BLOCK

PINS

TAILS

panels for the box front and back (M), and two panels for the box ends (N). (See the Technique article on page 112 for tips on gluing up panels.)

After the glue has dried, plane the panels flat and trim them to finished width and rough length *(Fig. 11)*.

Note: So the dovetail corners can be sanded flush after assembly, the measurements given are 1/16" longer than the final dimensions of the box.

DOVETAILS. Next, I laid out the dovetails *(Fig. 12)*, and cut them using a jig. (See the Shop Info box on the facing page.) Of course, if the spirit moves you, the joint can also be cut by hand. After the dovetails are cut, the box is glued together, and the corners are sanded flush. (See the Shop Tip above for a tip about clamping this joint.)

ASSEMBLY

Once the main box is glued together, the drawer carcase is attached to the bottom.

UPPER FRAME. The first step is to attach the upper frame of the drawer carcase to the box. To do this, flip the box over and center the upper frame on the bottom of

the box *(Fig. 13a)*. Then clamp the frame to the box, and drill shank and pilot holes 1⁷⁄₁₆" from the outside edges of the frame members. Now unclamp the frame, apply glue to the bottom edge, and screw the upper frame to the box.

APRONS. To assemble the rest of the drawer carcase, glue the three aprons in place (with splines in the mitered corners), then add the drawer dividers, and finally glue the lower frame in place.

Note: Be sure to keep the fronts of the drawer dividers and end aprons in line.

KICKBOARD. Once the glue has had a chance to dry, the kickboard is glued to the rabbet on the bottom of the lower drawer carcase frame *(Fig. 13)*.

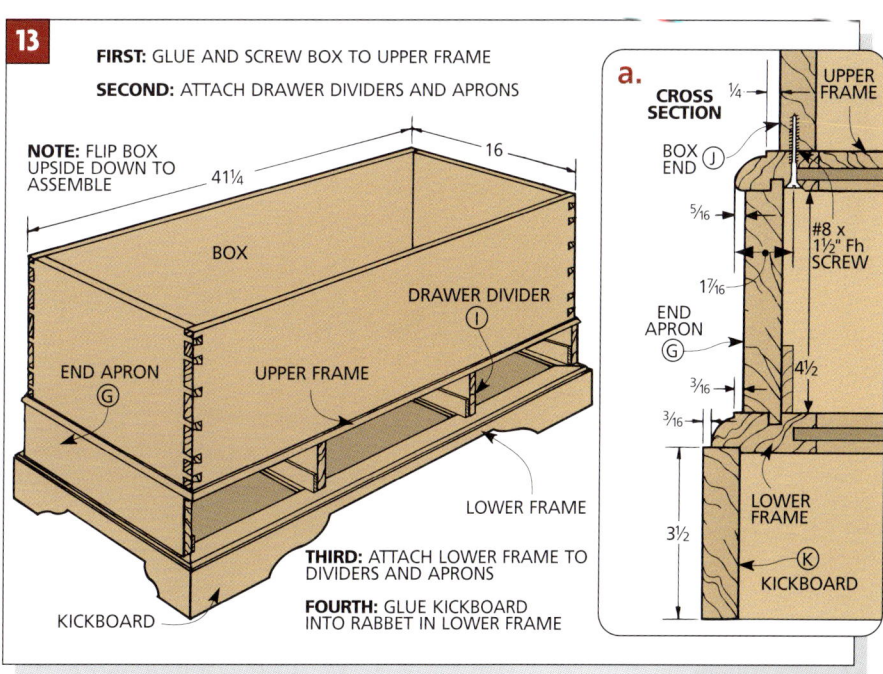

13

FIRST: GLUE AND SCREW BOX TO UPPER FRAME

SECOND: ATTACH DRAWER DIVIDERS AND APRONS

NOTE: FLIP BOX UPSIDE DOWN TO ASSEMBLE

41¼

16

BOX

DRAWER DIVIDER
Ⓘ

END APRON
Ⓖ

UPPER FRAME

LOWER FRAME

THIRD: ATTACH LOWER FRAME TO DIVIDERS AND APRONS

FOURTH: GLUE KICKBOARD INTO RABBET IN LOWER FRAME

KICKBOARD

a. **CROSS SECTION**

¼

UPPER FRAME

BOX END Ⓙ

#8 x 1½" Fh SCREW

5/16

1⁷⁄₁₆

END APRON Ⓖ

4½

3/16

3/16

3½

LOWER FRAME

Ⓚ KICKBOARD

LIP MOLDING

With the bottom section completed, I flipped the cabinet right side up and started on the lip molding for the top edge of the box.

To make the lip molding (O, P), rip enough stock $1^3/_4$" wide for all four sides of the box. Then cut a $^1/_8$"-deep groove $^5/_8$" from the outside edge of the molding *(Fig. 14)*. The width of this groove should fit the top edge of the box side.

Next, rout a $^1/_2$" cove on the bottom outside edge of the lip molding, and remove the sharp corners on the inside edges using a $^1/_4$" roundover bit set at a very shallow depth. Finally, miter the lip molding to fit the rim of the box, and glue it in place *(Fig. 15)*.

LID

The first thing most people notice on a blanket chest is its lid. For the lid on this chest, I decided to follow the design found on most traditional chests — a flat, solid-wood lid that's simple in design.

GLUING. Since the lid overhangs the lip molding $^3/_8$" on all four sides, the first step is to glue up enough stock to produce a panel that can be trimmed down to provide the $^3/_8$" overhang. Then the lid (Q) is planed flat, and trimmed to its final dimensions *(Fig. 16)*.

MOLDING. After the lid is trimmed to final size, its outside edge is molded with a $^1/_2$" roundover bit, leaving a $^1/_8$" shoulder *(Fig. 16a)*. Then on the bottom outside edge, the sharp edge is removed using a $^1/_4$" roundover bit, again set at a very shallow depth of cut.

CLEATS. To keep the lid from cupping, two $1^1/_4$"-wide cleats (R) are screwed to its bottom face *(Fig. 16)*.

DRAWERS

One of the things that makes this chest different from its modern counterparts is three drawers that actually work.

DRAWERS. The first step is to cut the $^1/_2$" drawer stock for the small drawer fronts and backs (S) and large drawer front and back (T). These pieces fit between the drawer runners with $^1/_{16}$" of clearance, and are $^1/_{16}$" narrower than the height of the openings *(Fig. 17)*.

Then cut the drawer sides (U) to the same width as the drawer fronts and 15" long. (This allows for a 1" clearance at the back of the drawer.)

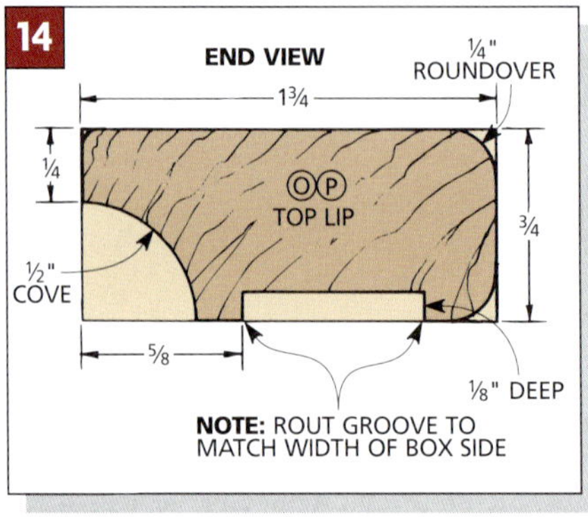
14
END VIEW
1/4" ROUNDOVER
1 3/4
1/4
O P
TOP LIP
3/4
1/2" COVE
5/8
1/8" DEEP
NOTE: ROUT GROOVE TO MATCH WIDTH OF BOX SIDE

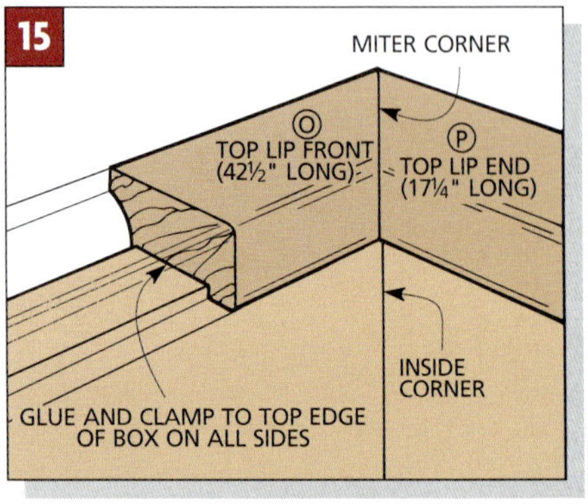
15
MITER CORNER
O
TOP LIP FRONT (42 1/2" LONG)
P
TOP LIP END (17 1/4" LONG)
INSIDE CORNER
GLUE AND CLAMP TO TOP EDGE OF BOX ON ALL SIDES

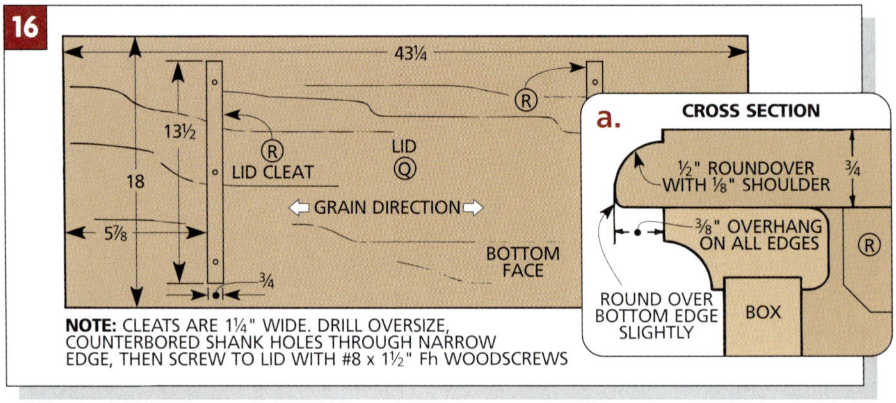
16
43 1/4
R
13 1/2
R
LID CLEAT
LID
Q
18
GRAIN DIRECTION
5 7/8
BOTTOM FACE
3/4
NOTE: CLEATS ARE 1 1/4" WIDE. DRILL OVERSIZE, COUNTERBORED SHANK HOLES THROUGH NARROW EDGE, THEN SCREW TO LID WITH #8 x 1 1/2" Fh WOODSCREWS
a.
CROSS SECTION
1/2" ROUNDOVER WITH 1/8" SHOULDER
3/4
3/8" OVERHANG ON ALL EDGES
R
ROUND OVER BOTTOM EDGE SLIGHTLY
BOX

JOINERY. To keep the joinery consistent, I used through dovetails on the drawers as well *(Fig. 17)*. (If you prefer, the drawers could also be joined with half-blind dovetails.)

After completing the corner joints, cut a $^1/_4$" groove for the drawer bottoms. The grooves are positioned $^3/_8$" from the bottom edge *(Fig. 17)*.

BOTTOM. Finally, dry-clamp the drawers together and measure between the grooves to find the sizes for the small and large drawer bottoms (V, W). Then cut the $^1/_4$" hardboard bottoms to size, and

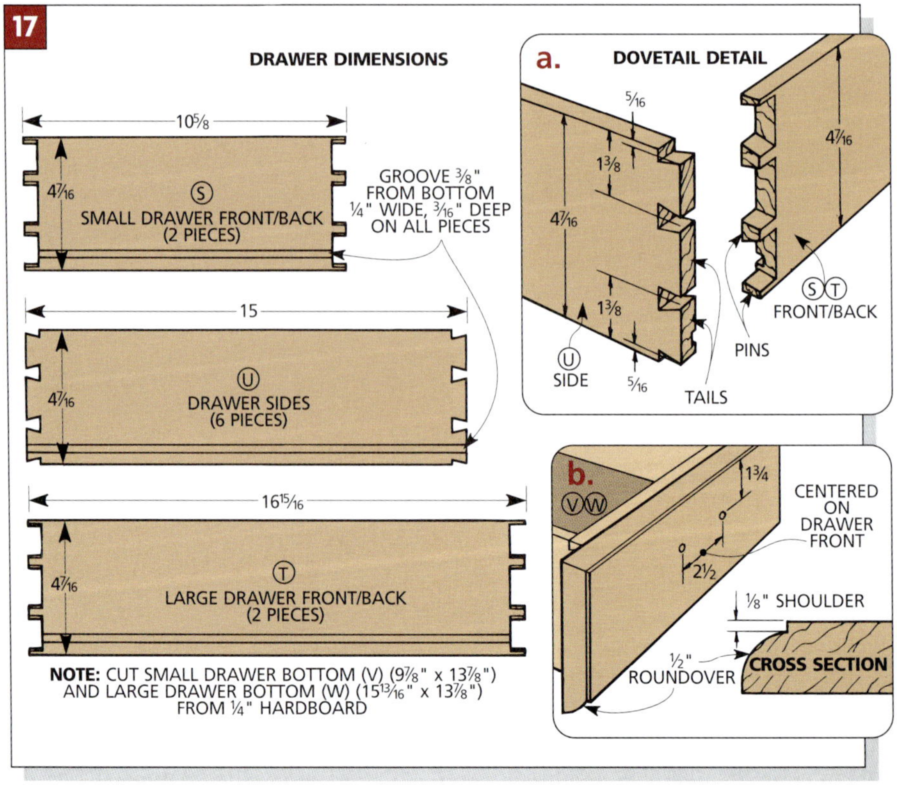
17
DRAWER DIMENSIONS
10 5/8
4 7/16
S
SMALL DRAWER FRONT/BACK (2 PIECES)
GROOVE 3/8" FROM BOTTOM 1/4" WIDE, 3/16" DEEP ON ALL PIECES
15
4 7/16
U
DRAWER SIDES (6 PIECES)
16 15/16
4 7/16
T
LARGE DRAWER FRONT/BACK (2 PIECES)
NOTE: CUT SMALL DRAWER BOTTOM (V) (9 7/8" x 13 7/8") AND LARGE DRAWER BOTTOM (W) (15 13/16" x 13 7/8") FROM 1/4" HARDBOARD
a.
DOVETAIL DETAIL
5/16
1 3/8
4 7/16
4 7/16
1 3/8
5/16
U
SIDE
TAILS
PINS
S T
FRONT/BACK
b.
V W
1 3/4
CENTERED ON DRAWER FRONT
2 1/2
1/8" SHOULDER
1/2" ROUNDOVER
CROSS SECTION

glue the drawers together with the drawer bottoms in place.

FALSE FRONTS. Each drawer has a $3/4$"-thick false front that's attached directly to the drawer front. To find the dimensions for the large false front (X), measure from center to center on the drawer dividers *(Fig. 18)*. Then cut the large false front to this length, and to the same height as the drawers.

To find the lengths for the small drawer fronts (Y), measure the distance from the center of the drawer dividers to the outside edge of the side apron, and subtract $1/16$" for clearance *(Fig. 18)*. Then cut the two small false fronts to length, and to the same height as the drawers.

MOLDING. The outside edges on the false fronts are routed with a $1/2$" roundover bit to match the edges on the cabinet *(Fig. 17b)*. Once the roundover was done, I attached the large false front centered on the large drawer, and the small false fronts so they were flush with the outside faces of the end aprons.

HARDWARE

All that remains is to mount the hardware and apply a finish.

DRAWER PULLS. The drawer pulls are centered on the width of the drawer fronts *(Fig. 17b)*. The screw holes for the pulls are counterbored from the back so the screw heads sit below the inside face of the drawer front.

LOCK. To mount the full-mortise lock, I drilled a series of $3/8$" holes centered 1" from the front edge of the lip molding *(Fig. 19)*. Square up the mortise with a chisel, then drill the keyhole, and mount the escutcheon over the keyhole.

HINGES. The hinges I used are designed for the extra overhang of the lid. (See page 126 for sources of these hinges.) They are mortised into both the lip molding and the lid *(Fig. 20a)*.

Cutting these mortises caused me the most tense moments on this project. After spending hours and hours on construction, a mere slip of the hand at this stage can ruin a project.

So I've come up with a few techniques that make this part of the process easier.

ALIGN HINGE. The first step is to use the hinge as a guide to lay out the mortise. To keep the hinge aligned properly as the mortise is marked out, I butt the edge of the hinge against the edge of a square. (Keeping the hinge square is necessary to eliminate binding during opening.)

SCORE OUTLINE. Then to mark the outline for the mortise, I score around the hinge with a razor knife, keeping the point of the blade tight against the bottom edge of the hinge. This scoring mark severs the fibers of the wood, reducing the chances of chipout at the edge of the mortise. The end result is a clean, accurate outline for the hinge mortise.

ROUT MORTISE. After the outline of the hinge was scored, I used a router to remove the waste from the mortise. Using a small ($1/8$" or $1/4$") straight bit eliminates the "pulling" you sometimes get with a router, and produces a mortise with a flat bottom and consistent depth.

However, trying to follow a shallow score mark by sighting through the collet opening in the base of a router can be difficult, if not impossible. So to make the outline of the mortise more visible, I outlined it with masking tape.

Then I set the router to cut about $1/32$" deep and slowly rout out the mortise within the masked-off borders. It usually takes several very light passes to reach the final depth of the mortise.

Note: Be sure to remove the masking tape on the last pass with the router so the depth of the mortise equals the exact thickness of the hinge flap.

SCREW HOLES. Now you can drill pilot holes for the hinge screws. Note that the rear holes on the bottom are shallow and the screws are cut off to avoid breaking through the molding *(Fig. 20a)*.

LID SUPPORT. Then I added an optional lid support to prevent the lid from slamming shut. (See Sources on page 126.)

FINISH. To highlight the wood, I applied two coats of tung oil sealer, and four coats of medium-luster tung oil.

Note: Don't apply finish to the cedar, or you'll reduce its aroma. ∎

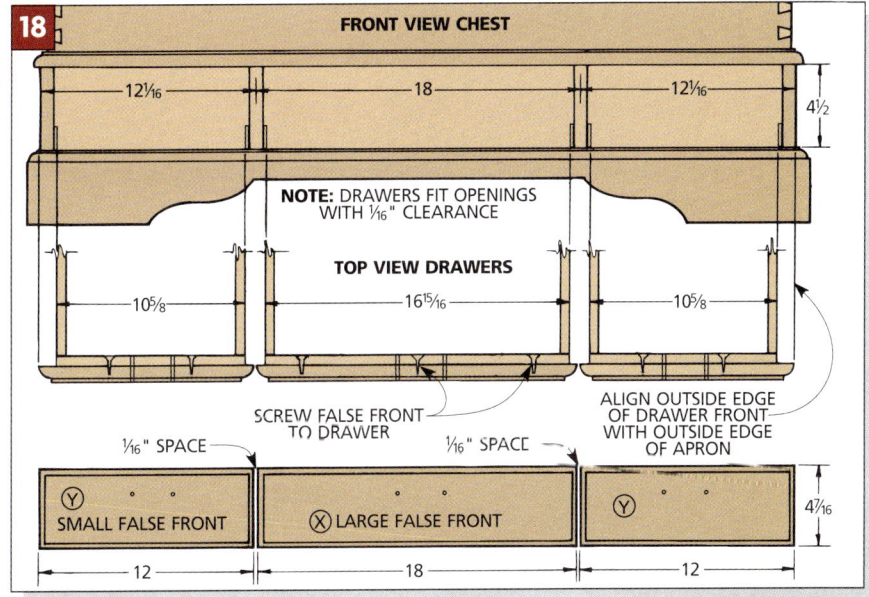

18 FRONT VIEW CHEST

12¹⁄₁₆ — 18 — 12¹⁄₁₆ 4½

NOTE: DRAWERS FIT OPENINGS WITH ¹⁄₁₆" CLEARANCE

TOP VIEW DRAWERS

10⁵⁄₈ — 16¹⁵⁄₁₆ — 10⁵⁄₈

SCREW FALSE FRONT TO DRAWER

ALIGN OUTSIDE EDGE OF DRAWER FRONT WITH OUTSIDE EDGE OF APRON

¹⁄₁₆" SPACE — ¹⁄₁₆" SPACE

(Y) SMALL FALSE FRONT — (X) LARGE FALSE FRONT — (Y)

4⁷⁄₁₆

12 — 18 — 12

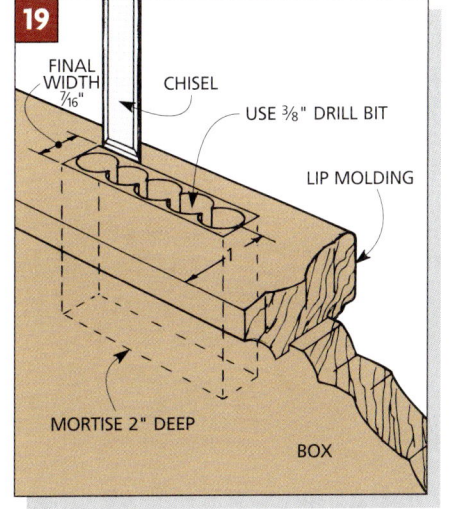

19

FINAL WIDTH ⁷⁄₁₆"

CHISEL

USE ⅜" DRILL BIT

LIP MOLDING

MORTISE 2" DEEP

1

BOX

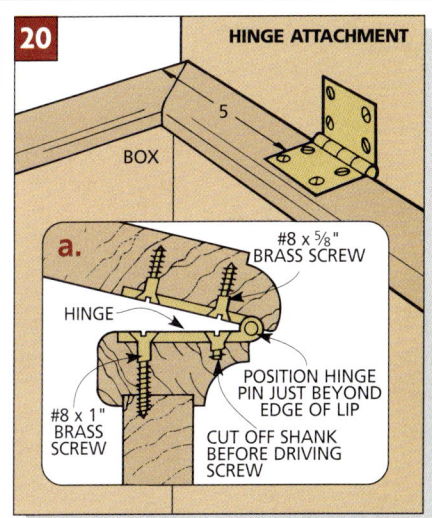

20 HINGE ATTACHMENT

BOX

5

a.

#8 x ⅝" BRASS SCREW

HINGE

#8 x 1" BRASS SCREW

POSITION HINGE PIN JUST BEYOND EDGE OF LIP

CUT OFF SHANK BEFORE DRIVING SCREW

DESIGNER'S NOTEBOOK

With frame and panel construction, there's no need to glue up panels for the front, back, and sides — or to cut dovetails. And there's a trick to making the panels look as good inside as outside.

CONSTRUCTION NOTES:

■ The drawer carcase and kickboard assembly are exactly the same as before.

The changes come when you begin building the box. Instead of solid-wood panels, it consists of four frame and panel assemblies. Each end assembly is a regular frame with two rails and two stiles surrounding a panel (refer to *Fig. 4*). The front and back assemblies have two rails and four stiles to create openings for three panels (similar to the upper and lower frames of the drawer carcase) *(Fig. 1)*. On these frames, though, instead of miters, I used stub tenons and grooves.

■ The stiles on the end frames have tongues that fit into grooves in the front and back frames. Since I find it easier to sneak up on the thickness of a tongue to fit a groove than to cut a groove to fit a tongue, I started construction with the front and back assemblies.

■ The first pieces to cut to size are the front and back rails (Z) and the front and back stiles (AA) *(Fig. 1)*.

■ Next, I cut the four center stiles (BB) to finished width and rough length.

■ For the panels, I used two pieces of ¼" plywood glued back to back so I ended up with two good faces *(Fig. 1a)*. The panels fit into centered grooves cut on the inside edges of the frame pieces.

To cut the groove, I set up a ¼" dado blade in my table saw and positioned the rip fence so the blade was just off center of the frame pieces. Then I cut a groove,

rotated the piece end for end, and made another pass to widen the groove *(Fig. 2)*. Next, I nudged the fence and repeated the procedure until two layers of plywood fit in the groove. Once the saw is set up, cut the grooves on all the frame pieces.

Note: The front and back rails and stiles (Z, AA) have grooves only on the inside edges. The center stiles (BB) have grooves on both edges *(Fig. 1)*.

■ Once the grooves are cut, you can move on to the stub tenons on the rails and center stiles. On the rails, this is pretty straightforward. Just sneak up on the thickness of each ³⁄₈"-long tongue until it fits the groove *(Fig. 3)*.

■ For the center stiles, you'll need to concentrate on the shoulder-to-shoulder dimension of the piece, instead of the length of the tongue. To determine this,

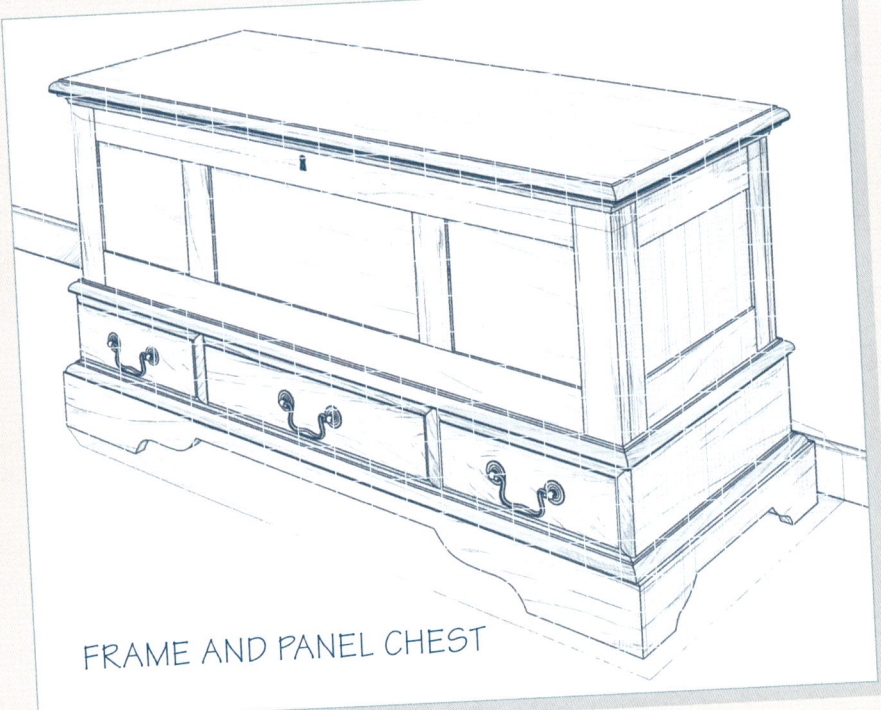

FRAME AND PANEL CHEST

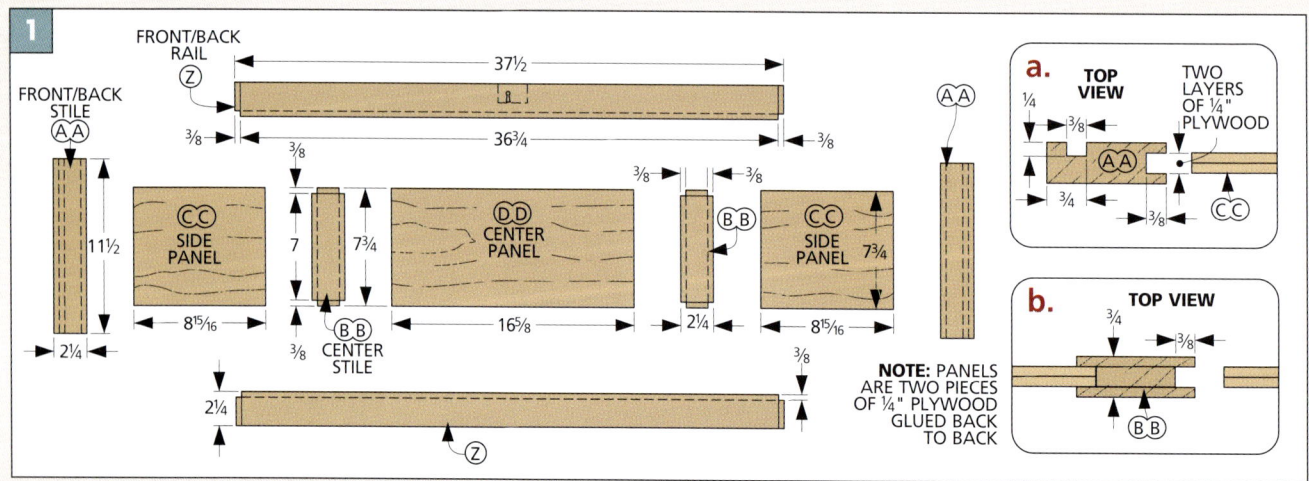

dry-assemble the rails (Z) and stiles (AA). The distance between the rails is the shoulder-to-shoulder length of the center stiles *(Fig. 1)*. I cut a tongue on each end of the center stiles, then trimmed the extra length off the tongues to bring the pieces to finished length *(Fig. 3)*.

■ Before the frames can be assembled, a groove needs to be cut on the inside face of each front and back stile (AA) to accept the tongue on the end assemblies. This is just a $\frac{3}{8}$"-wide groove, cut $\frac{1}{4}$" deep *(Fig. 1a)*. The groove is positioned so the end assembly will be flush with the outside edge of the stile.

■ The last pieces to cut for the front and back assemblies are the panels. I started by cutting eight side panels (CC) to rough size and laminating them together with the good faces out *(Fig. 1)*.

■ When the glue is dry, you can trim the panels to finished size.

■ To determine the size of the center panels, dry-assemble the rails, stiles and side panels. Then measure between the center stiles and add in the depth of the grooves ($\frac{3}{4}$"). Cut four center panels (DD) slightly larger than this size and glue them together, back to back.

■ Once the glue is dry, you can trim the panels to size and glue up each of the front and back assemblies.

Note: If you have trouble fitting a panel into a groove, try sanding or planing a slight chamfer around the edges so it will slip into the groove more easily.

■ Now you can work on the end assemblies. This is just like the front and back.

■ First, cut the end rails (EE) and end stiles (FF) to size *(Fig. 4)*. Note that the end stiles are narrower than those on the front and back assemblies. When the frames are glued together, the edges of the front stiles will help make the narrower end stiles look wider *(Fig. 6)*.

■ Next, cut the grooves in the edges.

■ As before, set up and cut the tenons on the rails to fit the grooves in the stiles.

■ Now you can cut the end panels (GG) to size and glue them together *(Fig. 4)*.

■ Finally, you can assemble the two end assemblies by gluing the frame pieces around the panel.

■ Before assembling the box, you'll need to cut $\frac{1}{4}$"-long tongues on the edges of the stiles to fit the grooves in the front and back assemblies *(Figs. 4a and 5)*.

■ Assembling the box is just a matter of gluing the end assemblies between the front and back assemblies *(Fig. 6)*.

■ The box is glued and screwed to the drawer carcase as before.

■ On this chest, I chose swan-neck bail pulls for the drawers (see drawing on opposite page). And because the overlay escutcheon is wider than the top rail, a press-in escutcheon is used *(Fig. 7)*.

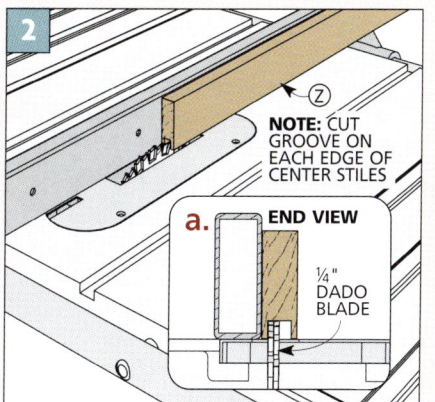

2

NOTE: CUT GROOVE ON EACH EDGE OF CENTER STILES

a. END VIEW

$\frac{1}{4}$" DADO BLADE

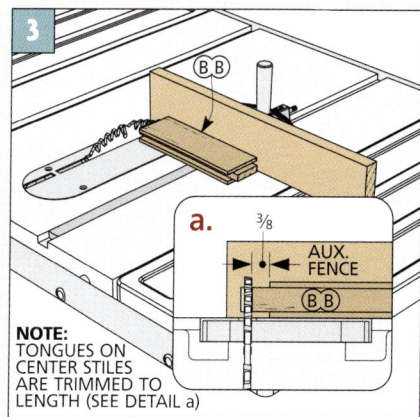

3

BB

a. $\frac{3}{8}$

AUX. FENCE

BB

NOTE: TONGUES ON CENTER STILES ARE TRIMMED TO LENGTH (SEE DETAIL a)

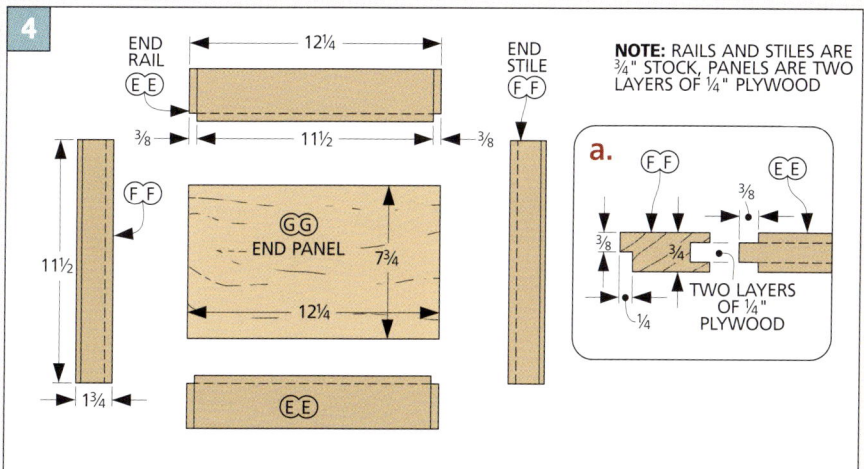

4

END RAIL EE — 12$\frac{1}{4}$

$\frac{3}{8}$ — 11$\frac{1}{2}$ — $\frac{3}{8}$

END STILE FF

NOTE: RAILS AND STILES ARE $\frac{3}{4}$" STOCK, PANELS ARE TWO LAYERS OF $\frac{1}{4}$" PLYWOOD

FF 11$\frac{1}{2}$

GG END PANEL 7$\frac{3}{4}$

12$\frac{1}{4}$

1$\frac{3}{4}$

EE

a. FF EE $\frac{3}{8}$

$\frac{3}{8}$ $\frac{3}{4}$

TWO LAYERS OF $\frac{1}{4}$" PLYWOOD

$\frac{1}{4}$

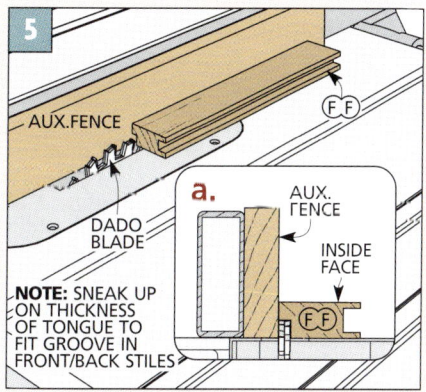

5

AUX. FENCE FF

a. AUX. FENCE

DADO BLADE

INSIDE FACE

FF

NOTE: SNEAK UP ON THICKNESS OF TONGUE TO FIT GROOVE IN FRONT/BACK STILES

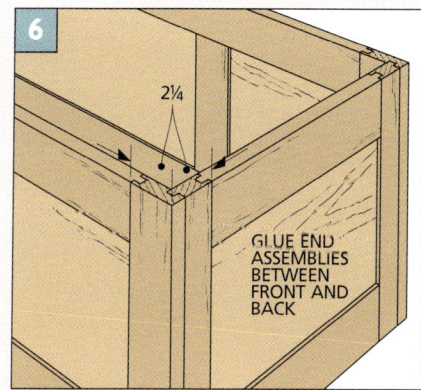

6

2$\frac{1}{4}$

GLUE END ASSEMBLIES BETWEEN FRONT AND BACK

7

TOP LIP

KEY HOLE

PRESS-IN ESCUTCHEON

FRONT RAIL

NOTE: RAIL IS TOO NARROW FOR OVERLAY ESCUTCHEON. USE PRESS-IN ESCUTCHEON

MATERIALS LIST

NEW PARTS

Z	Front/Back Rails (4)	$\frac{3}{4}$ x 2$\frac{1}{4}$ - 37$\frac{1}{2}$
AA	Front/Back Stiles (4)	$\frac{3}{4}$ x 2$\frac{1}{4}$ - 11$\frac{1}{2}$
BB	Center Stiles (4)	$\frac{3}{4}$ x 2$\frac{1}{4}$ - 7$\frac{3}{4}$
CC	Side Panels (8)	$\frac{1}{4}$ ply - 7$\frac{3}{4}$ x 8$\frac{15}{16}$
DD	Center Panels (4)	$\frac{1}{4}$ ply - 7$\frac{3}{4}$ x 16$\frac{5}{8}$
EE	End Rails (4)	$\frac{3}{4}$ x 2$\frac{1}{4}$ - 12$\frac{1}{4}$
FF	End Stiles (4)	$\frac{3}{4}$ x 1$\frac{3}{4}$ - 11$\frac{1}{2}$
GG	End Panels (4)	$\frac{1}{4}$ ply - 7$\frac{3}{4}$ x 12$\frac{1}{4}$

Note: Do not need parts M, N.

HARDWARE SUPPLIES
(1) Press-in brass escutcheon
(3) Brass swan-neck bail pulls

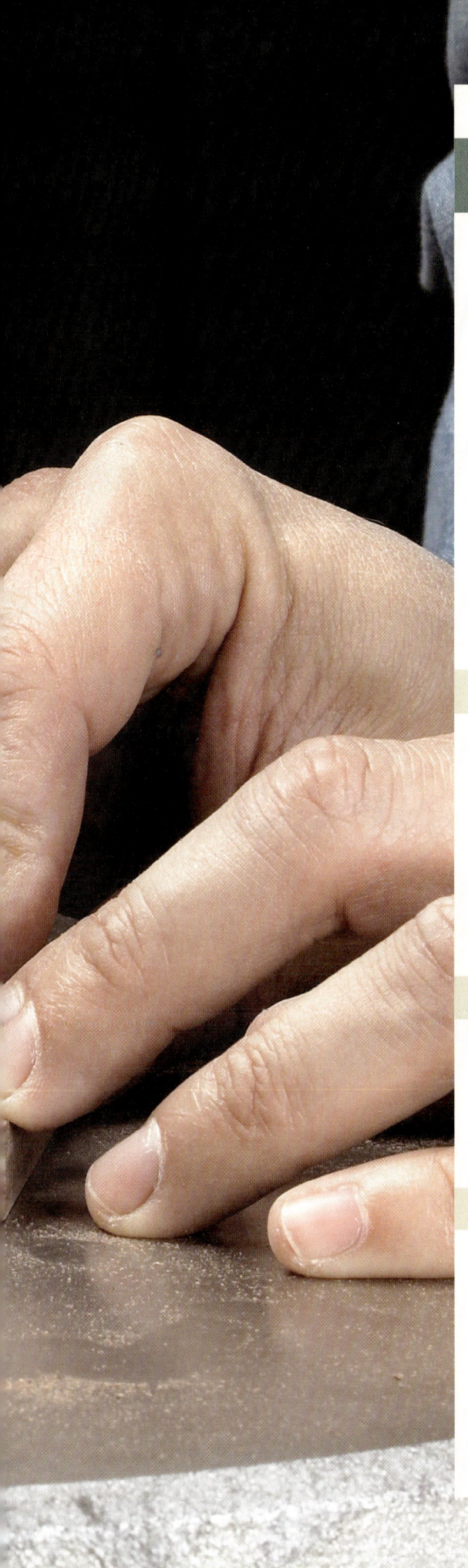

THE CLASSICS

What sets a classic piece of furniture apart from an ordinary, everyday piece? It's all in the details, whether they can be seen or not.

The obvious details on the formal hall table include scroll-sawn brackets and bead molding. Each is simple to make.

With the barrister's bookcases, some of the details are hidden from view. For example, why build a fully enclosed case when one stacks on top of another? Instead, each case has a set of interlocking cleats at the top and bottom.

Finally, the slant front desk uses sliding dovetail joinery, a feature that makes cross-grain joints between solid wood panels possible. An optional pigeonhole insert provides extra organization — and conceals a hidden compartment.

Formal Hall Table

The closer you look, the more details you begin to see: scroll-sawn brackets, tapered legs, and beaded drawers and aprons. But don't let the details fool you. There's nothing tricky about its construction.

There's an understated elegance about this small table. While the table is solid and built to withstand heavy use, the tapered legs and scroll-sawn brackets give it a light, graceful appearance. Narrow strips of bead molding highlight the drawer fronts and the bottom edges of the aprons. And the dark walnut adds to the "rich," formal feel of the project.

The nice thing is that none of these details are difficult to "pull off." The long, gradual tapers on the legs were cut with a shop-made jig. The scrollsawn brackets are small, fairly simple curves with only one "inside" cut to make. The table also

looks fine without these corner brackets (refer to the photo on page 94). And the different pieces of bead molding are simply routed with roundover bits.

JOINERY. The construction of the table is traditional, but the joinery can be cut with ordinary power tools. The legs are joined to the aprons with mortises and tenons. This joint is easy to make with a drill press and table saw. The table saw is all you need for the stub tenons and grooves on the web frames. These frames fit between the aprons to create the opening for the drawers. And the drawers feature half-blind dovetails routed with the help of a dovetail jig.

WOOD. If you can't find any walnut locally, cherry or mahogany would suit the style of this table. (The only plywood you'll need is $1/4$" maple.) These woods are often used for formal furniture, and their subtle grain won't detract from the overall elegance of the table.

FRAME AND PANEL TOP. The top of the table shown in the photo is a glued-up panel of solid wood. As an option, I've also included a design that uses a frame and panel top with three plywood panels. It gives the table a less formal look.

As part of this design option, I also made one wide drawer instead of three. Details about all of this are on page 95.

EXPLODED VIEW

OVERALL DIMENSIONS:
30W x 18D x 28H

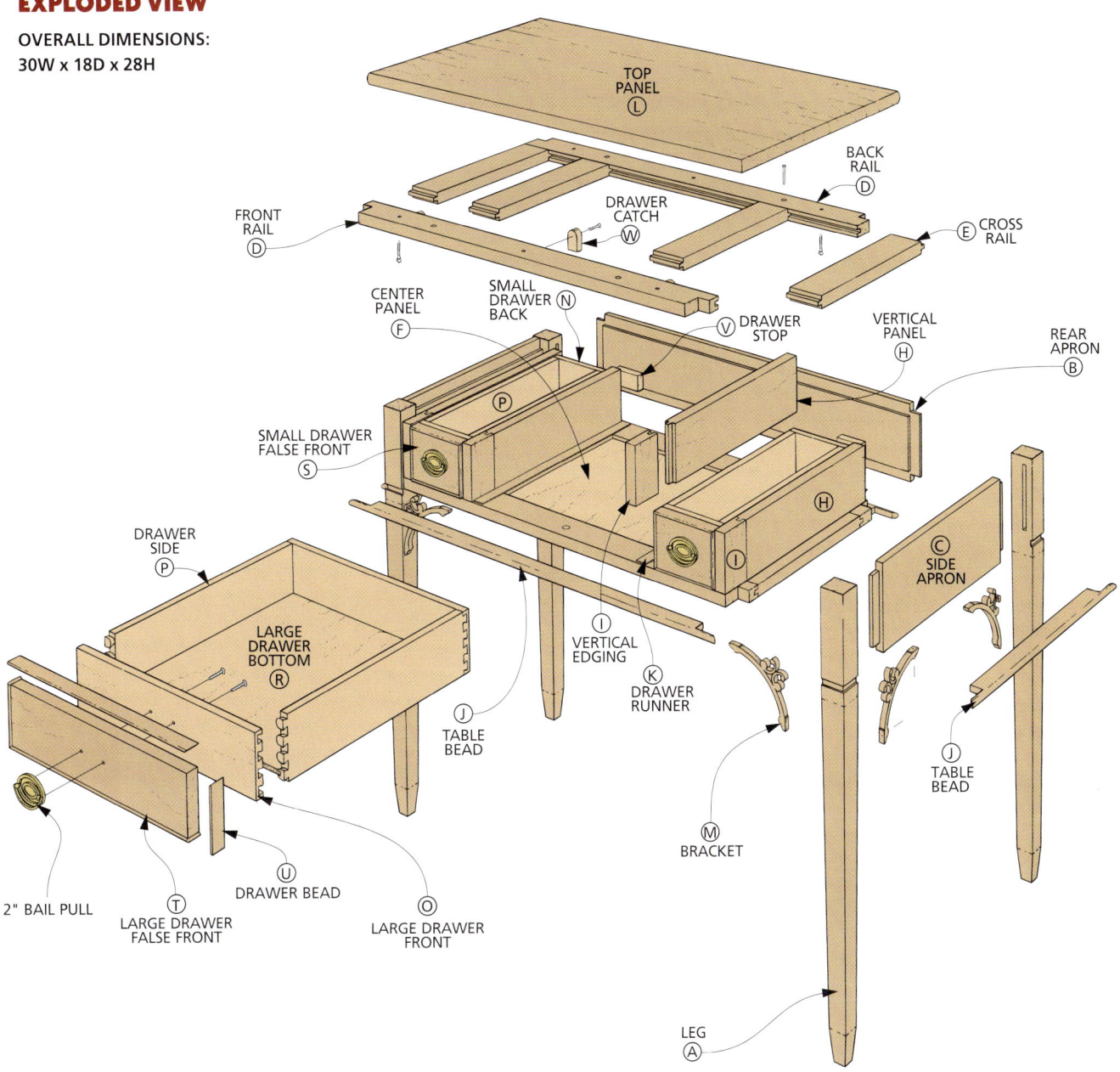

TOP PANEL (L)

BACK RAIL (D)

FRONT RAIL (D)

DRAWER CATCH (W)

CROSS RAIL (E)

CENTER PANEL (F)

SMALL DRAWER BACK (N)

DRAWER STOP (V)

VERTICAL PANEL (H)

REAR APRON (B)

SMALL DRAWER FALSE FRONT (S)

(P)

(H)

(I)

DRAWER SIDE (P)

LARGE DRAWER BOTTOM (R)

VERTICAL EDGING (I)

DRAWER RUNNER (K)

SIDE APRON (C)

TABLE BEAD (J)

TABLE BEAD (J)

2" BAIL PULL

LARGE DRAWER FALSE FRONT (T)

DRAWER BEAD (U)

LARGE DRAWER FRONT (O)

BRACKET (M)

LEG (A)

MATERIALS LIST

WOOD

A Legs (4)   1½ x 1½ - 27¼
B Rear Apron (1)   ¾ x 5⅛ - 27½
C Side Aprons (2)   ¾ x 5⅛ - 15½
D Fr./Bk. Rails (4)   ¾ x 2¼ - 27⅝
E Cross Rails (8)   ¾ x 2¼ - 12⅜
F Center Panel (1)   ¼ ply - 13¼ x 12⅜
G Side Panels (2)   ¼ ply - 3¹³/₁₆ x 12⅜
H Vertical Panels (4)   ¾ x 3⅝ - 14
I Vertical Edging (4)   ¾ x 2¼ - 3⅝
J Table Bead   ⅜ x 1⅛ - 104 ln. in.

K Drawer Runners (6)   ¾ x 1/₁₆ - 15¾
L Top Panel (1)   ¾ x 18 - 30
M Brackets (6)   ⅜ x 3 - 7 rough
N Sm. Drwr. Fr./Bk. (4)   ½ x 3½ - 4⅜
O Lg. Drwr Fr./Bk. (2)   ½ x 3½ - 13⅞
P Drawer Sides (6)   ½ x 3½ - 14¼
Q Sm. Drwr. Btm. (2)   ¼ ply - 3⅞ x 14⅛
R Lg. Drwr. Btm. (1)   ¼ ply - 13⅜ x 14⅛
S Sm. Drwr. False Fr. (2) ¾ x 3¼ - 4⅛
T Lg. Drwr. False Fr. (1) ¾ x 3¼ - 13⅝
U Drawer Bead   ⅛ x ⅞ - 80 ln. in.

V Drawer Stops (3)   ¾ x ½ - 2
W Drawer Catches (3)   ¾ x ½ - 1¼

HARDWARE SUPPLIES

(8) No. 8 x 1½" Fh woodscrews
(9) No. 8 x 1¼" Fh woodscrews
(12) No. 17 x ⅝" wire brads
(3) Hepplewhite-style 2" bail pulls
(6) 8-32 x 1¾" machine screws

CUTTING DIAGRAM

¾ x 5½ - 96 WALNUT (3.7 Bd. Ft.)

| B | C | C | L |

¾ x 5½ - 96 WALNUT (3.7 Bd. Ft.)

| L | L | L |

¾ x 5½ - 96 WALNUT (3.7 Bd. Ft.)

| D | D | E | E | E |
| D | D | E | E | E |

¾ x 5½ - 96 WALNUT (3.7 Bd. Ft.)

| E | H | H | H | H | S | S | I I I I |
| E | | | | | | | |

K

½ x 5½ - 96 MAPLE (3.7 Sq. Ft.)

| P | P | P | P | P | P |

1½ x 7½ - 36 WALNUT (3.75 Bd. Ft.)

| A |
| A |

¾ x 5½ - 48 WALNUT (1.8 Bd. Ft.)

| T | J |
V   W

½ x 5½ - 48 MAPLE (1.8 Sq. Ft.)

| N | N | N | N | O | O |

⅛ x 5½ - 48 WALNUT (3 Boards @ 1.8 Sq. Ft. Each)

| M | M | M | M | M | M |
U

ALSO NEED:
ONE 24" x 48" PIECE
OF ¼" MAPLE PLYWOOD
FOR PARTS F, G, Q, AND R

LEGS

This project started with a small challenge. The legs require 8/4 walnut (1¾" thick), while the aprons are 4/4 stock (¾" thick). This meant I had to work a little harder to find boards that matched in color and grain. Usually, though, it only takes a few extra minutes when going through the stacks, and the extra effort is worth it.

LEGS. With the stock picked out, you'll want to start with the legs (A) *(Fig. 1)*. They're planed and ripped down to 1½" square and then cut to final length.

After cutting the legs to size, the next thing to do is cut the mortises for the aprons *(Figs. 1b and 1c)*. I laid them out carefully on each leg and then double-checked the layout by standing them on end in their proper orientation. I also drilled them a little deeper (13⁄16") than I planned on cutting the tenons (¾") to allow for excess glue.

Next, I cut ⅛"-deep dadoes for the beading that will wrap around the table after it's assembled *(Figs. 1 and 1b)*. While the mortises were cut on the inside faces, these dadoes need to be cut across the outside faces. And to make sure they align perfectly with each other, I used a stop block clamped to an auxiliary miter gauge fence to position the legs as they were pushed across the dado blade.

Now that the dadoes are cut, the legs are ready to be tapered. There are two tapers here: a long, gradual taper and a short taper that creates a "foot" profile near the bottom.

The gradual taper starts ¾" below the dadoes. To taper all four sides of the leg, you'll need an adjustable jig (see the opposite page). One setting works for two adjacent faces, then the workpiece is repositioned for the last two tapers.

When the gradual tapers have been cut and sanded smooth, the short tapers on the bottoms of the legs can be cut. This time, though, instead of a taper jig, you'll use a band saw and a block plane. (See the Technique box on page 88.)

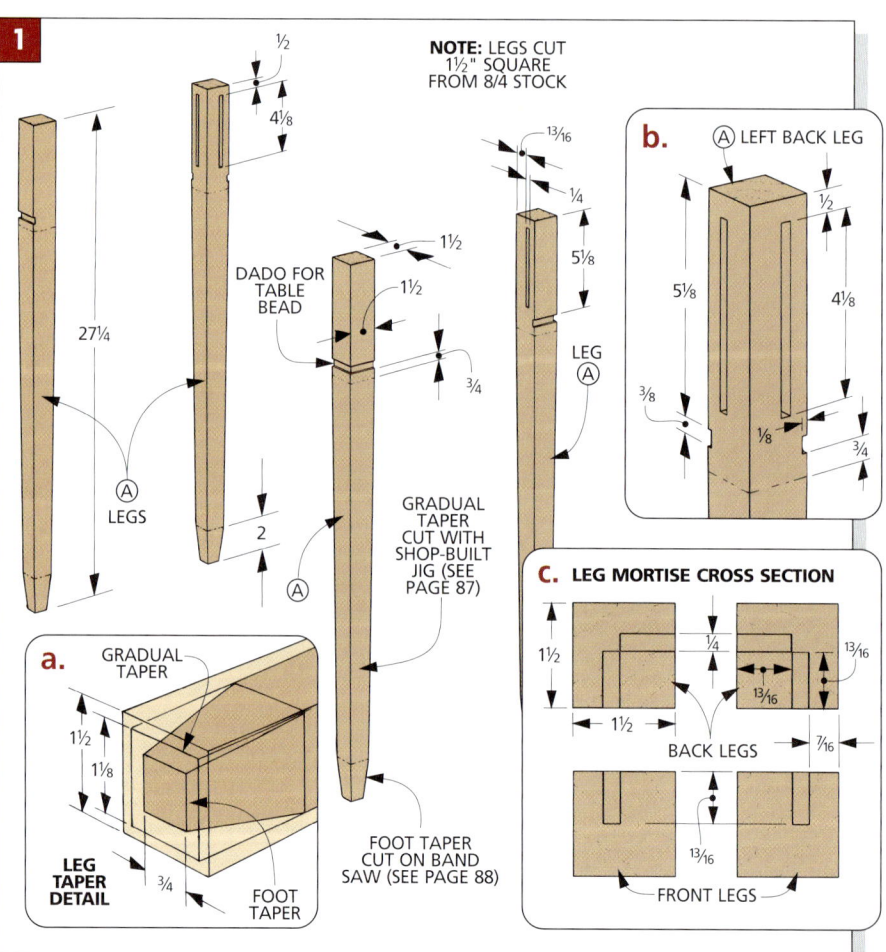

With four-sided tapers, you need an adjustable jig. That's because after two faces are tapered, the leg needs to be repositioned on the jig to allow for the tapers already cut. I came up with a sled that makes it easy to do this without a lot of fussing and measuring.

NOTCH. On this jig, the key to getting the same taper on all four faces is a notch in a stop block at the end of the fence *(Fig. 1a)*. When you taper the first two faces, the legs rest fully against the fence. When you cut the third and fourth faces, the notch positions the leg away from the fence the amount of the taper.

FENCE. The first piece to make is the fence *(Fig. 1)*. The important thing here is that it's a little longer than the taper.

HOLD-DOWNS. After cutting the fence to size, holes can be drilled and counterbored for the carriage bolts that support the hold-downs *(Fig. 1b)*. The hold-downs are just thin strips of maple. Nuts on two of the bolts support the back edge of each hold-down, while a star knob at the front applies clamping pressure.

The last thing to add to the fence is the notched stop block *(Fig. 1a)*. When you attach it, make sure the notch is positioned exactly $3/16$" from the fence.

BASE. Next, I cut a piece of $1/4$" hardboard to size to serve as the base of the jig *(Fig. 1)*. The leg rests on the base as the cuts are made.

Note: After ripping the hardboard to width, don't move the table saw fence.

ASSEMBLY. Assembling the jig is pretty easy. All you need is a leg with the taper laid out on one of the faces and the end.

To position the plywood fence on the base, place the leg so the layout lines are on the edge of the base *(Fig. 2)*. Then mark the base along the back of the leg. Finally, align the fence along this mark and screw it in place from the bottom.

CUT TAPERS. Now you're ready to cut the tapers. For the first two faces, the leg rests snug against the fence of the jig and against the stop block *(Fig. 3)*.

After the first pass, rotate the leg so the newly tapered face is up *(Fig. 3)*. (You may need to re-adjust the hold-downs.) Then make a second pass.

For the last two faces, the leg rests in the notch *(Fig. 4)*. After each pass, rotate the leg so the side you just cut faces up.

Then sand away any saw marks and blend each taper into the top of the leg.

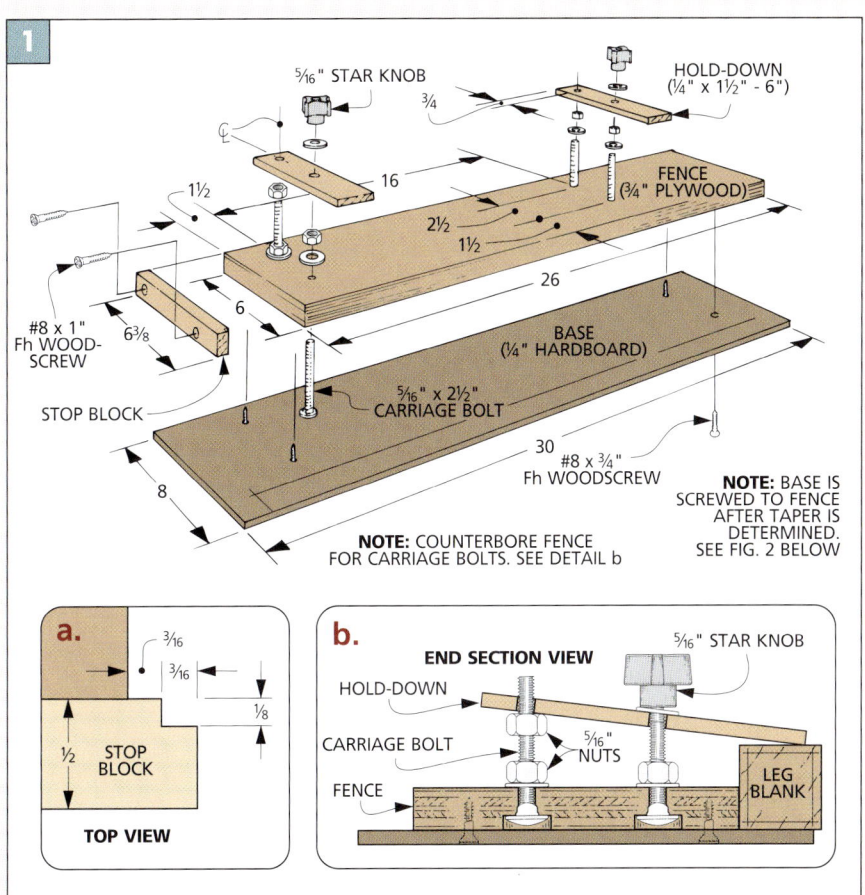

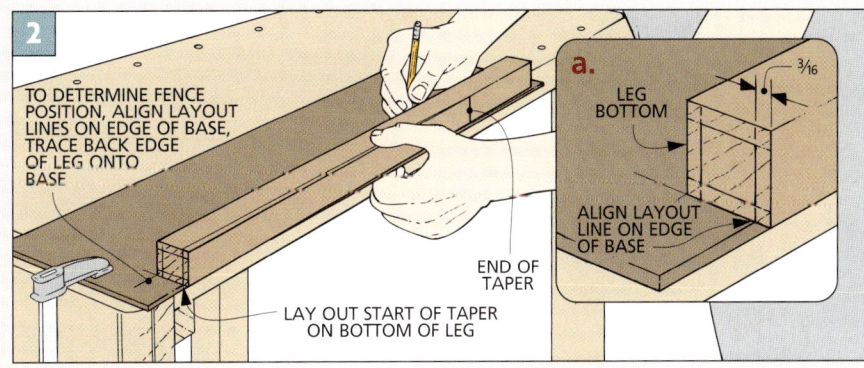

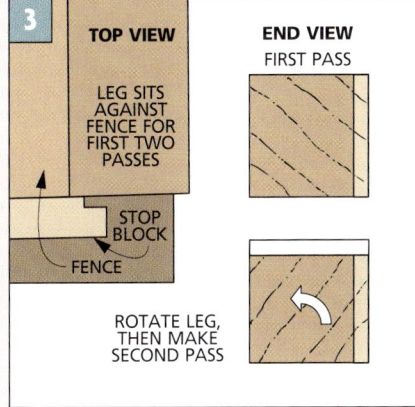

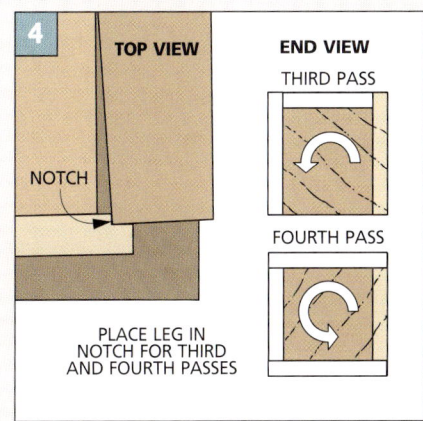

reating the tapers at the bottom of the legs doesn't require a special jig. A band saw (or even a hand saw) and a sharp block plane will do the trick.

LAYOUT AND CUT. Before cutting the taper, you'll need to lay it out on the end of the leg (refer to *Fig. 1a* on page 86). And don't try to cut right to the layout lines (*Step 1*). Stay to the waste side so you can clean up the tapers with a plane.

PLANE. When cleaning up the tapers (*Step 2*), work down to the end of the leg. Try to make the starting points of the tapers line up on all four faces. But even so, you'll still need to sand the faces to get a straight edge between the gradual tapers and foot tapers.

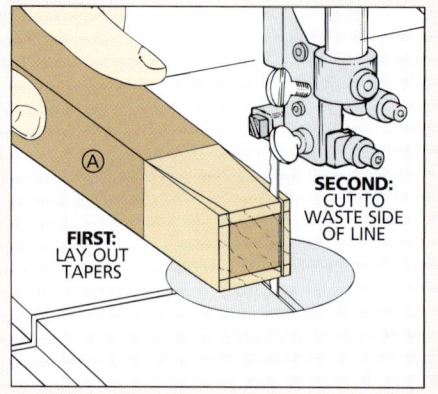

FIRST: LAY OUT TAPERS

SECOND: CUT TO WASTE SIDE OF LINE

1 *After laying out two tapers on one face of the leg, rough out the tapers with the band saw. Then rotate the leg and repeat for the remaining tapers.*

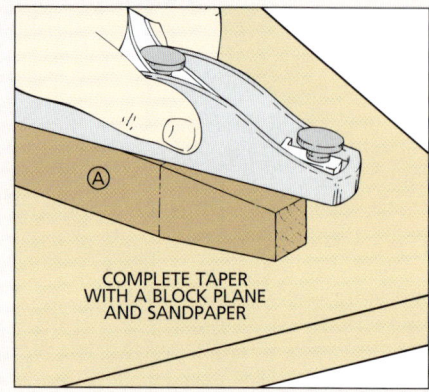

COMPLETE TAPER WITH A BLOCK PLANE AND SANDPAPER

2 *Keeping an eye on the layout lines on the faces and ends of the legs, clean up the tapers with a block plane. Finally, sand the faces as needed.*

APRONS

At this point the legs are complete. But you'll want to keep them nearby as you begin work on the three aprons (*Fig. 2*). That way you can check the fit and sizing of the pieces as you work.

The rear (B) and side aprons (C) are cut to size from $^3/_4$"-thick stock. But the critical thing here is the width (height) of the aprons. They need to be flush with the shoulder of the dadoes ($5^1/_8$") (*Fig. 2b*). Otherwise, you'll run into problems when adding the beading later.

With the aprons cut to size, tenons can be cut on the ends to fit the leg mortises (*Figs. 2a and 2b*). There are a number of ways to do this, but again, the important thing is that the aprons end up flush with the dadoes on the legs. (I cut the tenons using a dado blade in the table saw, because the same setup can be used again in the next step.)

The last thing to do to the aprons is to rabbet their inside edges (*Fig. 2c*). These rabbets will hold some web frames later, so the width of the rabbets should equal the thickness of your stock exactly.

ASSEMBLY. Now the legs and aprons are ready to be glued together. To keep the assembly square, I clamped a scrap spacer between the two front legs (see the Shop Tip on the opposite page). And it's a good idea to leave the spacer in place for support while you begin work on the web frames that fit into the rabbets. So cut the spacer to width so that it fits between the rabbets in the aprons.

WEB FRAMES

With the legs and aprons assembled, it's time to create the drawer openings. The first thing to do is make a pair of web frames (*Fig. 3*). These are identical except that the lower frame has dust panels. Then four vertical dividers can be built to fit between the frames.

FRONT AND BACK RAILS. The first frame pieces to work on are the front and back rails (D) (*Fig. 3*). The lower rails here will have the strips of bead glued to them later, so you want the thickness of these rails (and the four cross rails that

2 (Figure)

SIDE APRON Ⓒ

RABBETS FOR WEB FRAMES

NOTE: APRONS ARE $^3/_4$" THICK

27½

26

Ⓑ REAR APRON

14

15½

5⅛

NOTE: TABLE ASSEMBLED WITH HELP OF SCRAP SPACER (SEE SHOP TIP ON PAGE 89)

NOTE: TENONS CUT FIRST, THEN RABBETS

Ⓒ SIDE APRON

a.
¾
¾
½
3⅝
INSIDE FACE
¼
4⅛
ⒷⒸ SIDE/REAR APRON
¼

b.
½
CROSS SECTION
ⒷⒸ
4⅛
Ⓐ
½
APRON FLUSH WITH DADO ON LEG

c.
END VIEW
AUX. FENCE
ⒷⒸ
¾
¼

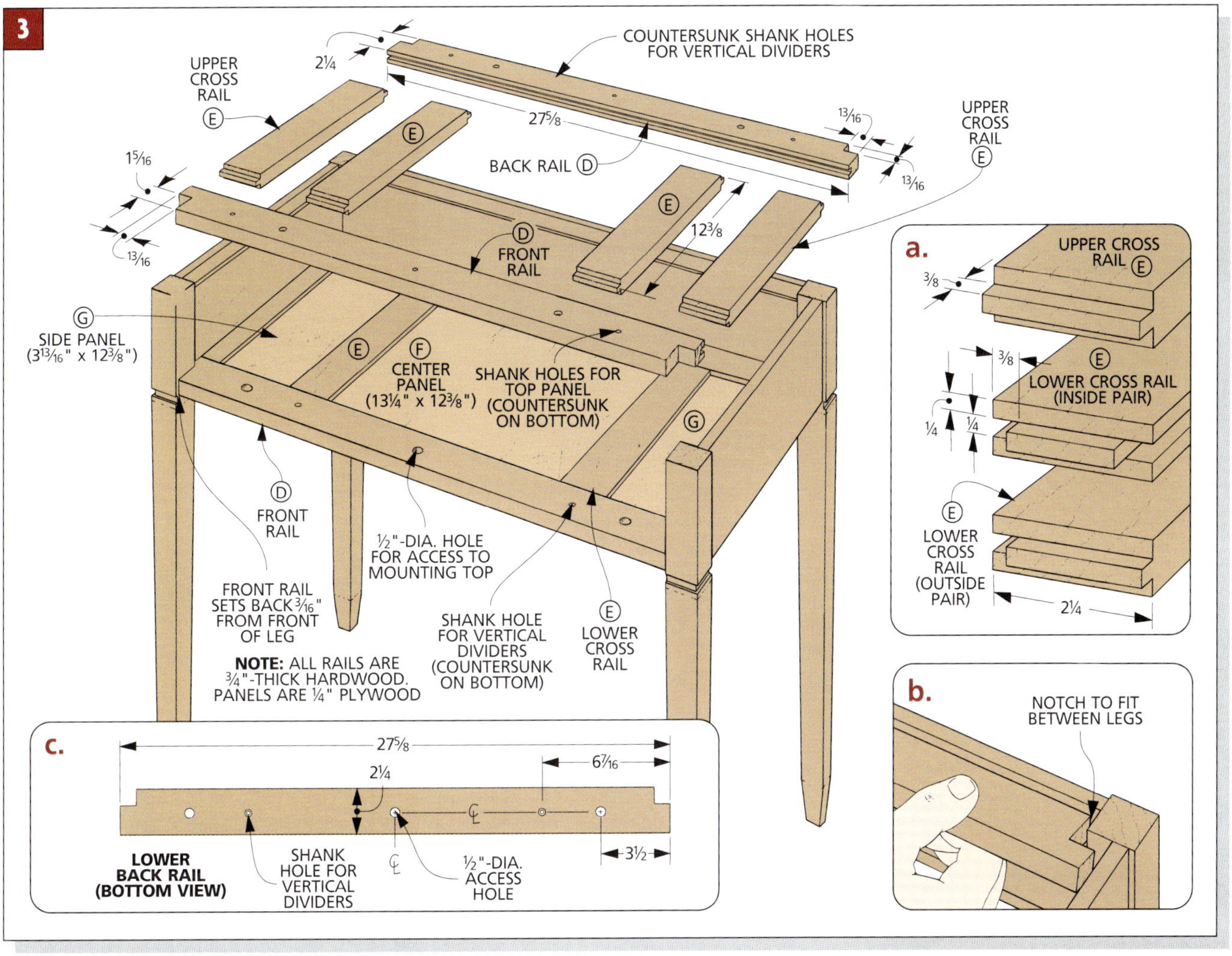

3

UPPER CROSS RAIL (E)

2¼

COUNTERSUNK SHANK HOLES FOR VERTICAL DIVIDERS

27⅝

13/16

UPPER CROSS RAIL (E)

13/16

1⁵/16

13/16

BACK RAIL (D)

(E)

12⅜

(D)
FRONT RAIL

(G)
SIDE PANEL
(3¹³/16" x 12⅜")

(E)   (F)
CENTER PANEL
(13¼" x 12⅜")

SHANK HOLES FOR TOP PANEL (COUNTERSUNK ON BOTTOM)

(G)

(D)
FRONT RAIL

½"-DIA. HOLE FOR ACCESS TO MOUNTING TOP

FRONT RAIL SETS BACK ³/16" FROM FRONT OF LEG

SHANK HOLE FOR VERTICAL DIVIDERS (COUNTERSUNK ON BOTTOM)

(E)
LOWER CROSS RAIL

NOTE: ALL RAILS ARE ¾"-THICK HARDWOOD. PANELS ARE ¼" PLYWOOD

a.

UPPER CROSS RAIL (E)

⅜

⅜

(E)
LOWER CROSS RAIL (INSIDE PAIR)

¼   ¼

(E)
LOWER CROSS RAIL (OUTSIDE PAIR)

2¼

b.

NOTCH TO FIT BETWEEN LEGS

c.

27⅝

2¼

6⁷/16

LOWER BACK RAIL (BOTTOM VIEW)

SHANK HOLE FOR VERTICAL DIVIDERS

½"-DIA. ACCESS HOLE

3½

---

connect them) to match the height of the rabbets in the aprons perfectly.

The front and back rails are cut to length to fit between the rabbets in the side aprons. Then they need to be notched to wrap around the legs. I did this by making multiple passes over my combination blade in the table saw *(Fig. 4)*.

The key dimension here is the shoulder-to-shoulder fit between the legs *(Fig. 3c)*. If there are any gaps at these points, they'll be fairly noticeable. So to get a clean, accurate cut, I used my rip fence as a stop, flipped each piece end for end between cuts, and nudged the rip fence to sneak up on the final fit.

As for the depth of these notches, the first thing to note is that they're not the same for the front and back rails. I raised the saw blade so the notch in the front rail would position the rail ³/16" from the front face of the legs *(Figs. 3 and 4a)*. The notch in the back rail is sized to fit into the rabbet in the rear apron *(Figs. 3b and 4b)*.

CROSS RAILS. When I was satisfied with the fit of the front and back rails, I dry-clamped the rails in position in the table and measured between them to find the shoulder-to-shoulder dimensions of the eight cross rails (E) *(Fig. 3)*. But when cutting these pieces to size, remember to add ¾" to that measurement to allow for the ⅜"-long stub tenon that will be cut on each end.

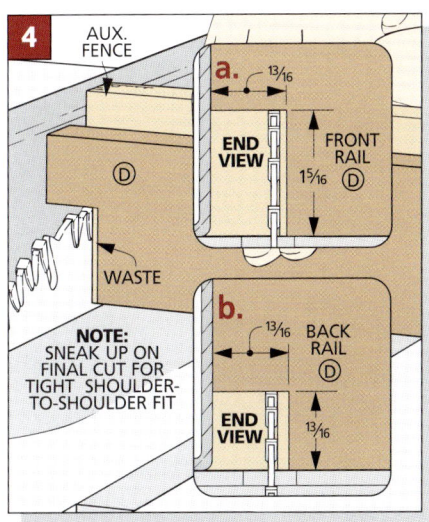

4

AUX. FENCE

a.   13/16

END VIEW

FRONT RAIL (D)

1⁵/16

(D)

WASTE

b.   13/16
BACK RAIL (D)

END VIEW   13/16

NOTE: SNEAK UP ON FINAL CUT FOR TIGHT SHOULDER-TO-SHOULDER FIT

SHOP TIP

Scrap Spacer

When it was time to assemble the legs and aprons, I needed a way to keep the three-sided assembly square. The solution was simple. I just cut a scrap spacer to match the distance between the rear legs and clamped it between the front legs.

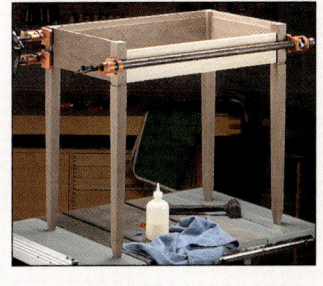

FRAME JOINERY. The frames are held together with stub tenon and groove joinery *(Figs. 5 and 6)*. The grooves are sized to hold a piece of ¼" plywood (for the dust panels in the lower frame), and they're cut in all four of the front and back rails. But to capture the panels, you'll also want to cut grooves on the *lower* cross rails only. Cut these grooves on one edge of the outside rails and on both edges of the inside rails (refer to *Fig. 3a* on the previous page).

With the grooves cut, the stub tenons can be cut on the cross rails *(Fig. 6a)*. Here, the important thing is that the shoulder-to-shoulder dimensions match the opening you measured earlier.

To do this, I dry-assembled the front and back rails in the table. Next, I attached an auxiliary fence to my table saw's rip fence, and made a cut on each end of a cross rail *(Fig. 6)*. Then I nudged the fence as necessary and made another cut until the shoulders of the cross rail fit between the front and back rails.

DUST PANELS. Next, I dry-assembled the lower frame and cut a center panel (F) and two side panels (G) to size from ¼" plywood (refer to *Fig. 3*). Then before gluing the frames together, I drilled mounting holes in the rails *(Fig. 3c)*. All four rails get two countersunk holes for attaching the center vertical dividers *(Fig. 7)*. Each upper rail has three more holes for attaching the top panel. But you won't be able to get a screwdriver to the screws unless you drill access holes in the lower rails *(Fig. 3c)*.

ASSEMBLY. All that's left for these frames is to glue them together. Then when the glue has dried, the frames can be added to the table.

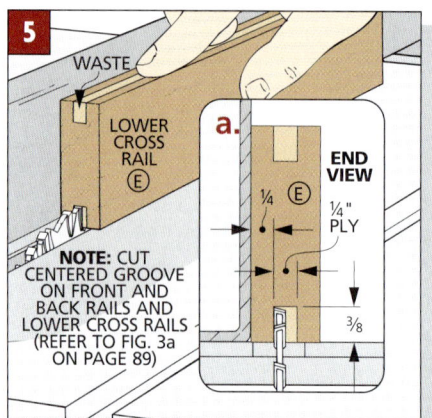

VERTICAL DIVIDERS

Now that the frames are in place, the four vertical dividers can be added *(Fig. 7)*. Each divider is a two-piece assembly: a vertical panel (H) and a breadboard vertical edging (I). These pieces will end up as tall as the drawer opening, but for now, it's a good idea to cut them oversized.

Each panel is joined to the edging with a stub tenon and groove joint *(Figs. 7 and 7a)*. The procedure for this joint is the same as for the frames, but this time, you can cut the grooves (and tenons) exactly ¼" wide since there is no plywood panel to hold.

Now the vertical edging can be glued to the panel, and when the glue is dry, the assembly can be cut to final size. The height of the vertical dividers should match the height of the openings. And they should be cut to length so they end up flush with the front rails *(Fig. 7a)*.

INSTALL DIVIDERS. To add these dividers to the table, the outside two can be glued to the legs *(Fig. 7)*. The inside two will be screwed in place, but you've

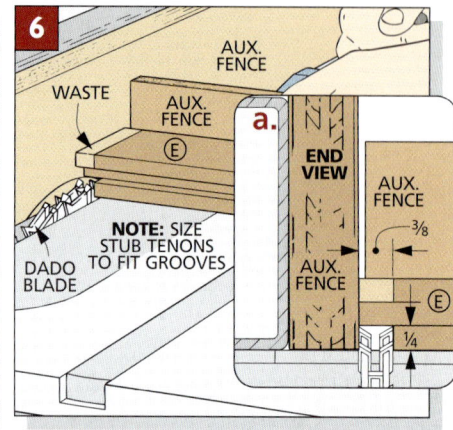

already drilled the shank holes for these. Still, I found that scrap spacers helped position them and hold them steady while I was screwing them in place.

BEAD & TOP PANEL

There are only a few parts left to add to the table before you can begin on the drawers. I built the bead molding and the drawer runners next *(Fig. 8)*. Then I made the solid wood panel for the top.

BEADING. The blanks for the table bead (J) need to be planed or resawn ⅜" thick to fit the dadoes in the legs *(Fig. 9)*. The bead will end up 1⅛" wide, but it would be better to leave the blanks extra wide for now so they're a little safer to work with on the router table.

With the pieces cut to rough size, the next thing to do is round over their front edges *(Fig. 9a)*. The workpieces are too thin to ride fully against the bearing on the bit, so use the router table fence to guide them during the cuts.

At this point, the bead can be ripped to final width. But before you miter the strips

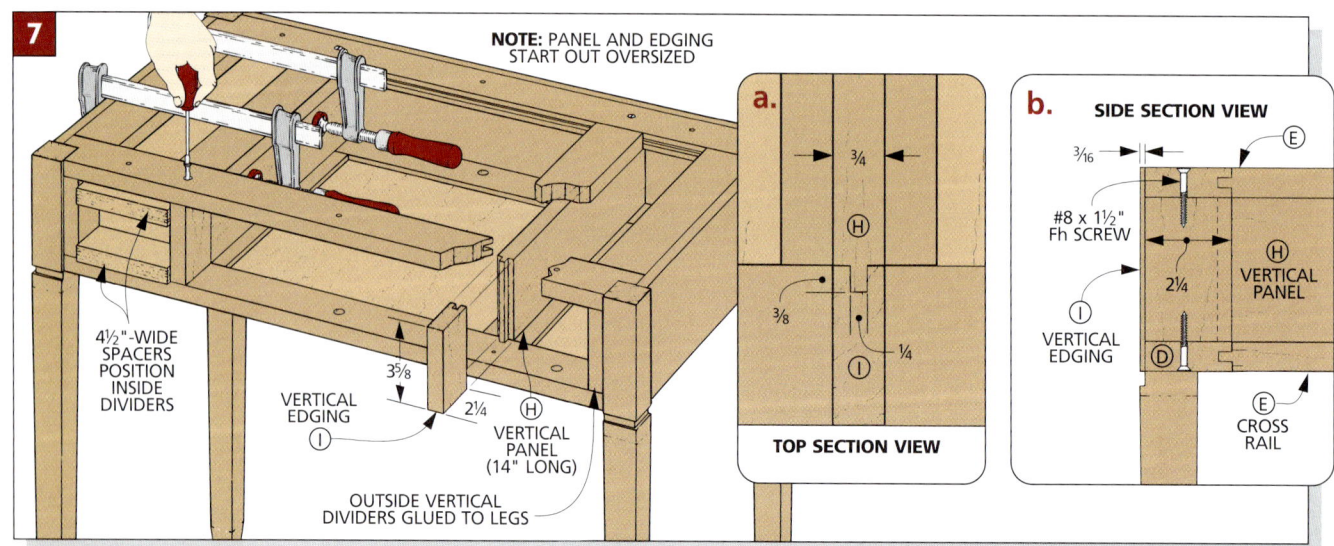

to final length, the ends need to be notched *(Fig. 9)*. This allows the strips to fit between the legs and also positions them so the bead stands proud $3/16$".

The procedure here is the same as the one you used to notch the front and back rails of the frame (refer to *Fig. 4* on page 89). And as before, you'll want to make sure the shoulder-to-shoulder fit between the legs is gap-free. Only this time, all the notches will be identical. Plus, you'll be able to use a dado blade (so there will be fewer passes to make).

After all the notches are cut, the last step is to miter the ends of the table bead so they will fit around the table. Since the inside corner of the joint will be hidden by the dado, fitting the miters is just a matter of sneaking up on each cut until the outside tips of the bead meet *(Fig. 8a)*. Then you can glue and clamp the strips to the table.

DRAWER RUNNERS. Before working on the top, I cut six drawer runners (K) and glued them to the cross rails inside the table *(Fig. 8)*. These $1/16$"-wide strips will create the gap under the drawers, keeping them from wearing on the lower front rail as they're opened and closed.

TOP PANEL. Now you're ready to work on the $3/4$"-thick solid-wood top panel (L) *(Fig. 8)*. After it's been planed or sanded flat, you can cut it to finished size so it overhangs the legs $1/2$" on each side.

Next, a bullnose profile needs to be routed around the edges. It's created with a $1/2$" roundover bit raised $5/16$" above the table *(Fig. 8b)*. And you'll want to start with the ends of the panel to reduce the chance of chipout.

You could screw the top in place now, but there are some good reasons to wait. For one thing, to add the drawer catches and stops later, you need easy access to the inside. But also, I like to apply finish to both the top and bottom faces so there's less chance the panel will warp. So I set the top panel aside until the table was almost completed.

CORNER BRACKETS. At this point, the optional corner brackets (M) can be added. The process of making the blanks and cutting them to shape is described in the Technique article on the next page.

Of course, these brackets can be left off if you prefer. To see what the table looks like without the brackets, see page 94.

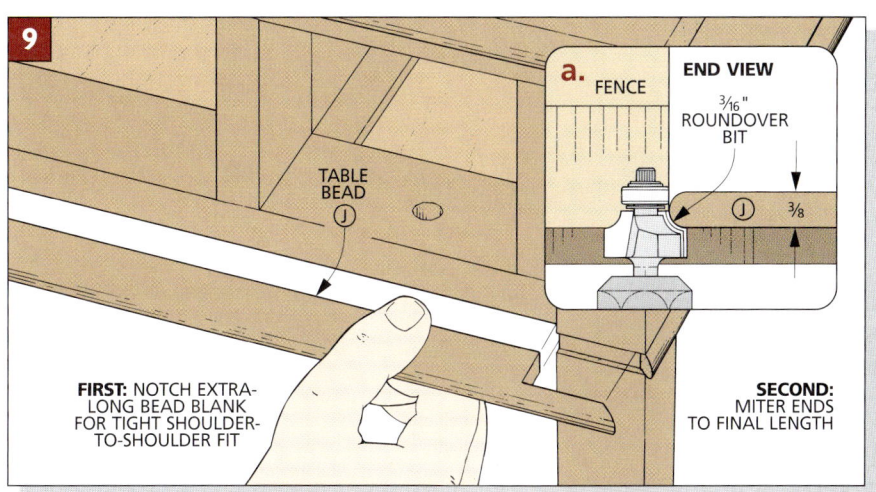

FIRST: NOTCH EXTRA-LONG BEAD BLANK FOR TIGHT SHOULDER-TO-SHOULDER FIT

TABLE BEAD (J)

a. FENCE   END VIEW
$3/16$" ROUNDOVER BIT
(J) $3/8$
SECOND: MITER ENDS TO FINAL LENGTH

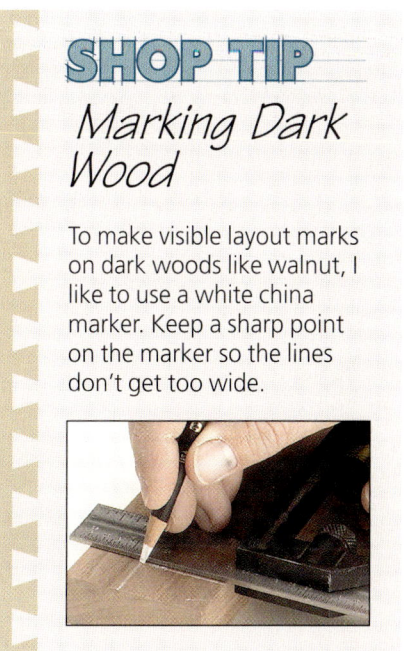

SHOP TIP
Marking Dark Wood

To make visible layout marks on dark woods like walnut, I like to use a white china marker. Keep a sharp point on the marker so the lines don't get too wide.

8

18   30

TOP PANEL ($3/4$" THICK)

BULLNOSE PROFILE

#8 x $1 1/4$" Fh SCREW

DRAWER RUNNERS ($3/4$" x $1/16$" - $15 3/4$")

NOTE: CUT NOTCHES ON BEAD TO FIT TIGHT BETWEEN LEGS

TABLE BEAD ($3/8$" x $1 1/8$")

$3/16$" ROUNDOVER

TABLE BEAD

a. TOP SECTION VIEW

$13/16$

BEAD STICKS $3/16$" PROUD OF LEG

DADO IN LEG

b. FENCE   END VIEW

TOP PANEL

$5/16$

$1/2$" ROUNDOVER BIT

When it came time to make the corner brackets for this table, I had one concern: the brackets look fragile. And the last thing I wanted was to have one break before (or after) it was in place. But there are two simple things you can do to make sure these pieces are plenty strong.

SHOP-BUILT PLYWOOD. First, instead of cutting the brackets out of solid wood, I made my own 3/8"-thick "plywood" from three layers of 1/8"-thick stock *(Fig. 1)*. This way, the grain direction of the center layer runs across the grain of the outside layers, so there are no weak spots.

Making your own plywood isn't as much work as it might sound. The plies are small *(Fig. 1)*, and you don't need special clamps or glue. (I used yellow glue and weighted the pieces while the glue dried.) The only thing to watch is the joint line of the two center pieces. It has to be tight, so there won't be any gaps when you cut out the bracket.

PATTERN ORIENTATION. Once you have your plywood in hand, you can lay out the shape for each bracket. The simple way to do it is to photocopy the pattern provided here at 112% *(Fig. 3)*.

I use rubber cement or a light coat of spray adhesive to fasten the patterns to the plywood. And here's the second thing you can do to make the brackets as strong as possible — position the patterns on the blanks so the long curve is oriented with the long grain *(Fig. 1)*. This means the pattern isn't tucked into the corner of the blank but runs along its length.

CUT TO SHAPE. After the pattern has been attached to the blank, you can begin cutting out the bracket *(Fig. 2)*. A scroll saw works best here, but you could easily make the triangle-shaped inside cut with a coping saw and cut the outside curves with a band saw and a narrow blade.

I started with the inside opening, drilling a starter hole and cutting the opening with a scroll saw *(Fig. 2)*. When that was done, I concentrated on the curved filigree in the center. Finally, I worked on the gentle curve that connects the leg to the table. What you want here is to make sure the inside and outside edges of the curve end up smooth and parallel.

FINISH AND FIT. Before you attach the brackets to the table, it's a good idea to apply a finish to them. However, you'll want to keep finish off the "pads" where the bracket attaches to the table. This is easy to do by covering each one with a strip of masking tape.

After the finish dries, you may have to sand these pads slightly to get a good, tight fit. Once you're satisfied with the fit, drill pilot holes through the brackets so there's no chance of splitting the thin stock. Then glue and nail them in place *(Fig. 3a* and photo).

1

SCROLL SAW PATTERN

1/8"-THICK PLIES

7

3

MIDDLE LAYER SHOULD HAVE TIGHT JOINT LINE

2

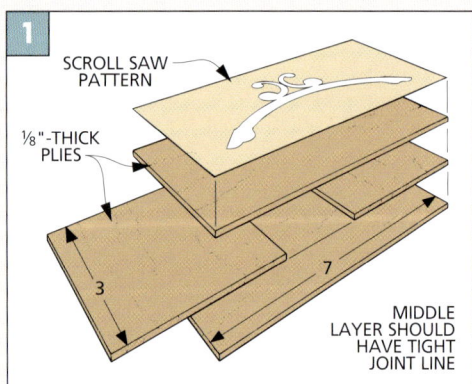

FIRST: DRILL STARTER HOLE

SECOND: CUT INSIDE OPENING

1/8"-THICK HARDBOARD USED AS BACKER TO PREVENT TEAROUT

THIRD: CUT OUT BRACKET AND SAND SMOOTH

3

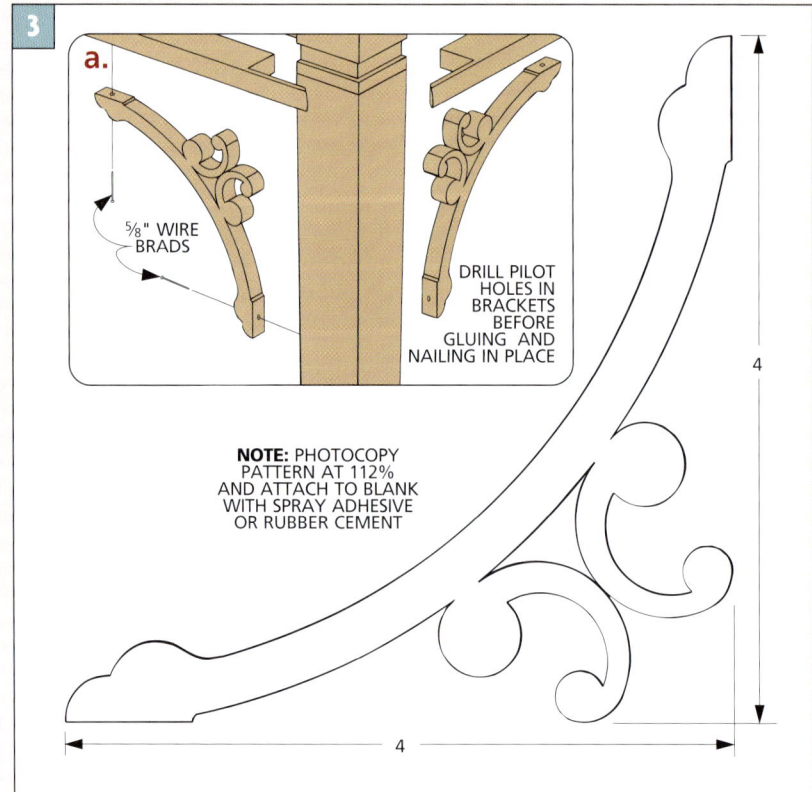

a.

5/8" WIRE BRADS

DRILL PILOT HOLES IN BRACKETS BEFORE GLUING AND NAILING IN PLACE

NOTE: PHOTOCOPY PATTERN AT 112% AND ATTACH TO BLANK WITH SPRAY ADHESIVE OR RUBBER CEMENT

4

4

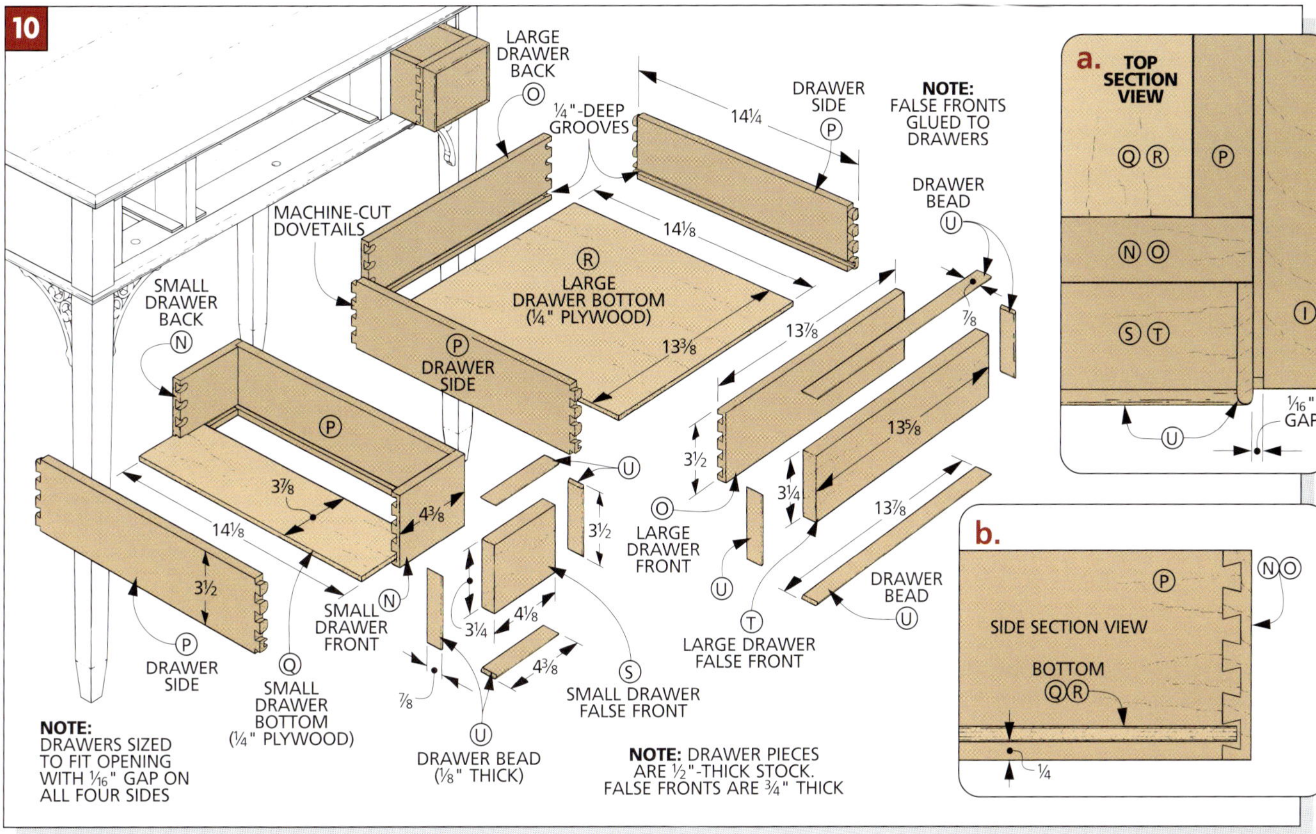

10

LARGE DRAWER BACK (O)
¼"-DEEP GROOVES
MACHINE-CUT DOVETAILS
SMALL DRAWER BACK (N)
14¼ — DRAWER SIDE (P)
14⅛ — LARGE DRAWER BOTTOM (¼" PLYWOOD) (R)
13⅜
DRAWER SIDE (P)
3⅞
14⅛ — DRAWER SIDE (P)
3½
SMALL DRAWER FRONT (N)
4⅜
3¼
4⅛
SMALL DRAWER BOTTOM (¼" PLYWOOD) (Q)
⅞
3¼
4⅜ (S)
DRAWER BEAD (⅛" THICK) (U)
SMALL DRAWER FALSE FRONT
3½
3½
LARGE DRAWER FRONT (O)
LARGE DRAWER FALSE FRONT (T)
DRAWER BEAD (U)
13⅜
13⅝
⅞
DRAWER BEAD (U)
DRAWER SIDE (P)

NOTE: FALSE FRONTS GLUED TO DRAWERS
DRAWER BEAD (U)

a. TOP SECTION VIEW
(Q)(R) (P)
(N)(O)
(S)(T)
(I)
(U)
1/16" GAP

b. SIDE SECTION VIEW
BOTTOM (Q)(R)
(P)
(N)(O)
¼

NOTE: DRAWERS SIZED TO FIT OPENING WITH 1/16" GAP ON ALL FOUR SIDES

NOTE: DRAWER PIECES ARE ½"-THICK STOCK. FALSE FRONTS ARE ¾" THICK

DRAWERS

One of the focal points of this project is the drawers. And one of the reasons they're so interesting is the small bead profile running around the edges of the false fronts *(Fig. 10)*. But other than this detail, they're really fairly typical.

CUT TO SIZE. I built the drawers with ½"-thick maple to provide a contrast with the walnut false fronts. The small and large fronts and backs (N, O) are sized to create a 1/16" gap on each side of the opening *(Fig. 10a)*. All the drawer sides (P) are identical and are cut to length so the drawers end up shallower than the full depth of the case. (A stop will be added behind each drawer later so the fronts will be flush with the front of the table.)

DOVETAILS. After the pieces have been cut to size, the next step is to rout half-blind dovetails on all the pieces. This is "business as usual" for the center drawer. But with my dovetail jig, the fronts and backs of the small drawers were a hair short for the jig to clamp them securely. My solution was to "extend" the pieces as shown in the Shop Tip at right. (More tips about routing half-blind dovetails can be found on page 70.)

DRAWER BOTTOMS. When the dovetails have been routed, the next step is to cut grooves centered on the bottom tails (and sockets) of the drawer pieces for the ¼" plywood bottoms *(Fig. 10b)*. Placing the groove here ensures that it's hidden when the drawer is assembled. Then you can cut the small and large bottoms (Q, R) to fit between these grooves.

ASSEMBLY. At this point, you can glue the drawer pieces together. (On page 77 there's a handy tip for clamping dovetail joints.) But before you start work on the false fronts, slide the drawers into the table to check their fit. You may need to sand or plane the drawers so they slide smoothly and have a consistent gap all the way around (1/16").

FALSE FRONTS. Typically, false fronts are larger than the drawers to which they're attached. Not here. The ¾"-thick small and large false fronts (S, T) are actually ¼" smaller than the drawer fronts in length and width *(Fig. 10)*. This allows for the ⅛"-thick bead that will be glued to each edge of the false fronts. (The completed false fronts should match the drawer fronts exactly.)

And here's one thing you may want to consider as you lay out the false fronts. Try to find a single board from which all three fronts can be cut. This way the wood grain will run continuously across the three drawers.

SHOP TIP

Dovetail Jig Extension

To fit my dovetail jig, I had to extend the small drawer fronts and backs with a simple platform. It's just a small piece the same width as the front and back pieces that's glued to a piece of ¼" hardboard. Add a piece of adhesive-backed sandpaper to the hardboard platform to help prevent the workpiece from shifting as you're routing the dovetails.

DRAWER BEAD. With the false fronts sized, you can make the $\frac{1}{8}$"-thick drawer bead (U) next. I started with oversize blanks. This makes them safer to work with while rounding over the edges on the router table *(Fig. 11)*.

After rounding over all four edges, the bead pieces can be ripped to final width ($\frac{7}{8}$"). Then you can begin the process of mitering them to length so they wrap around the false fronts. Just be sure to use a sparing amount of glue when gluing them in place — you don't want squeezeout on the front faces that could affect the finish.

Now the false fronts are ready to be attached to the drawers. These drawers are small enough that I simply glued them in place, taking care that the edges were flush *(Fig. 12a)*.

After the glue had dried, I laid out the locations for the machine screws for the pulls. Then the holes can be drilled and the pulls added *(Figs. 12 and 12a)*. (Sources of the Hepplewhite-style pulls I used can be found on page 126.)

One thing to note here — when you buy your pulls, make sure you get machine screws long enough to go through both the drawer fronts and false fronts. (Mine were $1\frac{3}{4}$" long.)

STOPS AND CATCHES. All that's left to do now is add a couple of pieces to the table so the drawers "work" better. At the back of each drawer opening, I glued a small stop (V). The thickness of these stops should position the *face* of the drawers (not the bead) flush with the

The brackets at each corner of the table are strictly ornamental. They can just as easily be left off if you want a slightly different appearance for your table.

rails and vertical dividers *(Fig. 12a)*. (My stops were $\frac{1}{2}$" thick.)

Finally, to prevent the drawers from being pulled out completely, I added a small catch (W) to the back edge of the front rail in each opening. It is screwed loosely in place and its top corners are

relieved so it can pivot up out of the way when you want to remove the drawer.

FINISH. To finish the table, I chose a tung-oil varnish and wiped on several coats. And remember to apply the finish to both faces of the top panel before screwing it to the table. ∎

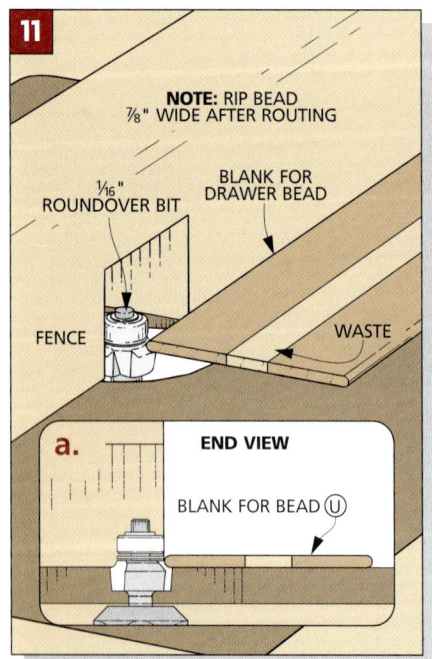

11

NOTE: RIP BEAD $\frac{7}{8}$" WIDE AFTER ROUTING

$\frac{1}{16}$" ROUNDOVER BIT

BLANK FOR DRAWER BEAD

FENCE

WASTE

a. END VIEW

BLANK FOR BEAD Ⓤ

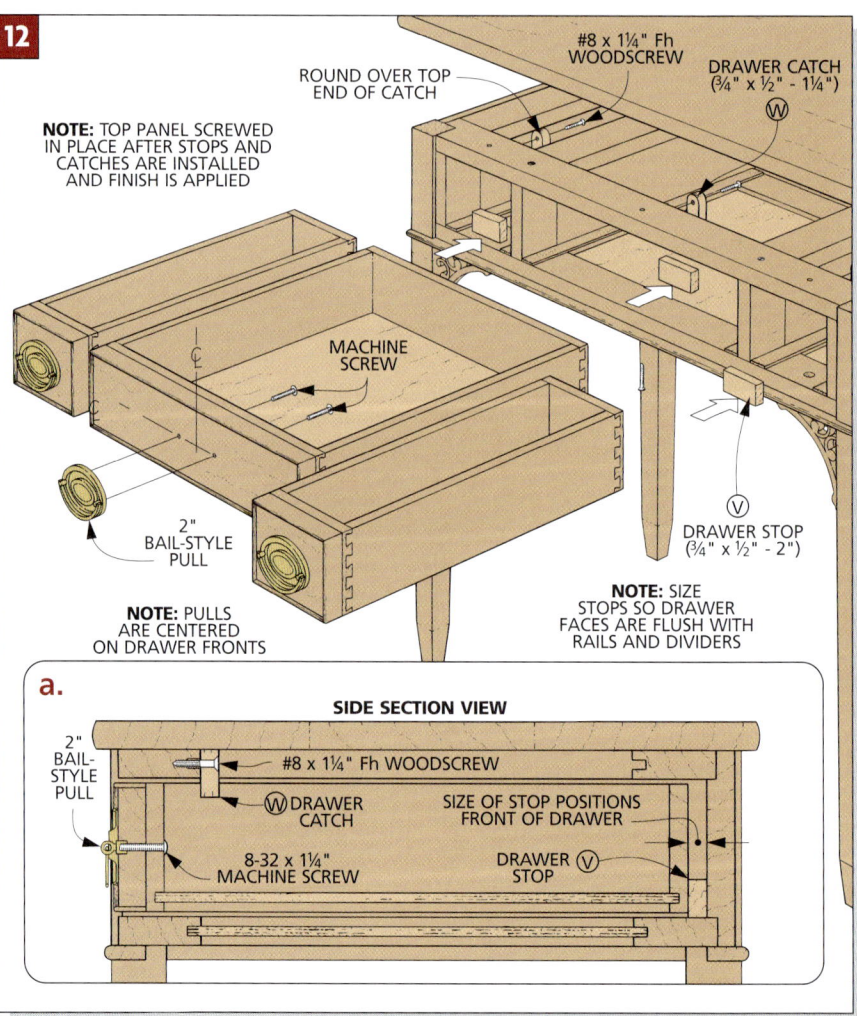

12

NOTE: TOP PANEL SCREWED IN PLACE AFTER STOPS AND CATCHES ARE INSTALLED AND FINISH IS APPLIED

ROUND OVER TOP END OF CATCH

#8 x 1¼" Fh WOODSCREW

DRAWER CATCH (¾" x ½" - 1¼") Ⓦ

MACHINE SCREW

2" BAIL-STYLE PULL

NOTE: PULLS ARE CENTERED ON DRAWER FRONTS

DRAWER STOP (¾" x ½" - 2") Ⓥ

NOTE: SIZE STOPS SO DRAWER FACES ARE FLUSH WITH RAILS AND DIVIDERS

a. SIDE SECTION VIEW

2" BAIL-STYLE PULL

#8 x 1¼" Fh WOODSCREW

Ⓦ DRAWER CATCH

SIZE OF STOP POSITIONS FRONT OF DRAWER

8-32 x 1¼" MACHINE SCREW

DRAWER Ⓥ STOP

DESIGNER'S NOTEBOOK

A single, wide drawer and a frame and panel top highlight this design option. The plywood panels in the top eliminate concerns about wood movement.

CONSTRUCTION NOTES:

■ The base of the table is built as before, but since there is just one drawer, you won't need the vertical panels (H) or the vertical edging (I).

■ The top panel (L) is replaced with a mitered frame that holds three plywood panels. First, cut the top front and back (X), top sides (Y), and top middle rails (Z) to rough length (see drawing).

■ Now cut a 1/4" groove on each piece centered on its thickness (detail 'a'). Note that the middle rails need a groove on each long edge.

■ The next step is to miter the front, back, and side pieces 1/2" longer than the outside dimensions of the table base (30" and 18" in my case) (see drawing).

■ Dry-assemble the pieces and measure between the front and back (X) to find the shoulder to shoulder dimension of the middle rails. Cut them 3/4" longer than this measurement.

■ Next, cut the top panels (AA) from 3/4" plywood (see drawing). Once that is done, you can set up a dado blade and begin work on the joinery.

■ First, cut tongues on the ends of the middle rails, centered on their thickness and sized to fit the grooves in the frame (details 'a' and 'b').

■ When cutting the tongues around the panels, note that they are not perfectly centered on the thickness of the panel (detail 'c'). You want the top face of the panel to end up flush with the top face of the frame pieces.

■ Construction of the drawer is the same as before, but the drawer is wider (24 3/8") (see drawing).

■ Next, attach the 3" swan-neck brass pulls to the drawer.

■ Finally, install two drawer stops at the back of the case (one below each cross rail) so the drawer will sit square in the opening of the table.

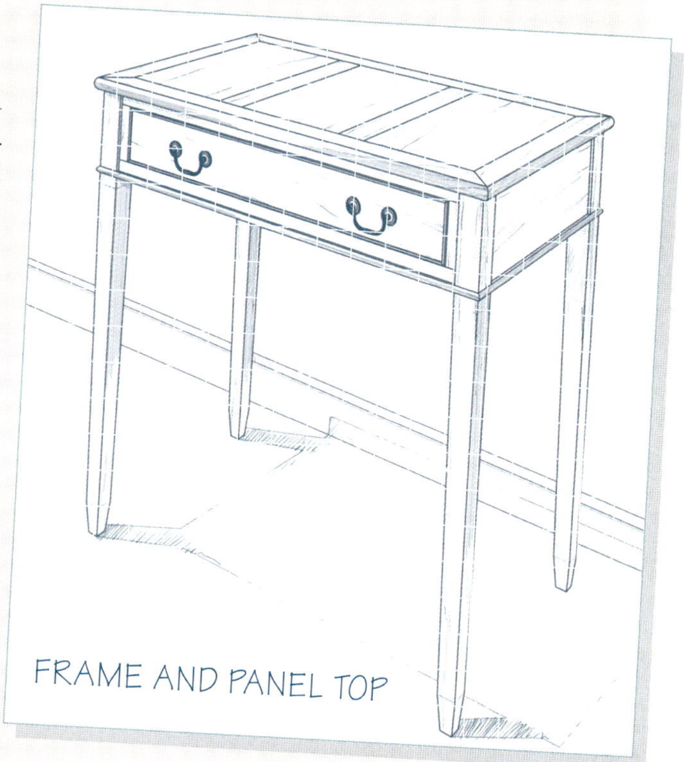
FRAME AND PANEL TOP

MATERIALS LIST

CHANGED PARTS

K	Drawer Runners (2)	3/4 x 1/16 - 15 3/4
O	Lg. Drwer Fr./Bk. (2)	1/2 x 3 1/2 - 24 3/8
P	Drawer Sides (2)	1/2 x 3 1/2 - 14 1/4
R	Lg. Drwr. Btm. (1)	1/4 ply - 14 x 23 7/8
T	Lg. Drwr. False Fr. (1)	3/4 x 3 1/4 - 24 1/8
U	Drawer Bead	1/8 x 7/8 - 60 ln. in.
V	Drawer Stops (2)	3/4 x 1/2 - 2
W	Drawer Catch (1)	3/4 x 1/2 - 1 1/4

NEW PARTS

X	Top Front/Back (2)	3/4 x 1 7/8 - 30
Y	Top Sides (2)	3/4 x 1 7/8 - 18
Z	Top Middle Rails (2)	3/4 x 1 7/8 - 15
AA	Top Panels (3)	3/4 ply - 8 1/4 x 15

HARDWARE SUPPLIES

(2) 3" brass swan-neck bail pulls w/ screws

Note: Don't need parts H, I, L, M, N, Q, S, Hepplewhite-style pulls and brads.

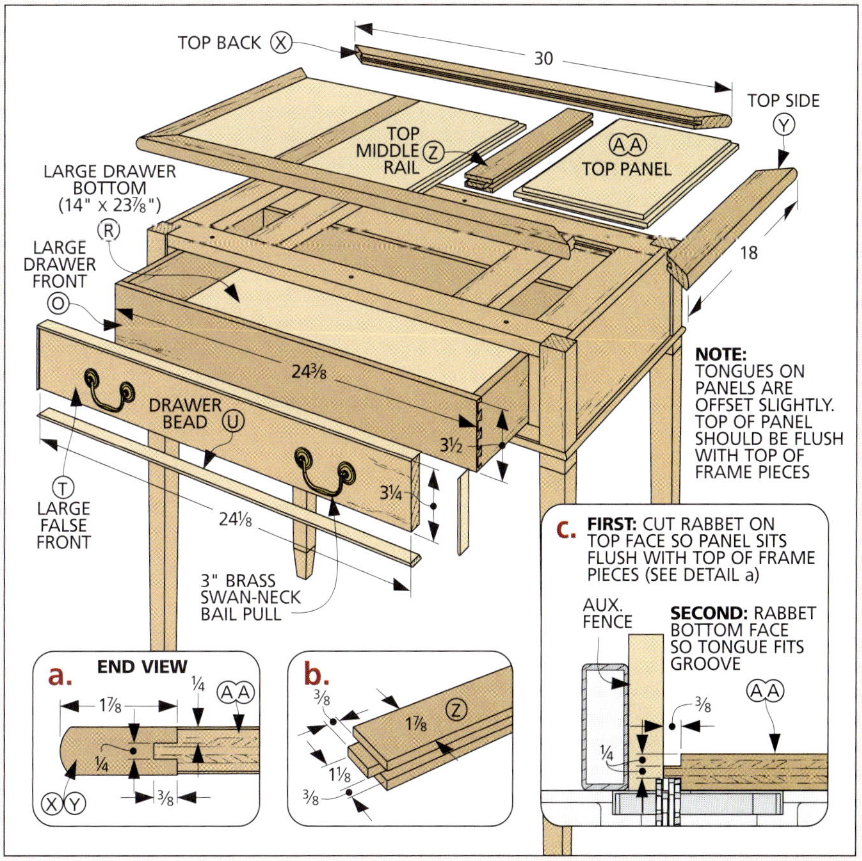
TOP BACK (X)
30
TOP SIDE (Y)
TOP MIDDLE (Z) RAIL
(AA) TOP PANEL
LARGE DRAWER BOTTOM (14" x 23 7/8") (R)
LARGE DRAWER FRONT (O)
18
24 3/8
NOTE: TONGUES ON PANELS ARE OFFSET SLIGHTLY. TOP OF PANEL SHOULD BE FLUSH WITH TOP OF FRAME PIECES
DRAWER BEAD (U)
3 1/2
3 1/4
(T) LARGE FALSE FRONT
24 1/8
3" BRASS SWAN-NECK BAIL PULL
c. FIRST: CUT RABBET ON TOP FACE SO PANEL SITS FLUSH WITH TOP OF FRAME PIECES (SEE DETAIL a)
AUX. FENCE
SECOND: RABBET BOTTOM FACE SO TONGUE FITS GROOVE
3/8
(AA)
1/4
a. END VIEW
1 7/8
1/4
(AA)
1/4
3/8
(X)(Y)
b.
3/8
1 7/8
(Z)
1 1/8
3/8

Barrister's Bookcase

It's easy to make a case for this modular bookcase — that's because the joinery is straightforward. The assemblies are small and easy to build, and you can customize it in several ways.

When you look at this Barrister's Bookcase, it would be easy to miss the fact that it's not built like most bookcases. Sure, there are framed glass doors that enclose the case, but more importantly, the bookcase isn't built as a single large unit — it's modular. And this particular Barrister's Bookcase is made up of three different sections: a cap, a base, and a case with about a 14" opening. (If you want you could also build a slightly shorter 12" version of the case that's featured in the Designer's Notebook on page 103.)

The nice thing about a modular design is that it allows for quite a bit of flexibility in the project. The modules can be mixed and matched in a variety of ways. (The taller version in the photo at left includes two tall and two short cases.) Plus, if at some point you end up with more books than shelf space, you can always build another section or two. (Which may be why these bookcases were so popular with lawyers, or barristers, to begin with.)

CLEAT SYSTEM. Of course, these individual sections have to connect securely. And here, instead of coming up with a "high tech" solution, I borrowed a simple cleat system I'd seen on some antiques. Two cleats in the bottom of one section straddle a single cleat in the top of the section below it. There's no hardware or tricky joinery to mess with. So when you move the bookcase out of the shop, these small sections won't strain your back.

There's one other thing to mention. Before building the bookcase, it's a good idea to decide how many modules you want to make (including the bases, caps, and shorter cases). Since all the sections have identical parts, you'll find the project will go together quicker (and fit together better) if you can build all the identical parts at the same time.

SOLID WOOD PANEL FRONT. For a different look, try the solid wood panel for the case fronts. This is featured in the Designer's Notebook on page 107.

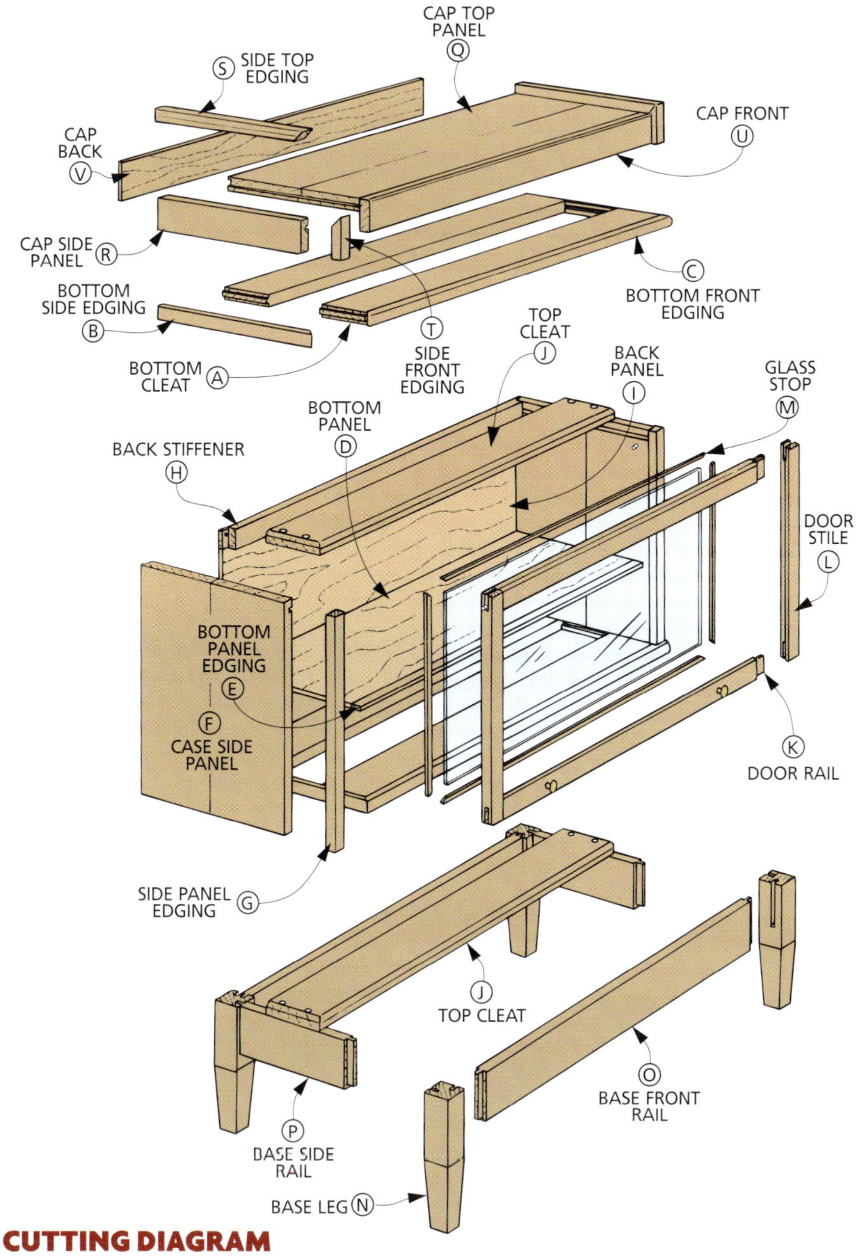

CAP TOP PANEL Q
S SIDE TOP EDGING
CAP BACK V
CAP FRONT U
CAP SIDE PANEL R
BOTTOM SIDE EDGING B
BOTTOM FRONT EDGING C
BOTTOM CLEAT A
T SIDE FRONT EDGING
TOP CLEAT J
BACK PANEL I
GLASS STOP M
BOTTOM PANEL D
BACK STIFFENER H
DOOR STILE L
BOTTOM PANEL EDGING E
F CASE SIDE PANEL
K DOOR RAIL
SIDE PANEL EDGING G
J TOP CLEAT
O BASE FRONT RAIL
P BASE SIDE RAIL
BASE LEG N

EXPLODED VIEW

OVERALL DIMENSIONS:
$34\frac{1}{2}$W x $12\frac{1}{8}$D x $25\frac{1}{2}$H
(ONE CASE, CAP, AND BASE)

MATERIALS LIST

WOOD

A	Bottom Cleats (4)	$\frac{3}{4}$ x $3\frac{1}{2}$ - $33\frac{3}{4}$
B	Btm. Side Edging (4)	$\frac{3}{4}$ x $\frac{5}{8}$ - $12\frac{1}{8}$
C	Btm. Fr. Edging (2)	$\frac{3}{4}$ x $\frac{5}{8}$ - $34\frac{1}{2}$
D	Bottom Panel (1)	$\frac{1}{4}$ ply - $9\frac{3}{4}$ x 32
E	Btm. Pnl. Edging (1)	$\frac{1}{4}$ x $\frac{3}{4}$ - 32
F	Case Side Panels (2)	$\frac{3}{4}$ x $11\frac{1}{4}$ - 14
G	Side Pnl. Edging (2)	$\frac{3}{4}$ x 1 - 14
H	Back Stiffener (1)	$\frac{3}{4}$ x $1\frac{1}{4}$ - $33\frac{1}{4}$
I	Back Panel (1)	$\frac{1}{4}$ ply - $13\frac{5}{8}$ x $33\frac{1}{4}$
J	Top Cleats (2)	$\frac{3}{4}$ x $4\frac{1}{4}$ - $33\frac{1}{4}$
K	Door Rails (2)	$\frac{3}{4}$ x $1\frac{1}{4}$ - $31\frac{7}{8}$
L	Door Stiles (2)	$\frac{3}{4}$ x $1\frac{1}{4}$ - $13\frac{7}{8}$
M	Glass Stop	$\frac{1}{4}$ x $\frac{3}{8}$ - 84 rough
N	Base Legs (4)	$1\frac{3}{4}$ x $1\frac{3}{4}$ - 8
O	Base Fr./Bk. Rails (2)	$\frac{3}{4}$ x 3 - $31\frac{1}{2}$
P	Base Side Rails (2)	$\frac{3}{4}$ x 3 - $9\frac{3}{8}$
Q	Cap Top Panel (1)	$\frac{3}{4}$ x $10\frac{3}{4}$ - $32\frac{1}{2}$
R	Cap Side Panels (2)	$\frac{3}{4}$ x $11\frac{1}{4}$ - $1\frac{7}{8}$
S	Side Top Edging (2)	$\frac{3}{4}$ x 1 - $11\frac{7}{8}$
T	Side Fr. Edging (2)	$\frac{3}{4}$ x 1 - $2\frac{1}{2}$
U	Cap Front (1)	$\frac{3}{4}$ x 2 - 32
V	Cap Back (1)	$\frac{1}{4}$ ply - $2\frac{5}{8}$ x $33\frac{1}{4}$

Note: Quantities above are for one case, cap, and base only.

HARDWARE SUPPLIES

(16) No. 8 x 2" Fh woodscrews
(10) No. 6 x 1" Fh woodscrews
(4) $\frac{1}{4}$"-dia. steel pins (1" long)
(2 pkg.) 18-gauge brads ($\frac{1}{2}$" long)
(2) $\frac{5}{8}$"-dia. brass knobs w/ screws
(1 pc.) $\frac{1}{8}$"-thick glass ($11\frac{3}{4}$" x $29\frac{3}{4}$" rgh.)

CUTTING DIAGRAM

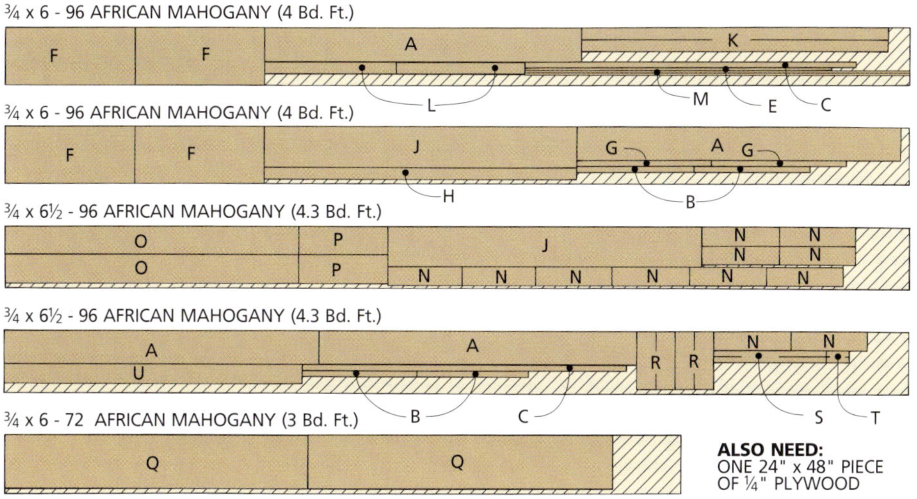

3/4 x 6 - 96 AFRICAN MAHOGANY (4 Bd. Ft.)
F F A K L M E C
3/4 x 6 - 96 AFRICAN MAHOGANY (4 Bd. Ft.)
F F J G A G H B
3/4 x 6½ - 96 AFRICAN MAHOGANY (4.3 Bd. Ft.)
O O P P J N N N N N N N N N N
3/4 x 6½ - 96 AFRICAN MAHOGANY (4.3 Bd. Ft.)
A U A R R N N B C S T
3/4 x 6 - 72 AFRICAN MAHOGANY (3 Bd. Ft.)
Q Q

ALSO NEED:
ONE 24" x 48" PIECE
OF $\frac{1}{4}$" PLYWOOD

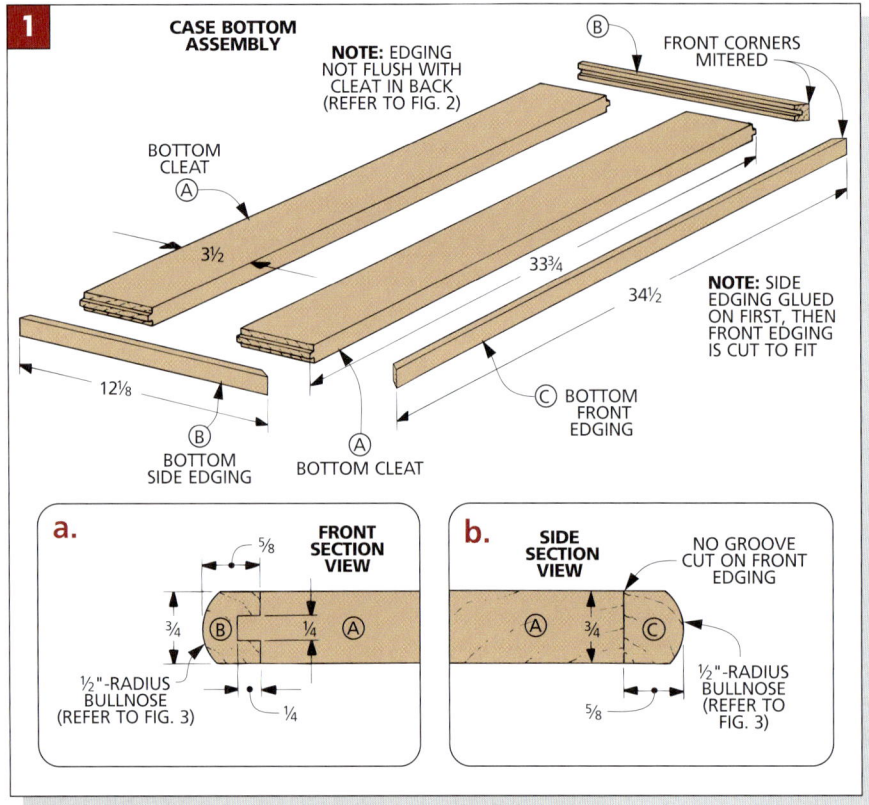

1 CASE BOTTOM ASSEMBLY

NOTE: EDGING NOT FLUSH WITH CLEAT IN BACK (REFER TO FIG. 2)

FRONT CORNERS MITERED — B

BOTTOM CLEAT — A

3½

33¾

34½

NOTE: SIDE EDGING GLUED ON FIRST, THEN FRONT EDGING IS CUT TO FIT

C BOTTOM FRONT EDGING

12⅛

B BOTTOM SIDE EDGING

A BOTTOM CLEAT

a. FRONT SECTION VIEW

⅝ · ¾ · B · ¼ · A

½"-RADIUS BULLNOSE (REFER TO FIG. 3)

¼

b. SIDE SECTION VIEW

NO GROOVE CUT ON FRONT EDGING

A · ¾ · C

½"-RADIUS BULLNOSE (REFER TO FIG. 3)

⅝

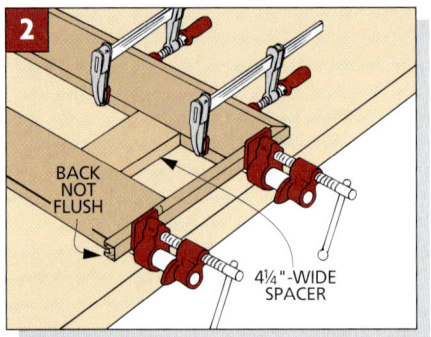

2 BACK NOT FLUSH

4¼"-WIDE SPACER

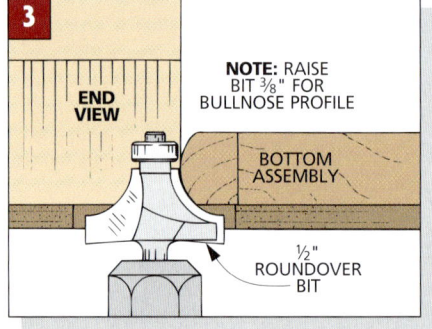

3 END VIEW

NOTE: RAISE BIT ⅜" FOR BULLNOSE PROFILE

BOTTOM ASSEMBLY

½" ROUNDOVER BIT

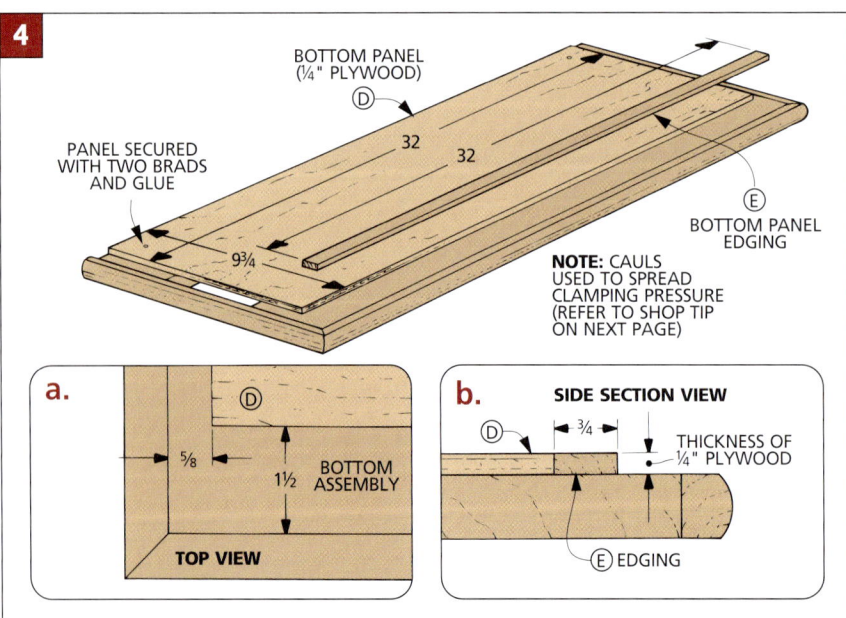

4 BOTTOM PANEL (¼" PLYWOOD) — D

PANEL SECURED WITH TWO BRADS AND GLUE

32 · 32

9¾

E BOTTOM PANEL EDGING

NOTE: CAULS USED TO SPREAD CLAMPING PRESSURE (REFER TO SHOP TIP ON NEXT PAGE)

a. ⅝ · D · 1½ · BOTTOM ASSEMBLY

TOP VIEW

b. SIDE SECTION VIEW

D · ¾ · THICKNESS OF ¼" PLYWOOD

E EDGING

CASE BOTTOM & SIDES

When building the Barrister's Bookcase, the best place to start is with the case. Depending on how many cases you decide to build, you'll save yourself some setup time if you can build all of them (as well as all the identical pieces for the base and cap) at the same time.

Note: The dimensions listed here are for the tall case. If you decide to build a short case or two, except for their heights, they are essentially the same. (Refer to the Designer's Notebook on page 103 for more on the small case.)

BOTTOM ASSEMBLY. To build the case, the place to start is with the bottom assembly *(Fig. 1)*. It's made up of two cleats with pieces of bullnose edging that wrap around the front and sides.

The bottom cleats (A) are made of ¾"-thick hardwood and building them is fairly straightforward. Once you've cut them to final size, the only thing left to do is cut the ¼"-thick, ¼"-long stub tenons on each end *(Fig. 1a)*.

These tenons help strengthen the hold of two pieces of bottom side edging (B) that connect the cleats *(Fig. 1)*. This edging starts out oversized so you can safely cut the grooves that are sized to hold the stub tenons on the cleats.

When the grooves have been cut, the side edging and the longer bottom front edging (C) can be ripped to final width (⅝"). Then you can miter the front corner of each side edging piece so it's 12⅛" long. (I left the back end of the side edging square. Later, when the back of the case is added, the edging helps hide the exposed plywood edge.)

At this point, the cleats and side edging can be glued together *(Fig. 2)*. When doing this, keep an eye on the front corners. The inside points of the edging should line up with the front of the cleat. Also, to get the cleats exactly 4¼" apart, I used two spacers. (Note that the edging will stick past the cleat ¼" in back.)

Now you can miter the front edging so it fits between the side pieces, then glue it and clamp it in place *(Fig. 2)*.

BULLNOSE PROFILE. I haven't forgotten about the bullnose profile on the edging. I decided it was best to wait until there was a larger assembly to work with. Now that the edging has been attached, the bullnose can be routed easily with a ½" roundover bit. To get the profile I wanted, I raised the bit ⅜" above the top of the router table *(Fig. 3)*.

Note: To guide the bottom assembly, you will need to use the router table fence.

BOTTOM PANEL AND EDGING. To create a "solid" bottom for the heavy books to sit on, I added a ¼" plywood bottom panel (D) that spans across the two cleats *(Fig. 4)*. To keep it from shifting when gluing it in place, I put two small brads in the back. Then to spread out the pressure, I used some clamping cauls (see the Shop Tip below).

To complete the bottom assembly, I glued a thin strip of bottom panel edging (E) to the front of the plywood bottom panel *(Fig. 4b)*. This strip has a dual purpose. It covers the exposed edge of the panel and also acts as a door stop later on.

SIDE PANELS. At this point, you can begin work on the side panels *(Fig. 5)*.

Note: The tall case side panel is 11¼" wide and 14" long *(Fig. 5)*. The shorter case side panel is only 12" long (refer to the Designer's Notebook on page 103).

The first thing to do is cut a dado to hold a steel pin that will guide and support the doors *(Fig. 5a)*. Then a hole is drilled near the front edge for a second steel pin *(Fig. 5b)*. (The door will rest on this steel pin when it's slid back into the case.)

EDGING. Next, I cut the side panel edging (G) to size *(Fig. 5)*. This edging acts as a stop at the front end of the dado, so the door can't be pulled out from the front. And other than rounding over its edges, the only thing to do to it is cut a rabbet so it wraps around the outside edge of the side panels *(Figs. 6 and 6a)*.

After the edging has been glued to the side panels, the last thing to do is cut a rabbet for the back panel that will be added later *(Fig. 5c)*.

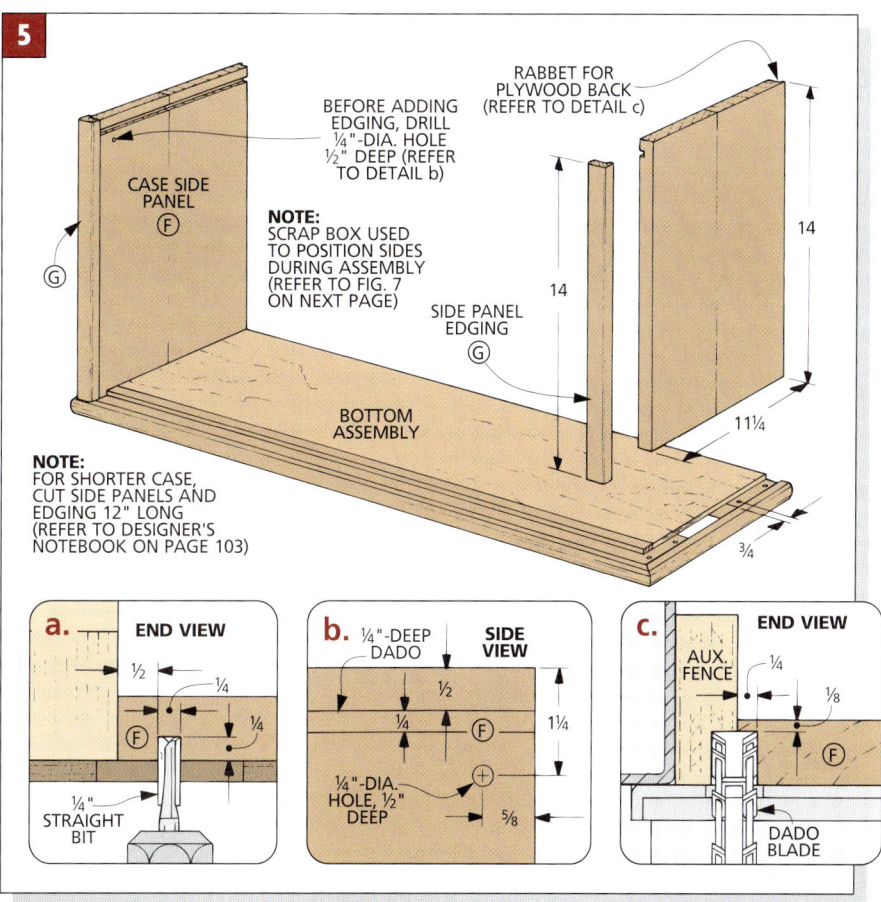

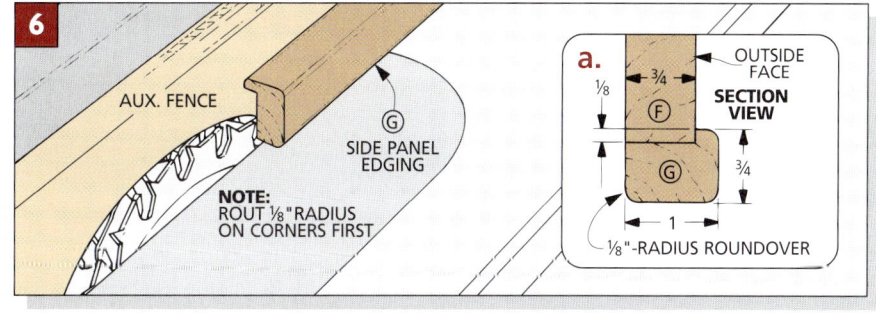

SHOP TIP . . . . *Clamping Cauls*

To spread the clamping pressure along the entire length of the bottom panel, I used a couple of 2x4s on edge as clamping cauls.

There's nothing fancy here. Just use 2x4s that are close in length (or slightly longer) than the surface of the pieces being glued up (see photo).

Pressure applied by the clamps at the end of the cauls is distributed along their length, ensuring a bond between the workpieces as the glue sets up.

ASSEMBLY. Now the side panels can be screwed to the bottom assembly. The plywood bottom helps position the sides, but to keep the assembly square, I built a box to act as a form *(Fig. 7)*.

Note: This form is easy to make with just four pieces of $^3/_4$" plywood. Be sure the top two spacers and the two side spacers are equal lengths so the case assembly ends up square.

As you work, there are two things to watch for. First, the back edges of the bottom side edging and the case side panels should be flush *(Fig. 7b)*. To make this easier to check (and adjust), I assembled the case upside down *(Fig. 7)*.

Also, you want the edges of the panels tight against the plywood bottom. So after adding screws at the back (to hold it flush) and front, I made sure the joint was tight before adding the other two screws.

CASE BACK & TOP CLEAT

With the side panels screwed in place, the next pieces to work on are the back and the cleat on top *(Fig. 8)*.

CASE BACK. There are actually two pieces that make up the back of the case: a hardwood stiffener and a $^1/_4$" plywood back panel *(Fig. 8)*. The back stiffener (H) is meant to be removable so you can add the door to the case later. It's cut to fit between the rabbets in the case sides. To get the stiffener to end up flush with the back of the case, you'll need to create a tongue on each end by cutting a small

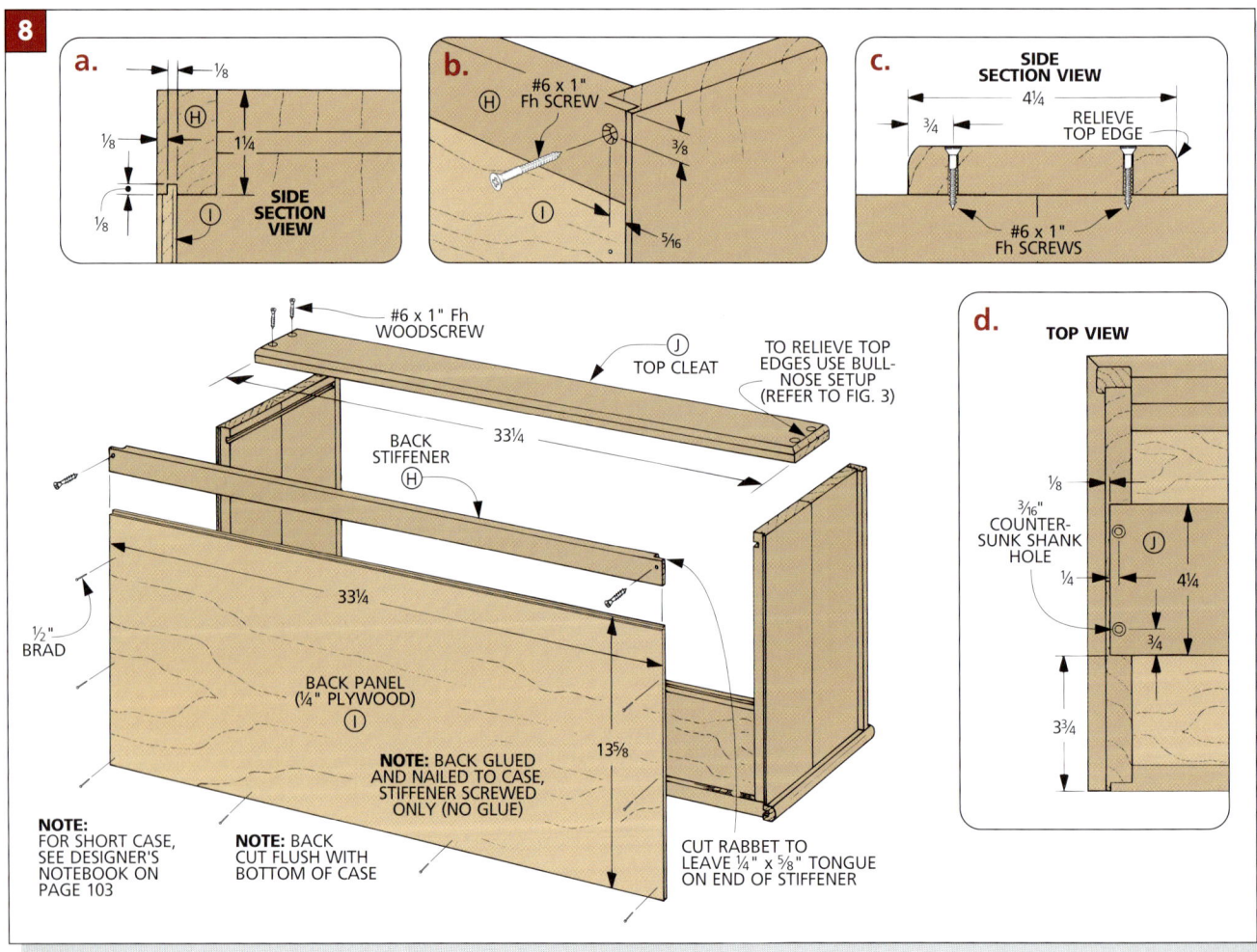

7

BOTTOM ASSEMBLY

SIDE PANEL ASSEMBLY

BACK CORNERS FLUSH (REFER TO DETAIL b)

ASSEMBLY BOX

a. SECTION VIEW — $^7/_8$ — #8 x 2" Fh WOODSCREW — ASSEMBLY BOX

b. TOP VIEW — (F) (D) (B) — BACK EDGES FLUSH

8

a. SIDE SECTION VIEW — $^1/_8$ — (H) $1^1/_4$ — $^1/_8$ — $^1/_8$ — (I)

b. #6 x 1" Fh SCREW — (H) — $^3/_8$ — (I) — $^5/_{16}$

c. SIDE SECTION VIEW — $4^1/_4$ — $^3/_4$ — RELIEVE TOP EDGE — #6 x 1" Fh SCREWS

#6 x 1" Fh WOODSCREW

(J) TOP CLEAT

TO RELIEVE TOP EDGES USE BULL-NOSE SETUP (REFER TO FIG. 3)

$33^1/_4$

BACK STIFFENER (H)

$33^1/_4$

$^1/_2$" BRAD

BACK PANEL ($^1/_4$" PLYWOOD) (I)

$13^5/_8$

NOTE: BACK GLUED AND NAILED TO CASE, STIFFENER SCREWED ONLY (NO GLUE)

NOTE: FOR SHORT CASE, SEE DESIGNER'S NOTEBOOK ON PAGE 103

NOTE: BACK CUT FLUSH WITH BOTTOM OF CASE

CUT RABBET TO LEAVE $^1/_4$" x $^5/_8$" TONGUE ON END OF STIFFENER

d. TOP VIEW — $^1/_8$ — $^3/_{16}$" COUNTER-SUNK SHANK HOLE — (J) — $^1/_4$ — $4^1/_4$ — $^3/_4$ — $3^3/_4$

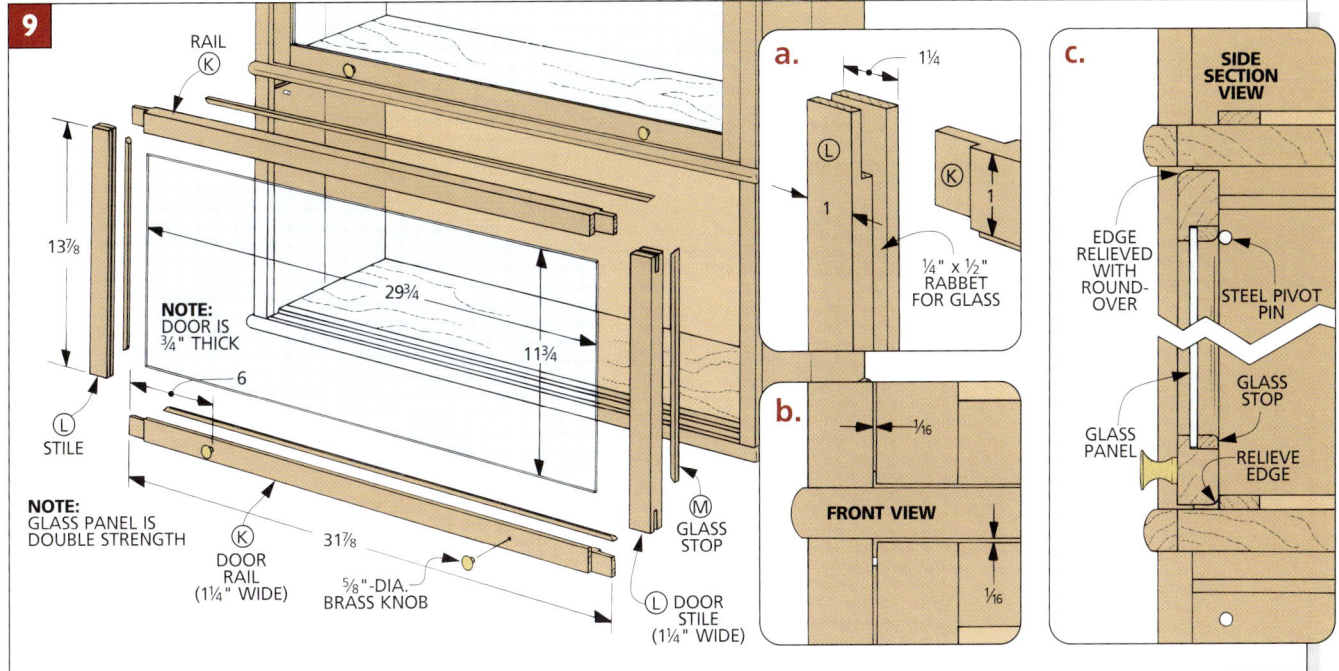

9

RAIL
K

13⅞

NOTE:
DOOR IS
¾" THICK

29¾

11¾

6

L
STILE

NOTE:
GLASS PANEL IS
DOUBLE STRENGTH

K

DOOR
RAIL
(1¼" WIDE)

31⅞

⅝"-DIA.
BRASS KNOB

M
GLASS
STOP

L DOOR
STILE
(1¼" WIDE)

a.

1¼

L

1

K

1

¼" x ½"
RABBET
FOR GLASS

b.

1/16

FRONT VIEW

1/16

c.

SIDE
SECTION
VIEW

EDGE
RELIEVED
WITH
ROUND-
OVER

STEEL PIVOT
PIN

GLASS
STOP

GLASS
PANEL

RELIEVE
EDGE

rabbet *(Fig. 8b)*. Then before screwing the stiffener in place, cut a small groove along the bottom edge to hold a tongue on the top of the back panel *(Fig. 8a)*.

The back panel (I) is made from ¼" plywood. Cut a small rabbet on top to create a tongue that fits into the groove you cut in the stiffener. Then simply glue and nail the back panel in place. (Don't get glue on the stiffener; it's removed later so you can install the door.)

TOP CLEAT. The next piece to add is the top cleat (J). It's sized to fit between the cleats in the bottom assembly. To relieve the top edges *(Fig. 8c)*, I used the same setup that was used for the bullnose profile earlier (refer to *Fig. 3* on page 98).

This cleat needs to be positioned so it will interlock with the bottom cleats on the other section (see the middle photo below). To make installing it easier, I cut a long spacer (3¾" wide) and clamped it flush with the back of the case before screwing the cleat in place.

DOOR

The door on the Barrister's Bookcase isn't like most cabinet doors. For one thing, to open it, the door swings up and slides into the case (see the far right photo below). So instead of hinges, steel pins (and a narrow dado cut in the sides) guide and support the door.

The joinery for this door is also unusual. Since this is a fairly narrow frame, I made the mortises and tenons as large as possible for added strength. So they're cut across the full width of the pieces *(Fig. 9a)*. (This joinery technique is called a "bridle joint.") The nice thing about this is that a bridle joint can be cut entirely on the table saw (more about that in a moment).

CUT TO SIZE. Another nice thing about the bridle joint is that the lengths of the door rails (K) and stiles (L) are easy to figure out. Both are cut ⅛" shorter than the case opening to create a 1/16" gap on each side of the door as well as at the top and bottom *(Fig. 9b)*.

Since it's never really seen, there's no top to the bookcase — just a top cleat that interlocks with the two bottom cleats in the case section above it.

The cleat system is very sturdy. These interlocking cleats create a quick, secure connection so the sections can be assembled in a matter of seconds.

So you can use both hands when removing a heavy book, the doors lift and slide back. Steel pins and a dado are all that's needed for a smooth-sliding fit.

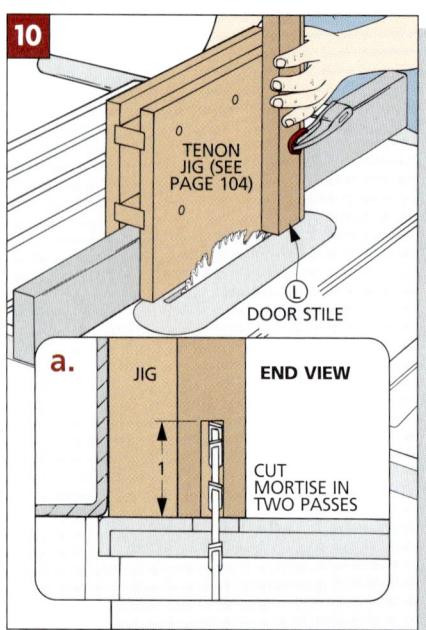

10

TENON JIG (SEE PAGE 104)

DOOR STILE

Ⓛ

a.

JIG

END VIEW

1

CUT MORTISE IN TWO PASSES

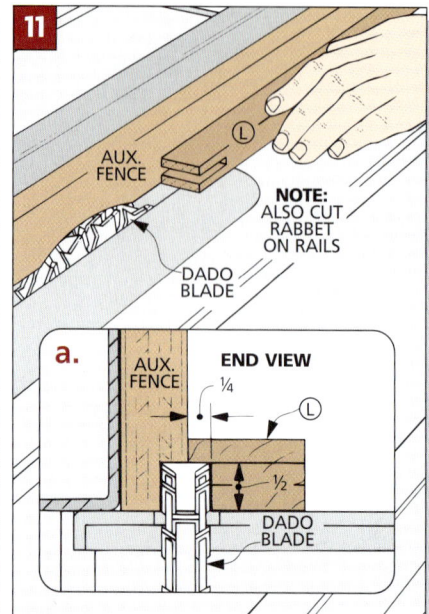

11

AUX. FENCE

Ⓛ

NOTE: ALSO CUT RABBET ON RAILS

DADO BLADE

a.

AUX. FENCE

¼

END VIEW

Ⓛ

½

DADO BLADE

12

AUX. FENCE

DOOR RAIL Ⓚ

DADO BLADE

NOTE: INSIDE CHEEK IS SHORTER TO ACCOUNT FOR RABBET FOR GLASS (SEE DETAIL a)

a.

1¼

1

END VIEW

INSIDE FACE

¼

¼ Ⓚ

¼

DADO BLADE

MORTISES. To cut the bridle joints, I started with the mortises *(Fig. 10)*. These are actually cut with a tenon jig that holds each stile vertically as it passes over the blade. (For tips on how to build a tenon jig, see the Shop Jig article on page 104.) To center each mortise, I made two passes across the blade, rotating the stile between each pass. (A combination or rip blade that cuts a flat-bottomed kerf will give you the cleanest fit.)

RABBETS. Before cutting the tenons, I cut a rabbet on all the frame pieces to hold the glass panel in place *(Fig. 11)*. Unfortunately, this will require an extra setup later for the tenons, but the door will end up a bit stronger.

TENONS. Even though a tenon jig was used for the mortises, I cut the tenons with the rails lying down *(Fig. 12)*. This way, you can more easily sneak up on the final position of each shoulder

(Fig. 12a). (To test the fit, just set the tenon across the outside face of the stile to see if the tenon is the right length.)

After cutting both cheeks of each tenon on the rails, the door frame can be glued together. Be sure to take it slow here. A bridle joint doesn't automatically lock into position like a regular mortise and tenon, so you'll want to check that the frame ends up square.

Even though you've got the door frame assembled now, there are still a few things to do before it will be ready to be added to the case.

ROUND OVER EDGES. The first thing to do is relieve a couple of the door's edges. Otherwise, they'll "catch" on the case as the door is opened and closed. You want to relieve the outside edge at the top of each door and the inside edge at the bottom *(Fig. 13)*. To do this, I routed these edges at the router table, using the

same setup that created the bullnose profile on the edging pieces (a ½" roundover bit raised ⅜" above the table).

ADD STEEL PINS. The next task is to add the steel pins that guide and support the door *(Figs. 14 and 15)*. I bought a ¼"-dia. steel rod from a local hardware store and cut it into 1"-long pins with a hack saw. (You'll need four pins for each case: two for the door and two for the case.)

Drilling the holes for the pins in the ends of the doors might seem a bit tricky *(Fig. 14)*. But they're not very deep — just ¾". (I used a piece of tape as a quick depth stop.) So as long as you work carefully, there shouldn't be any problems.

After the holes have been drilled, the pins can be glued into the door and case. (I used epoxy to do this.)

GLASS AND GLASS STOP. Now is a good time to order the glass panels to fit into the rabbets in the frame *(Fig. 14a)*.

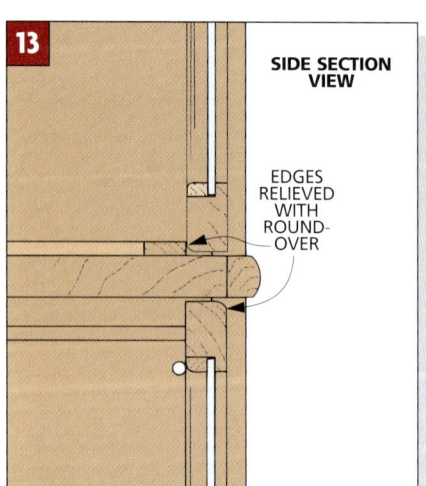

13

SIDE SECTION VIEW

EDGES RELIEVED WITH ROUND-OVER

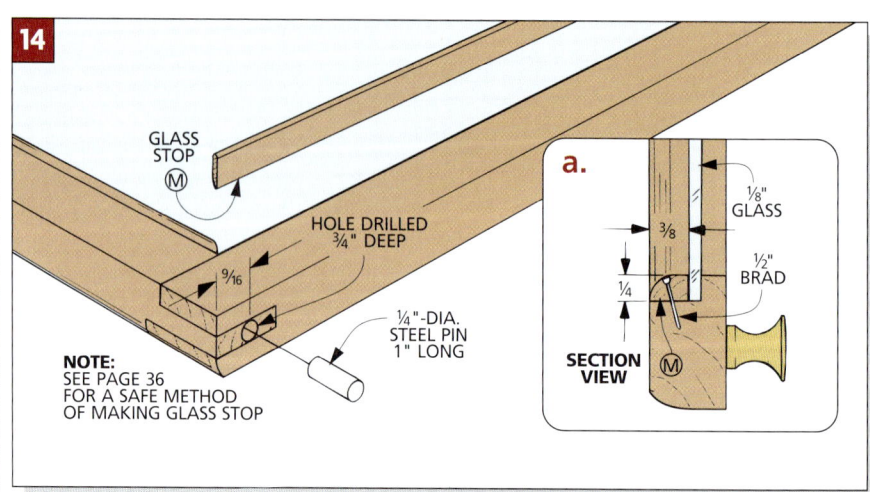

14

GLASS STOP Ⓜ

HOLE DRILLED ¾" DEEP

9⁄16

¼"-DIA. STEEL PIN 1" LONG

NOTE: SEE PAGE 36 FOR A SAFE METHOD OF MAKING GLASS STOP

a.

⅛" GLASS

⅜

½" BRAD

¼

SECTION VIEW Ⓜ

I ordered double-strength glass (about $\frac{1}{8}$" thick), and to make sure the glass fit comfortably, I had the panels cut $\frac{1}{8}$" less than the opening in both directions.

To hold the panels in place, I used strips of glass stop (M) (*Fig. 14*). These are rather small pieces to work with, but you can make them safely and quickly on the table saw. (For more on this, see the Shop Tip on page 36.) Then the strips can be mitered to fit into the rabbet.

The glass stop will be nailed into place with small brads, but I waited to do this until after the bookcase was finished. Then I added the glass and brass knobs. Finally, to install the doors, I removed the back stiffener and slid the pins into the dadoes in the case sides (*Fig. 15*).

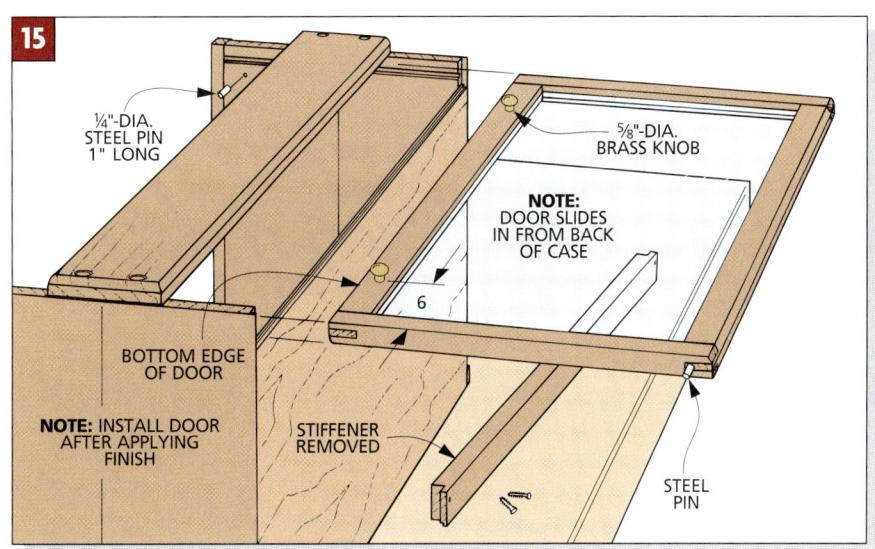

15

$\frac{1}{4}$"-DIA. STEEL PIN 1" LONG

$\frac{5}{8}$"-DIA. BRASS KNOB

NOTE: DOOR SLIDES IN FROM BACK OF CASE

BOTTOM EDGE OF DOOR

NOTE: INSTALL DOOR AFTER APPLYING FINISH

STIFFENER REMOVED

6

STEEL PIN

DESIGNER'S NOTEBOOK

With a shorter case or two placed on the top, a tall stack of bookcases won't look top heavy.

CONSTRUCTION NOTES:

■ A lot of Barrister's Bookcases had two case sizes — the ones at the top were a little shorter than those on the bottom. Frankly, I wasn't too sure what this would involve. Often a small change in size can really affect a lot of dimensions. But that definitely wasn't the case with this project.

■ To make a shorter version of the case (with a 12" opening), only a few pieces change: the side panels and their edging, the back, and the door stiles. (You'll also need to adjust the size of the glass to fit the door.) Other than that, the procedure and dimensions for building the short case are identical to the taller version.

MATERIALS LIST

CHANGED PARTS

F	Case Side Panel (2)	$\frac{3}{4}$ x $11\frac{1}{4}$ - 12
G	Side Panel Edging (2)	$\frac{3}{4}$ x 1 - 12
I	Back Panel (1)	$\frac{1}{4}$ ply - $11\frac{5}{8}$ x $33\frac{1}{4}$
L	Door Stiles (2)	$\frac{3}{4}$ x $1\frac{1}{4}$ - $11\frac{7}{8}$
M	Glass Stops	$\frac{1}{4}$ x $\frac{3}{8}$ - 80 rough

HARDWARE SUPPLIES

(1 pc.) $\frac{1}{8}$"-thick glass ($9\frac{3}{4}$" x $29\frac{3}{4}$" rough)

SHORT CASE

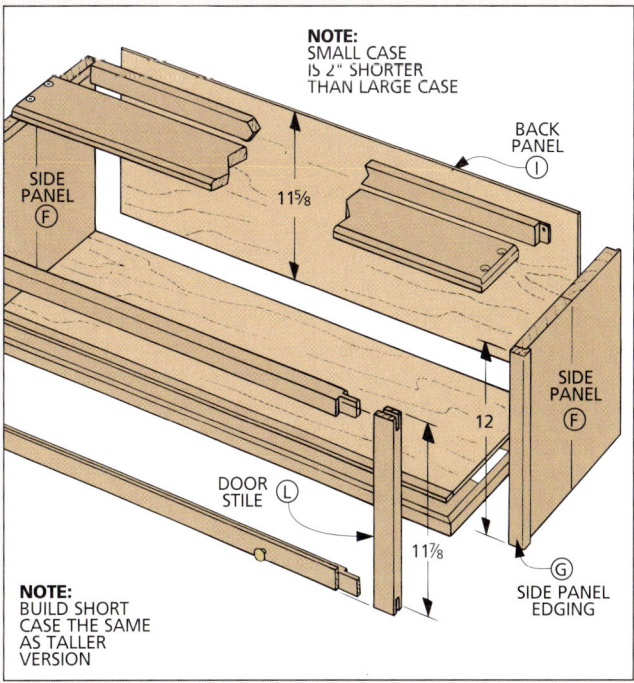

NOTE: SMALL CASE IS 2" SHORTER THAN LARGE CASE

BACK PANEL (I)

SIDE PANEL (F)

$11\frac{5}{8}$

DOOR STILE (L)

12

SIDE PANEL (F)

$11\frac{7}{8}$

SIDE PANEL EDGING (G)

NOTE: BUILD SHORT CASE THE SAME AS TALLER VERSION

SHOP JIG ...... *Tenon Jig for Mortises*

The stiles and rails for the doors I built to fit the Barrister's Bookcase are held together with a bridle joint (refer to *Fig. 9* on page 101). Since the pieces used to construct the door are fairly narrow, the mortises and tenons are larger to add strength.

A bridle joint is just another name for a mortise and tenon joint where the mortise is cut right through the end of the stile. In effect, a tenon cut on the stiles slips into the open mouth of the mortise cut on the rails, somewhat like a bridle fits into a horse's mouth.

It may seem odd, but an easy way to cut this type of mortise on the long rails is with a simple tenoning jig on the table saw. Instead of cutting away the outside faces of the workpiece to leave a tenon, you cut away the middle to leave an open mortise (see photo).

The jig is just a tall vertical carriage that straddles the rip fence (see the drawing below). And to keep the workpiece 90° to the table, there's a vertical hardwood stop that backs it up.

FACE PIECES. To make this jig, I started by cutting two identical face pieces out of

¾" medium-density fiberboard (MDF). (Mine were 8" tall and 12" long.)

Next, a couple of ¼"-deep dadoes are cut on the inside face of each piece (see detail 'a'). These are sized to hold the MDF crosspieces added later.

Note: Be sure to position the bottom dado so the crosspiece will clear the top of your table saw's rip fence and any exposed bolt heads on the fence (see the photo and detail 'a').

I also cut one more dado for the vertical stop. Since this stop will get chewed up with repeated passes over the blade, the dado allows you to easily replace the vertical stop in the correct position.

CROSSPIECES. With the two face pieces complete, they're connected with a couple of crosspieces (see drawing and detail 'a'). These crosspieces are sized so the face pieces fit just snug against the rip fence. The goal here is to allow the jig to slide easily — but without any "slop."

When the crosspieces are cut to final size, the jig can be screwed together. I also waxed the inside faces of the jig (below the bottom crosspiece) so it would slide back and forth easily along the rip fence.

VERTICAL STOP. Finally, the hardwood vertical stop is cut to fit the dado and screwed in place. To avoid damaging the blade, place the bottom screw above the highest blade setting.

USING THE JIG. Cutting the mortise is just a matter of raising the blade to the correct height (equal to the width of the mortise), butting the workpiece against the stop, and clamping it in place.

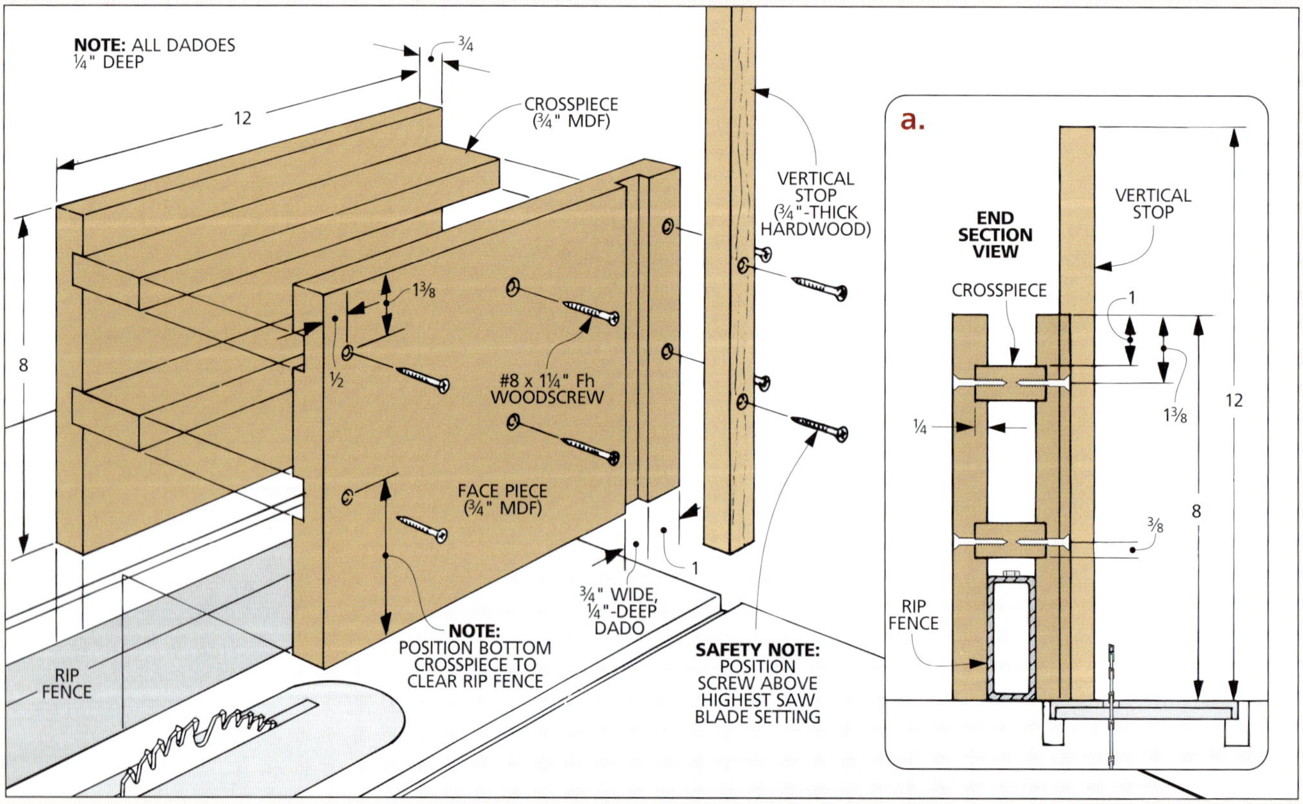

NOTE: ALL DADOES ¼" DEEP

¾

12

CROSSPIECE (¾" MDF)

VERTICAL STOP (¾"-THICK HARDWOOD)

8

1⅜

½

#8 x 1¼" Fh WOODSCREW

FACE PIECE (¾" MDF)

RIP FENCE

NOTE: POSITION BOTTOM CROSSPIECE TO CLEAR RIP FENCE

¾" WIDE, ¼"-DEEP DADO

1

SAFETY NOTE: POSITION SCREW ABOVE HIGHEST SAW BLADE SETTING

a.

END SECTION VIEW

CROSSPIECE

VERTICAL STOP

1

1⅜

¼

12

8

⅜

RIP FENCE

BASE

With the cases complete, it's time to build a base for everything to sit on. The base here is like a small table with four legs and rails *(Fig. 16)*. But instead of a table top, there's a single cleat that interlocks with the cleats in the case above it.

LEGS. The first things I worked on were the legs (N). These will end up $1\frac{3}{4}$" square and can be made with 8/4 stock. Or you can glue a $\frac{3}{4}$"-thick piece between two $\frac{1}{2}$"-thick pieces *(Fig. 16a)*. (This way, you won't cut through the joint lines when tapering the legs later.)

The first thing to do to the legs is cut the open mortises at the top.

Note: If the legs were glued up, lay out the mortises carefully, so the joint lines all face the same direction.

I cut them at the router table with a $\frac{1}{4}$" straight bit and a stop block clamped to the fence *(Fig. 17)*. I took two passes for each mortise, and since they're not centered, I reset the fence for the second mortise *(Fig. 17b)*.

The next thing to do is cut the tapers on the legs *(Fig. 18)*. All four faces are tapered, and to do this safely on these short pieces, I made a simple sled with an L-shaped fence *(Fig. 18b)*. To position the fence pieces on the jig, I laid out a taper on one of the leg blanks and set it on the base of the jig so the layout line was flush with the edge of the base.

After tapering each leg, the last thing to do is soften the edges with some roundovers *(Fig. 16c)*. I also softened the bottom end to prevent chipping.

RAILS. The legs are connected with front and back rails (O) and side rails (P). When sizing these pieces, you want the outside faces of the legs aligned with the edging on the side panels (not the case bottom assembly).

The only thing to do to these rails is to cut tenons to fit the mortises in the legs *(Fig. 16b)*. Just keep in mind there isn't a shoulder on the top of these tenons.

TOP CLEAT. With the tenons cut, the legs and rails can be glued together. Then the last piece to add is a top cleat (J) that's just screwed in place *(Fig. 16a)*.

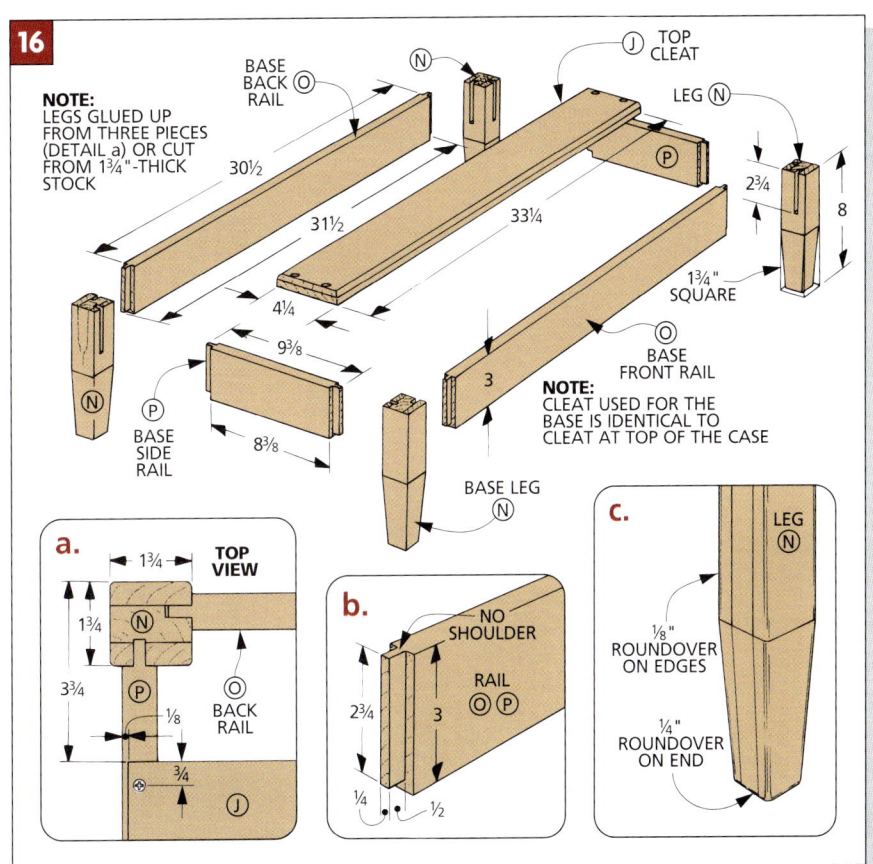

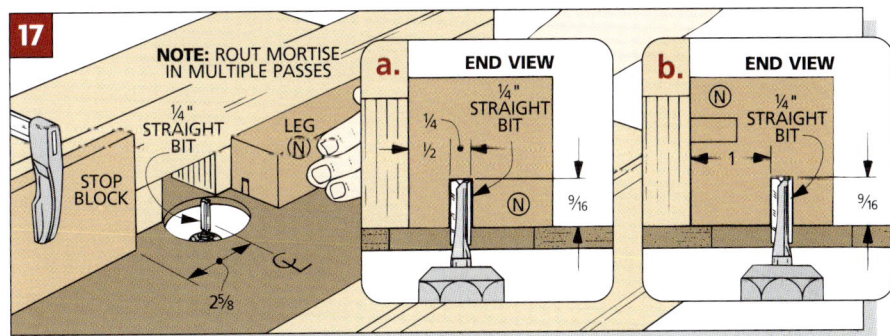

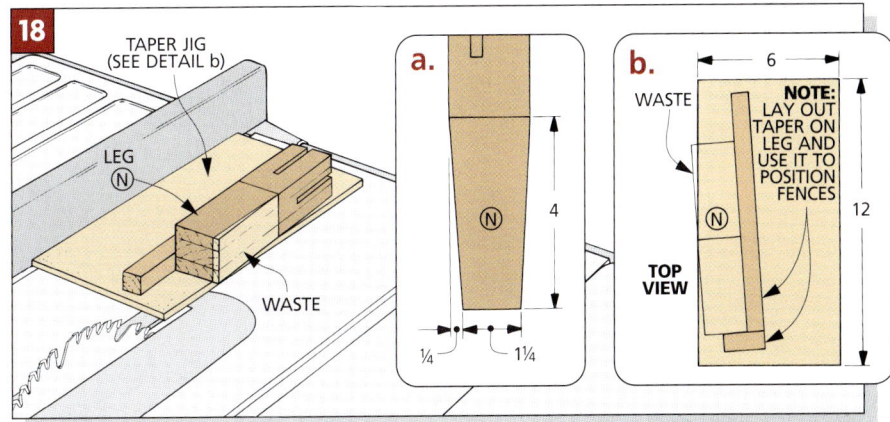

The cases are supported by a simple base. It interlocks securely with the same cleat system that "ties" the cases together.

The final section to build is the cap. The main purpose of the cap is to give the top of the bookcase a finished look. But if you're building a shorter version of the bookcase, then this section also serves as a shelf to set things on.

BOTTOM ASSEMBLY. To make the cap look like an integral part of the bookcase, I used some of the components that were used for the other sections. So the first thing to do is make a bottom assembly (if you haven't built it already). You can turn to page 98 for this since the construction and the size of the cap bottom assembly are identical to those of the case bottom assembly.

TOP AND SIDE PANELS. Next I worked on the top assembly, starting with a cap top panel (Q) and the two cap side panels (R) *(Fig. 19)*. The two side panels are short ($1\frac{7}{8}$" tall) versions of the case side panels. (You'll even add the same edging.) The top panel is slightly narrower. But all three panels get rabbeted to hold a $\frac{1}{4}$" plywood back *(Fig. 20)*.

To join the top and side panels, the first step is to cut a $\frac{1}{4}$" dado across the sides *(Fig. 21)*. Then you can cut a rabbet along each end of the top, sneaking up on the size of the remaining tongue until it fits the dadoes *(Fig. 19a)*.

EDGING. It's time to add edging to the side panels *(Fig. 22)*. The profile here is identical to the side panel edging on the cases *(Fig. 19a)*. (Refer to page 98 to make these pieces.) The edging doesn't cover just the side panel fronts. It also wraps around to cover the top edge. The side top edging (S) and side front edging (T) join at the front corner with a miter. But since the top edging runs across the grain of the side panel, I only applied glue to the front half so the panel could expand and contract *(Fig. 22)*.

CAP ASSEMBLY. Now that the side panels are complete, the top panel can be

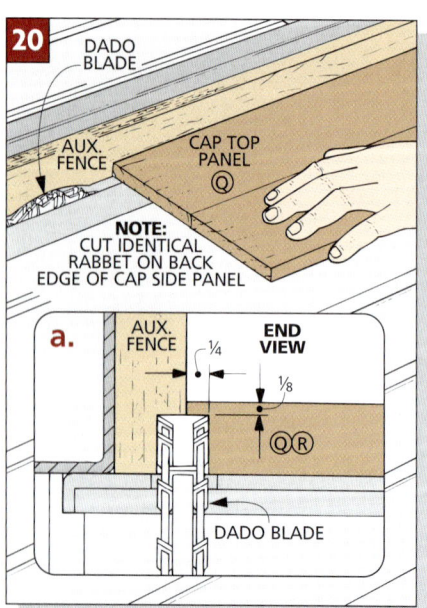

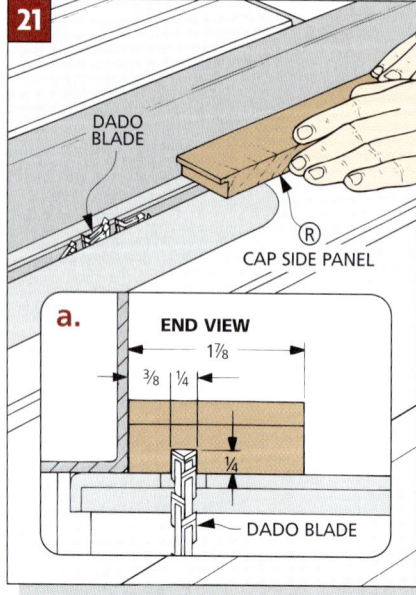

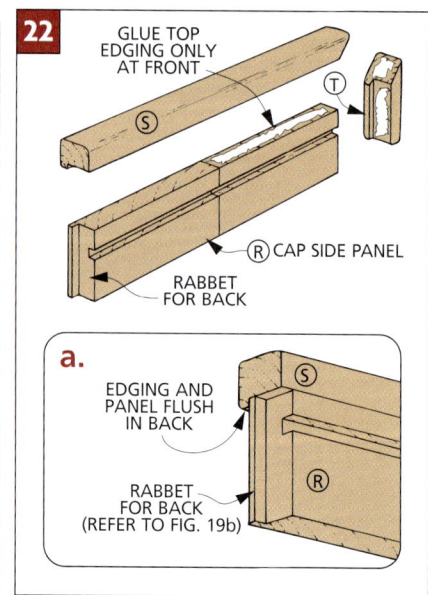

glued between the two side panels. (Their back edges should be flush.) Then this cap assembly can be centered over the bottom assembly and attached with woodscrews and glue (*Fig. 23*).

BACK AND FRONT. All that's left to add to the cap assembly are the front and back pieces. The cap front (U) is made from ³/₄"-thick hardwood and is cut to fit between the two sides (*Fig. 24*). Its top is flush with the top panel, and before gluing it in place, I softened the top front edge with a ¹/₈" roundover.

The last piece, the ¹/₄" plywood cap back (V), is just as easy to make as the front. It's sized to fit in the rabbets in the cap back (*Fig. 25*). Then it's simply glued and nailed in place.

FINISH. Before applying a few light coats of varnish, I mixed up a stain to give the mahogany a deeper reddish color. The stain I chose was three parts of a mahogany stain blended with one part of an antique cherry stain. ■

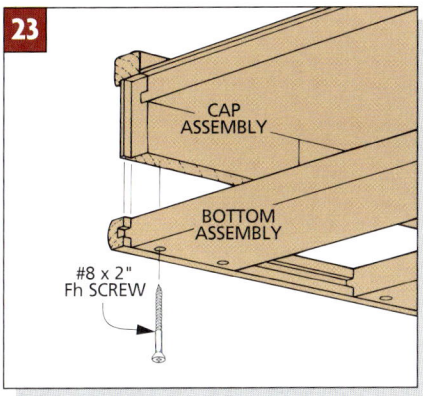

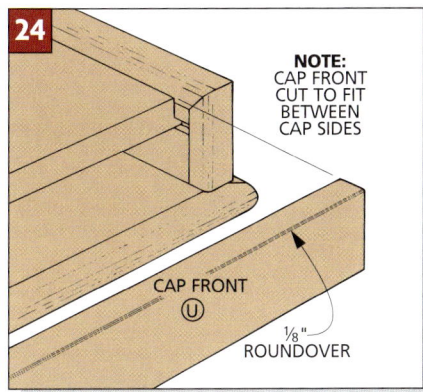

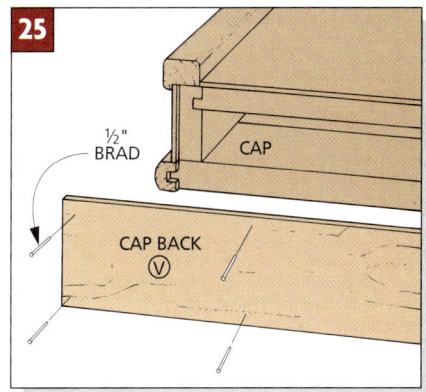

DESIGNER'S NOTEBOOK

A solid-wood raised panel would be an attractive alternative to the glass fronts on a few of your cases.

WOOD DOOR PANEL

■ A glass door panel is a great way to display items. But you could substitute a solid-wood panel instead of glass.

■ The door frames aren't built any differently than when they hold glass panels. That's because after the raised field is cut on the front of the panel, you can cut a rabbet on the back to leave a "tongue" that fits into the rabbet (see detail 'a').

■ To build the wood door panel (W), first glue up a couple of pieces of ¹/₂"-thick hardwood stock and then cut the panel to size (11³/₄" x 29³/₄") (see drawing).

■ To cut the raised field, tilt the saw blade

to 10° and support the panel with a tall auxiliary fence clamped to the rip fence. (I used a sanding block with a bevel cut on one edge to square up the shoulder of the raised field.)

■ Then to make room to add the glass stop, use a dado blade in the table saw to cut a ³/₁₆"-deep rabbet around the back.

MATERIALS LIST

NEW PART

W   Wood Door Panel (1)   ¹/₂ x 11³/₄ - 29³/₄

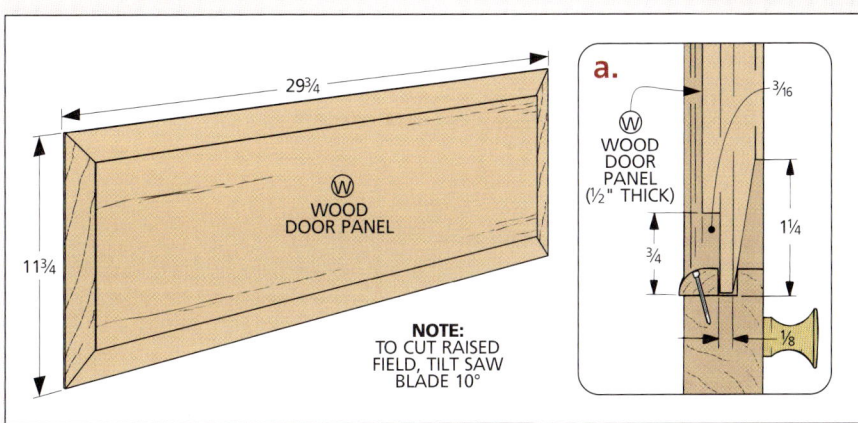

Slant Front Desk

When joining solid wood to solid wood, you have to allow for expansion and contraction of the pieces. The solution used on this desk is a sliding dovetail joint. It locks the parts together without glue.

Wood movement. It's a real concern with projects that feature solid-wood panels. That's because as solid wood expands and contracts with seasonal changes in humidity, joints can pop and boards may warp or split. Since this desk is built with a number of wide panels made of glued-up cherry, it requires special joinery to deal with the problems of wood movement.

JOINERY. Sliding dovetails are one answer. With this joinery technique, the wide side panels are free to move independently of the frames that hold the panels together. (For more on this, see the Joinery box on page 116. Plus, you'll find a complete description on gluing up wide panels in the Technique article starting on page 112.)

But the drop-down door that is such a prominent feature of slant front desks required another answer. To keep this panel flat I used "breadboard" ends.

EXTRAS. Ogee bracket feet complete the case and raise it off the ground. The feet on this desk look just like traditional feet from an early-American craftsman, but they're made with a table saw and a band saw instead of the hand tools of old. I've included separate step-by-step instructions on how to make them in the Technique article starting on page 122.

WOOD AND FINISH. All the visible parts of this desk are solid, $^3/_4$"-thick black cherry. Only the drawer sides — and some other parts that aren't visible — are different. For these I used $^1/_2$"-thick maple for more wear and less expense.

Finally, I went an extra step for the finish — four top coats of satin oil finish.

OPTIONS. This desk can really be enhanced by adding an insert behind the door. So you'll find plans for a pigeonhole unit in the Designer's Notebook on page 120. It features an egg-crate system with two areas that have vertical dividers. But the best parts are the drawers — and an easy-to-build hidden compartment.

EXPLODED VIEW

OVERALL DIMENSIONS:
42 1/2W x 22 1/2D x 42H

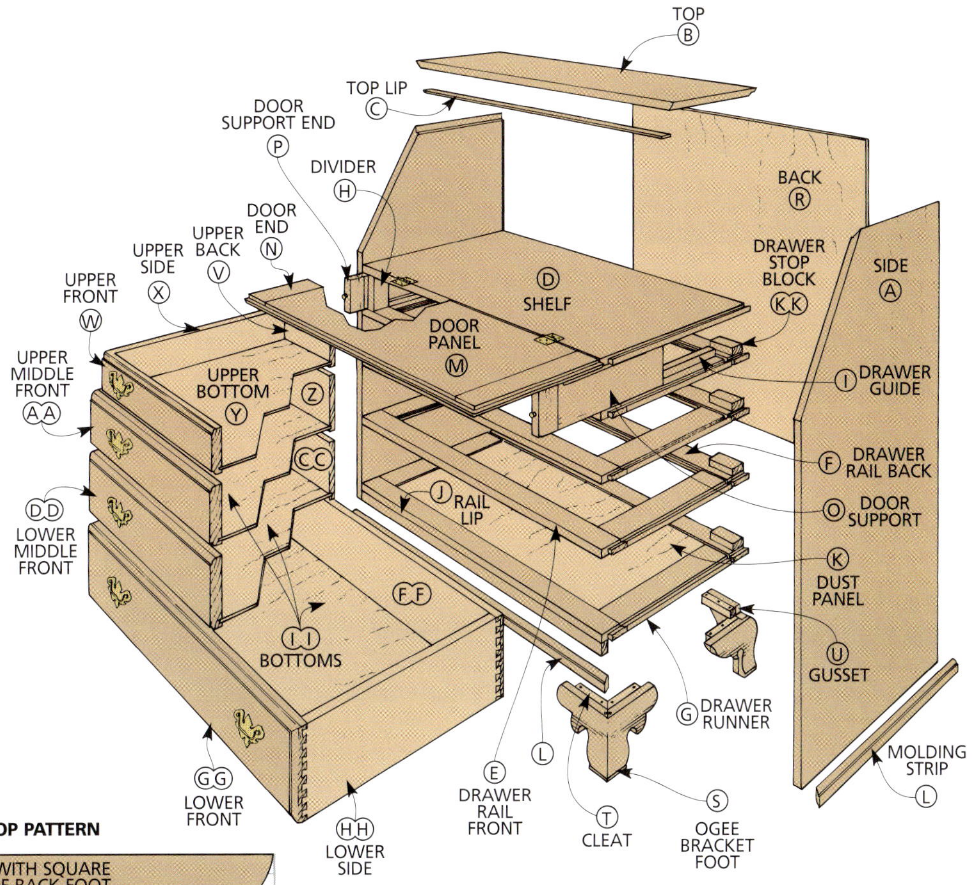

TOP B
TOP LIP C
DOOR SUPPORT END P
DIVIDER H
BACK R
DOOR END N
UPPER BACK V
UPPER SIDE X
UPPER FRONT W
SHELF D
DRAWER STOP BLOCK KK
SIDE A
DOOR PANEL M
UPPER MIDDLE FRONT AA
UPPER BOTTOM Y
Z
CC
DRAWER GUIDE I
DRAWER RAIL BACK F
RAIL LIP J
DOOR SUPPORT O
DD LOWER MIDDLE FRONT
DUST PANEL K
FF
II BOTTOMS
GUSSET U
DRAWER RUNNER G
L
MOLDING STRIP L
GG LOWER FRONT
HH LOWER SIDE
E DRAWER RAIL FRONT
T CLEAT
S OGEE BRACKET FOOT

SCALLOP PATTERN

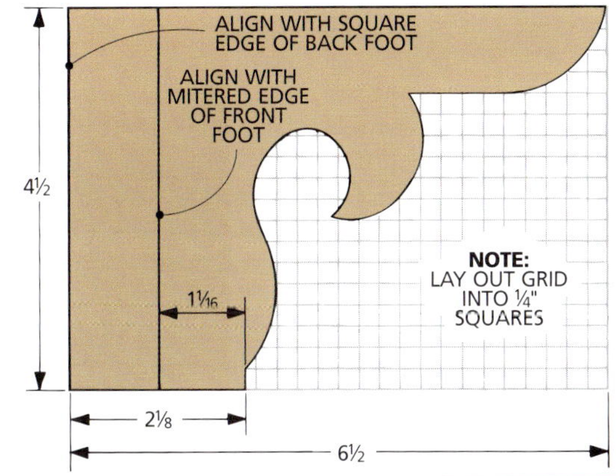

ALIGN WITH SQUARE EDGE OF BACK FOOT
ALIGN WITH MITERED EDGE OF FRONT FOOT
4 1/2
1 1/16
NOTE: LAY OUT GRID INTO 1/4" SQUARES
2 1/8
6 1/2

NOTE: CUT TEMPLATES FROM POSTERBOARD

OGEE PROFILE

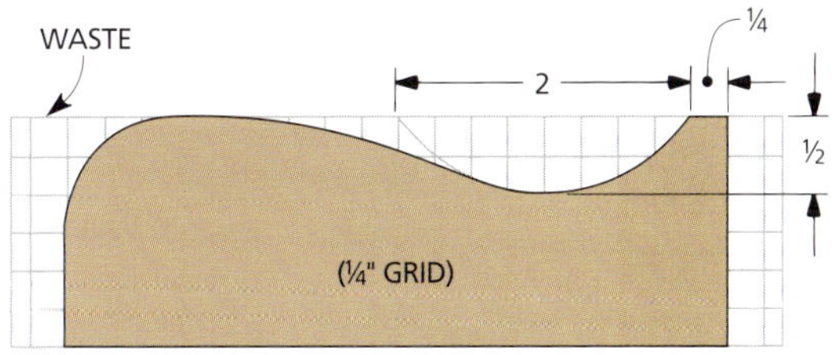

WASTE
2
1/4
1/2
(1/4" GRID)

MATERIALS LIST

CASE

A	Sides (2)	3/4 x 21 - 37 1/2
B	Top (1)	3/4 x 12 15/16 - 40
C	Top Lip (1)	3/16 x 7/8 - 38 1/2
D	Shelf (1)	3/4 x 20 3/4 - 39 1/4
E	Drwr. Rail Front (4)	3/4 x 2 - 39 1/4
F	Drwr. Rail Back (4)	3/4 x 2 - 39 1/4
G	Drwr. Runners (8)	3/4 x 2 3/4 - 17 1/2
H	Dividers (2)	3/4 x 2 - 4 1/8
I	Drawer Guides (2)	3/4 x 3/4 - 16 1/2
J	Rail Lip (1)	3/4 x 1 - 38 1/2
K	Dust Panel (1)	1/4 ply - 34 3/4 x 17 1/2
L	Molding Strip (1)	1/2 x 1 1/16 - 96 rgh.
M	Door Panel (1)	3/4 x 15 - 35 1/8
N	Door Ends (2)	3/4 x 2 1/2 - 15
O	Door Supports (2)	3/4 x 3 9/16 - 18
P	Door Sppt. Ends (2)	3/4 x 2 - 3 9/16
Q	Door Sppt. Stops (2)	1/2 dowel - 1
R	Back (1)	1/4 ply - 36 1/8 x 39 1/4
S	Ogee Ft. Blanks (3)	1 1/2 x 5 1/4 - 16
T	Cleats (8)	3/4 x 3/4 - 4 1/2
U	Gussets (2)	3/4 x 5 1/2 - 5 1/2

DRAWERS

V	Upper Back (1)	1/2 x 3 1/2 - 35 1/4
W	Upper Front (1)	3/4 x 3 1/2 - 35 1/4
X	Upper Sides (2)	1/2 x 3 1/2 - 19 3/8
Y	Upper Bottom (1)	1/4 ply - 34 3/4 x 19 1/4
Z	Upper Mdl. Back (1)	1/2 x 4 3/8 - 38 3/8
AA	Upper Middle Fr. (1)	3/4 x 4 3/8 - 38 3/8
BB	Upper Mdl. Sides (2)	1/2 x 4 3/8 - 19 3/8
CC	Lwr. Mdl. Back (1)	1/2 x 5 1/4 - 38 3/8
DD	Lwr. Middle Fr. (1)	3/4 x 5 1/4 - 38 3/8
EE	Lwr. Mdl. Sides (2)	1/2 x 5 1/4 - 19 3/8
FF	Lower Back (1)	1/2 x 6 1/8 - 38 3/8
GG	Lower Front (1)	3/4 x 6 1/8 - 38 3/8
HH	Lower Sides (2)	1/2 x 6 1/8 - 19 3/8
II	Bottoms (3)	1/4 ply - 37 7/8 x 19 1/4
JJ	Drawer Glides	3/4 x 1/16 - 148 ln. in.
KK	Drwr. Stop Blks. (8)	3/4 x 1 1/4 - 3

HARDWARE SUPPLIES

- (14) No. 4 x 3/4" Fh woodscrews
- (2) 2" x 3 1/16" brass hinges w/ screws
- (8) 3 1/8" brass pulls w/ screws
- (1) Brass escutcheon plate w/ screws
- (3) 1/2" x 1/2" brass knobs w/ screws

CUTTING DIAGRAM

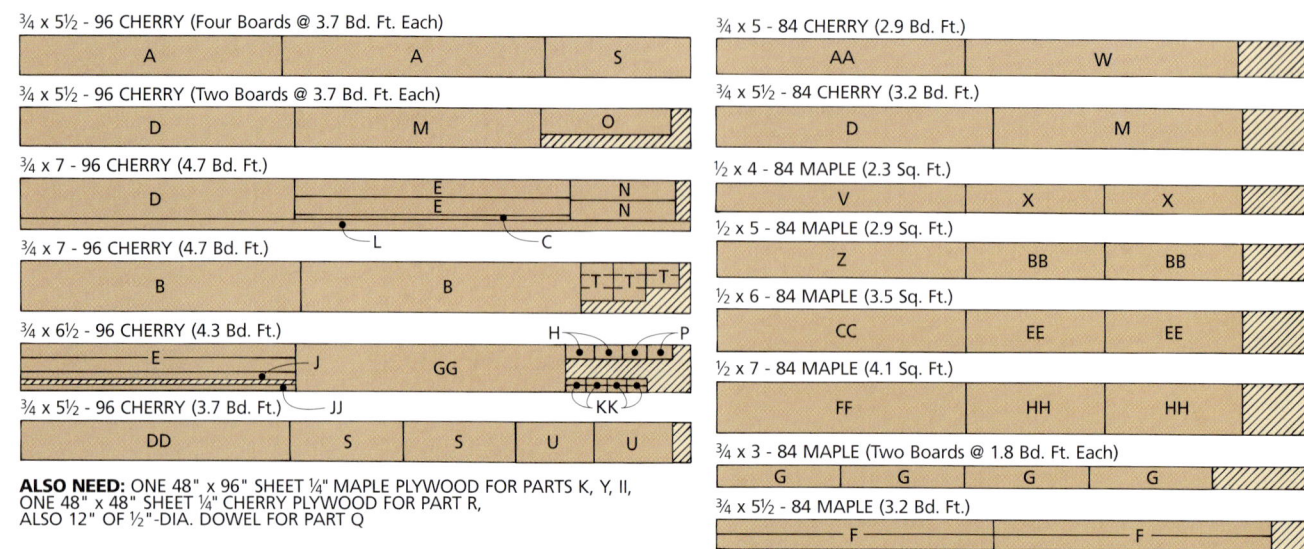

¾ x 5½ - 96 CHERRY (Four Boards @ 3.7 Bd. Ft. Each)

A	A	S

¾ x 5½ - 96 CHERRY (Two Boards @ 3.7 Bd. Ft. Each)

D	M	O

¾ x 7 - 96 CHERRY (4.7 Bd. Ft.)

D / E E / N N / L / C

¾ x 7 - 96 CHERRY (4.7 Bd. Ft.)

B	B	T T T

¾ x 6½ - 96 CHERRY (4.3 Bd. Ft.)

E / J / GG / H / P

¾ x 5½ - 96 CHERRY (3.7 Bd. Ft.) / JJ / KK

DD	S	S	U	U

ALSO NEED: ONE 48" x 96" SHEET ¼" MAPLE PLYWOOD FOR PARTS K, Y, II,
ONE 48" x 48" SHEET ¼" CHERRY PLYWOOD FOR PART R,
ALSO 12" OF ½"-DIA. DOWEL FOR PART Q

¾ x 5 - 84 CHERRY (2.9 Bd. Ft.)

AA	W	

¾ x 5½ - 84 CHERRY (3.2 Bd. Ft.)

D	M	

½ x 4 - 84 MAPLE (2.3 Sq. Ft.)

V	X	X	

½ x 5 - 84 MAPLE (2.9 Sq. Ft.)

Z	BB	BB	

½ x 6 - 84 MAPLE (3.5 Sq. Ft.)

CC	EE	EE	

½ x 7 - 84 MAPLE (4.1 Sq. Ft.)

FF	HH	HH	

¾ x 3 - 84 MAPLE (Two Boards @ 1.8 Bd. Ft. Each)

G	G	G	G	

¾ x 5½ - 84 MAPLE (3.2 Bd. Ft.)

F	F	

CASE CONSTRUCTION

I started work on the desk by building three solid panels for the outside case *(Fig. 1)*. But building a project with solid wood panels calls for some planning. Since each of the panels must be glued up from several boards, it's important to select these boards from stock that looks like it came from the same tree.

(For tips on how to glue up large panels, see the Technique on page 112.)

CUT TO ROUGH SIZE. After gluing enough boards together for three over-size blanks (two for the sides and one for the top), cut the sides (A) to finished width and rough length (39") *(Fig. 1)*.

Note: You'll be cutting the sides to finished length later, after the rabbeted miter joint is cut across the top.

Then cut the top (B) to rough width (13¾") but finished length (40") *(Fig. 1)*.

Note: The top end of the sides and the front edge of the top should be finish-quality cuts. That is, flat, smooth, and square to their adjacent edges.

RABBET MITER JOINT. In order to hide the end grain where the case sides meet the top, I used a variation on a miter joint. A common miter joint would work, but by rabbeting the miter, the joint is stronger and assembly is easier. (The workpieces won't shift around when they're glued and clamped together.)

The rabbeted miter joint is cut on both ends of the top and the top end of the sides (see the Joinery box at right).

SIDES

After cutting the rabbeted miter joint, cut the sides to finished length *(Fig. 1)*. Do this by cutting off the bottom ends square to the edges.

The sides of the case are held together by a shelf and web frames that are built later (refer to *Fig. 8* on page 114). To hold the shelf and web frames in place (and allow the solid wood sides to move), I used sliding dovetail joints.

This joint involves using a dovetail tongue on the ends of the shelf (and web frames) that locks in a dovetail groove on the insides of the case sides. (Refer to the Joinery box on page 116.)

LAY OUT DOVETAIL GROOVES. The frames that fit in the grooves do more than hold the sides of the case together. The web frames also support the drawers

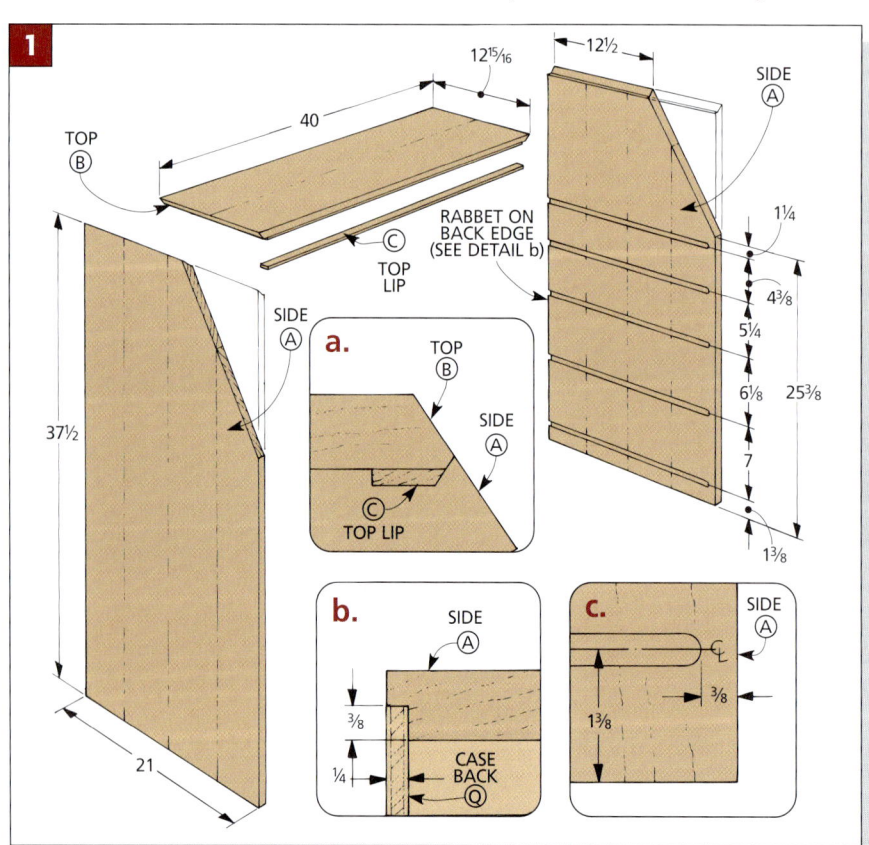

Miters are commonly used to hide the end grain on the ends of two pieces that are being joined. It's a joint that works well for picture frames and small boxes that won't get a lot of handling. But by itself, a miter joint isn't that strong. So I decided to use a variation of a miter joint — a rabbeted miter — to join the side panels of the case to the top.

SMALL KERF. First, a narrow kerf is cut across the inside face of all three pieces *(Step 1)*. Then the miter is cut with the blade aligned with the kerf *(Step 2)*. (A hardboard rub strip helps to align the blade to the top of the kerf.)

RABBET. Lastly, the rabbet is cut on both ends of the top piece only *(Step 3)*. Again, the rub strip helps line up the cut.

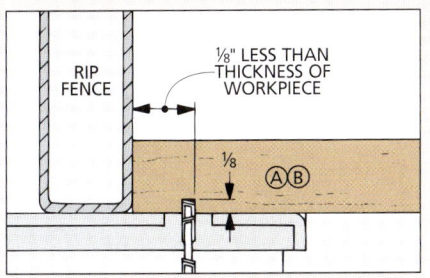

1 The rabbeted miter joint starts out the same on the top and side panels. First, cut a 1/8"-deep kerf with a regular saw blade across the inside face of all three pieces.

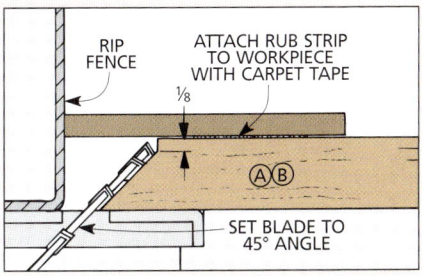

2 Now, cut the miter. Be sure the blade is aligned to the kerf. To help, stick a piece of hardboard to the workpiece. Then adjust the fence and sneak up on the cut.

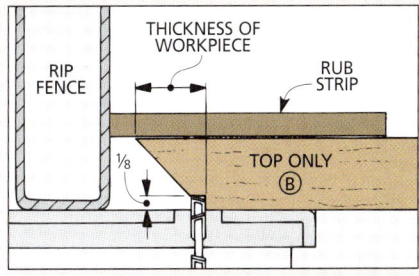

3 Finally, the last cut is a rabbet on the top piece only. Again use a hardboard rub strip, but this time to help position the blade in relation to the long point of miter.

inside the case. And since all four of the drawers are different heights, the grooves need to be spaced at different distances.

So to lay out the position of the dovetail grooves, measure up from the bottom of the case sides *(Fig. 1)*. Then draw a line across the inside face of each side to indicate the center of the dovetail grooves.

Note: Since the sliding dovetail joints are to be hidden on the front of the case, these grooves stop 3/8" from the front edge *(Figs. 1 and 1c)*.

ROUT DOVETAIL GROOVES. Now you can rout the dovetail grooves. To do this, I used a dovetail bit and guided the router along a straightedge clamped to the workpiece *(Figs. 2 and 2a)*.

RABBET. Next, cut a rabbet along the back edge of the side panels to accept a plywood back panel *(Figs. 1b and 2)*.

ANGLED CORNERS. With the dovetail grooves and the rabbets routed in the case sides, the next thing you'll do is cut off the front corners at a 35° angle to make the slant front *(Fig. 3)*.

To do this, first lay out the angle on both of the case sides *(Fig. 1)*. Then the angle is cut in two steps. First, cut to within about 1/16" of the line. (Make the cut on both side panels.)

The side panels could be clamped together and hand-planed to the mark to get the same angle on both panels. But I did something different. After the rough

cut, I clamped a straightedge on the pencil line (on the right-hand panel) and used a flush trim bit in the router to complete the cut and smooth the edge *(Fig. 3)*.

Note: To avoid chipout, rout from left to right, starting at the lower corner and finishing at the upper corner.

Finally, to cut the second (left-hand) panel identical to the first, I clamped them together so they were flush along the top, back, and bottom edges. Then I ran the bearing of the flush trim bit along the smooth edge of the first panel to trim a matching edge on the second panel.

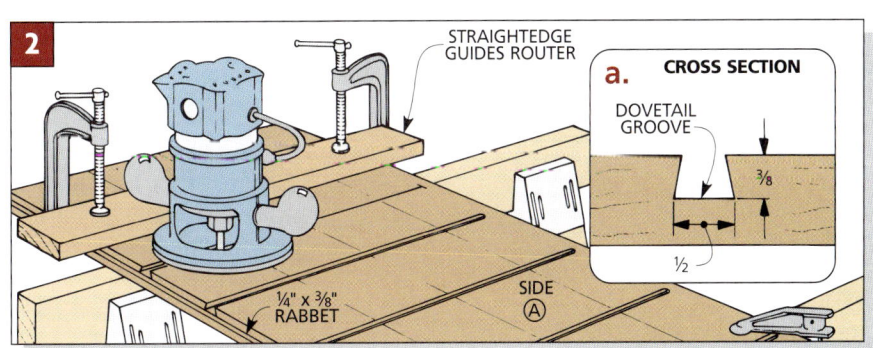

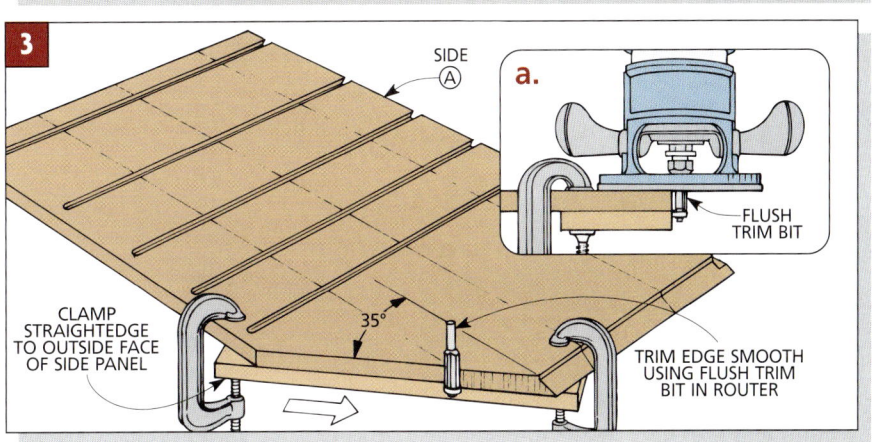

It's not easy to make a bunch of boards look like a single, flat piece of wood. But when making a solid, edge-glued panel, that's exactly the goal.

The colors should match. The grain of one piece should merge into the grain of the next. The joint lines should be practically invisible. If the panel looks like a bunch of boards slapped together, it will draw attention to itself — detracting from the appearance of the entire project.

And that's only half the battle. While an edge-glued panel should look like one wide piece of wood, it better not act like it. You want to avoid cupping or bowing and warping with changes in humidity. But if the pieces are arranged and prepared properly, this can be avoided.

SELECT & ARRANGE BOARDS

When edge-gluing, the easiest step to rush through is selecting the wood. But choosing and arranging the boards into a panel are essential for good results.

CHOOSING LUMBER. Straight boards make clamping much easier. Some warp is unavoidable and can be corrected. For example, a cupped board can be ripped in two, and a slightly bowed piece can be forced flat while clamping. But a twisted board is difficult to "untwist."

After selecting the lumber, I arrange the boards on the benchtop as they will appear in the panel.

APPEARANCE. First, I match the color. Then I try to fit the pieces together, turning and flipping, until the grain patterns match (see photo above). Your panel will look best if the straight grain is next to straight grain and the curved grain merges with curved grain.

While appearance is the most important consideration, it isn't the only one.

GRAIN DIRECTION. After the panel is glued up, you'll need to smooth it. This usually means planing it by hand or with a power planer. If the grain on the boards runs in varying directions, some pieces will probably chip or tear out.

To determine the grain direction, look at the *edge* of the board *(Fig. 1)*. Grain that's consistently curving the same way makes the job easy. But often, you simply have to pick the direction it curves the most and hope for the best.

Note: I like to draw an arrow on each face to note the direction of the grain *(Fig. 1)*. This way it will be easier to arrange the boards later.

END GRAIN. There's one more thing to consider before you go on — how will the panel cup with changes in humidity?

For a large glued-up panel that's anchored (such as a table top screwed to

aprons), cupping is rarely a problem. But a panel that's not secured (a chest lid, for example) can cup pretty badly.

For those panels that won't be anchored, I like to alternate the end grain from board to board *(Fig. 2)*. This way, by varying the growth rings, the whole panel won't cup in one direction. This is because each board cups in the opposite direction of the boards on either side of it.

MARK ORDER. Once the boards are arranged into a panel, I chalk Roman numerals (or a cabinetmaker's triangle) across the joints *(Fig. 3)*. I prefer the Roman numerals to help prevent the boards from getting mixed up, especially if I'm gluing up a number of panels.

Okay, so which item is most important: appearance, grain direction, or end grain? For me, it's appearance. I usually try to get the grain direction and end grain arranged correctly as well, but often, it's a compromise.

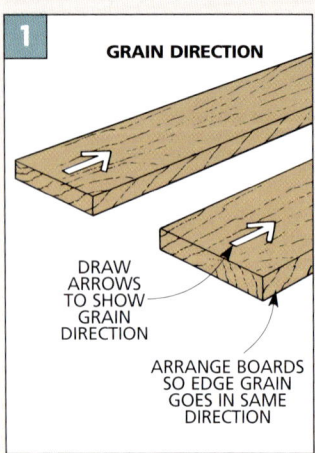

1 **GRAIN DIRECTION**

DRAW ARROWS TO SHOW GRAIN DIRECTION

ARRANGE BOARDS SO EDGE GRAIN GOES IN SAME DIRECTION

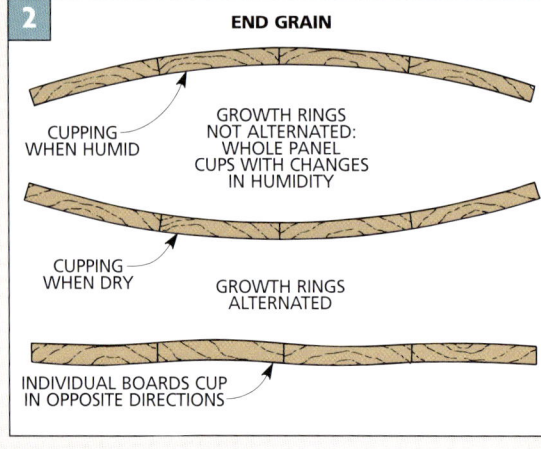

2 **END GRAIN**

GROWTH RINGS NOT ALTERNATED: WHOLE PANEL CUPS WITH CHANGES IN HUMIDITY

CUPPING WHEN HUMID

CUPPING WHEN DRY

GROWTH RINGS ALTERNATED

INDIVIDUAL BOARDS CUP IN OPPOSITE DIRECTIONS

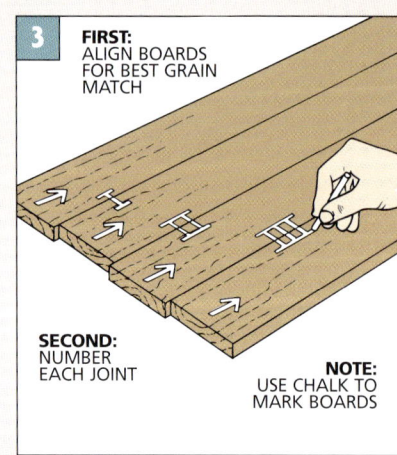

3 **FIRST:** ALIGN BOARDS FOR BEST GRAIN MATCH

SECOND: NUMBER EACH JOINT

NOTE: USE CHALK TO MARK BOARDS

JOINTING EDGES

Jointing the edges requires precision. If the edges aren't smooth, straight, and square to the faces, you'll have problems when gluing or clamping. Either the glue won't bond properly, or the whole panel can cup across its width.

CUPPED PANELS. A strong joint is as easy as cutting smooth, straight edges. But if the edges aren't square to the face of the board, the panel will cup as it's clamped together *(Figs. 4 and 4a)*. To prevent this, make sure your machine is set up correctly. With a jointer, set the *fence* exactly 90° to the table. And with a table saw, set the *blade* 90° to the table.

JOINTER. A jointer takes a uniform amount off each board, and you don't have to adjust the fence with every pass *(Fig. 5)*. I slowly feed the workpiece with the grain *(Fig. 5a)*. After a few light passes, the board has a smooth edge that's ready to be glued.

TABLE SAW. If you don't have a jointer, you can joint edges with a table saw and a good combination blade *(Fig. 6)*.

I use a double-cut method. Begin by ripping the boards straight. Then repeat the cut, this time only removing about half the thickness of the saw blade *(Fig. 6a)*. This second, lighter cut results in a smooth surface with virtually no saw marks or burning.

GLUING

I don't like to take chances when I'm gluing, so to make sure there aren't any surprises, I always dry-assemble a panel before glue up. And then when adding glue, I make sure there's enough on both edges for a good bond.

APPLYING GLUE. Some woodworkers put glue on only one edge of each board and don't bother to spread it out. This does have some advantages. It's quick, and the glue doesn't set up quite so fast. But I want to know that there's a thin, even film on both edges, so I spread the glue on with a brush (see photo below).

CLAMPING

When I'm ready to assemble the panel, I like to use ³/₄" pipe clamps. I space them 6" to 8" apart and alternate them above and below the panel to equalize the pressure and prevent cupping *(Fig. 7)*.

CLAMPING PRESSURE. After the boards are flush, tighten the clamps until tiny beads of glue appear along each joint line. The clamps should be tight, but the important thing is to equalize the pressure along the joint line — not "cranking" down on the clamps as hard as you can. Then add more clamps to the sections where there isn't any glue oozing out.

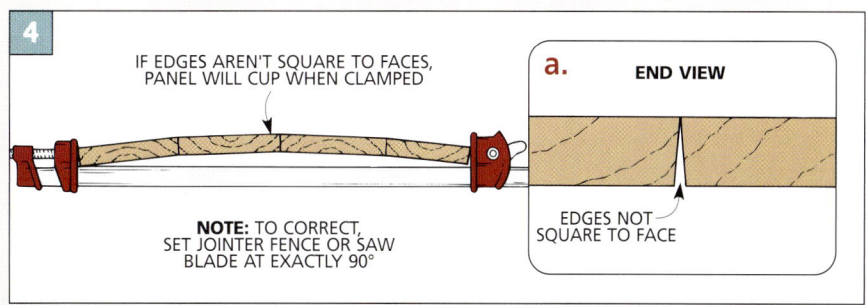

4 IF EDGES AREN'T SQUARE TO FACES, PANEL WILL CUP WHEN CLAMPED

NOTE: TO CORRECT, SET JOINTER FENCE OR SAW BLADE AT EXACTLY 90°

a. END VIEW

EDGES NOT SQUARE TO FACE

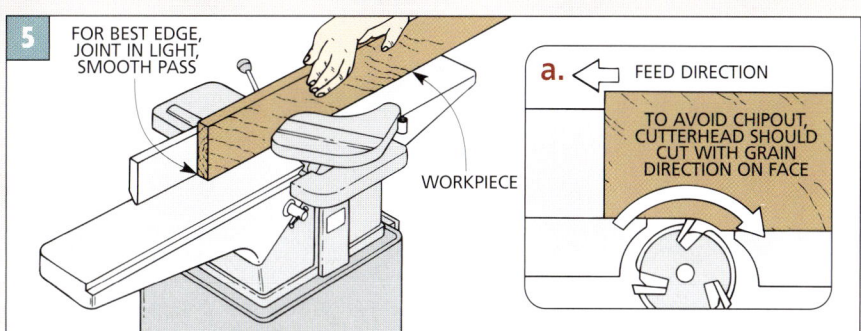

5 FOR BEST EDGE, JOINT IN LIGHT, SMOOTH PASS

WORKPIECE

a. FEED DIRECTION

TO AVOID CHIPOUT, CUTTERHEAD SHOULD CUT WITH GRAIN DIRECTION ON FACE

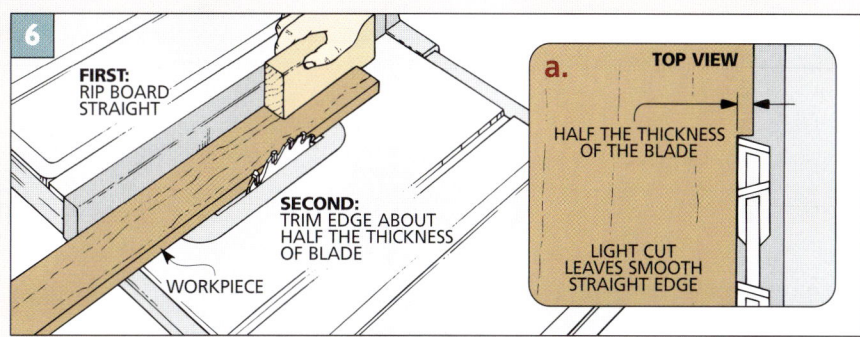

6 FIRST: RIP BOARD STRAIGHT

SECOND: TRIM EDGE ABOUT HALF THE THICKNESS OF BLADE

WORKPIECE

a. TOP VIEW

HALF THE THICKNESS OF THE BLADE

LIGHT CUT LEAVES SMOOTH STRAIGHT EDGE

Glue Joint Strength. *A strong joint requires a thin, consistent layer of glue. For a good bond, apply it to both edges and spread it out with a brush.*

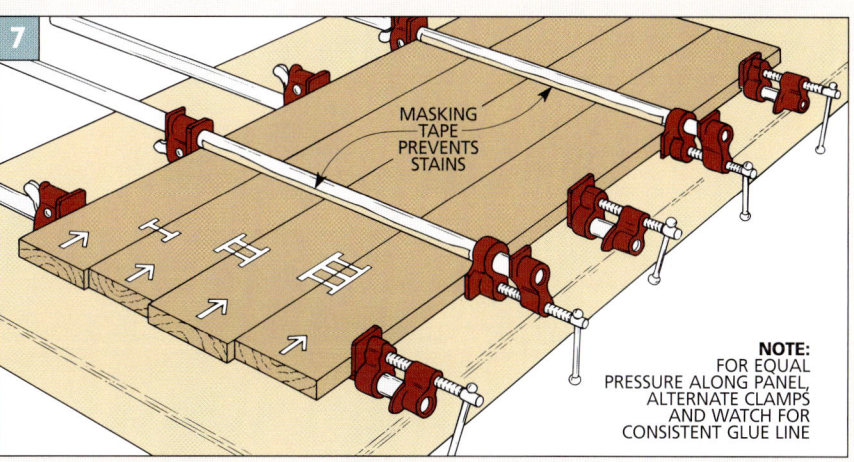

7 MASKING TAPE PREVENTS STAINS

NOTE: FOR EQUAL PRESSURE ALONG PANEL, ALTERNATE CLAMPS AND WATCH FOR CONSISTENT GLUE LINE

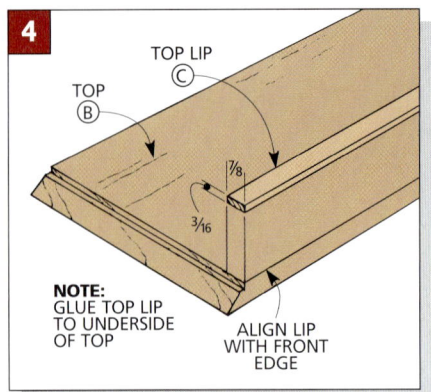

4

TOP LIP
C

TOP
B

7/8

3/16

NOTE:
GLUE TOP LIP
TO UNDERSIDE
OF TOP

ALIGN LIP
WITH FRONT
EDGE

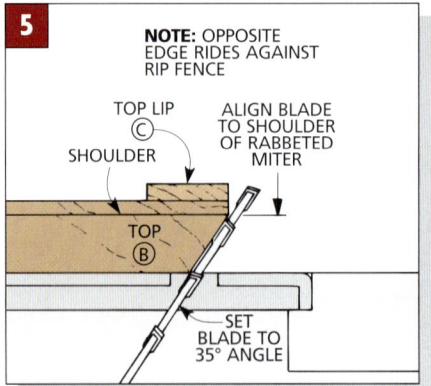

5

NOTE: OPPOSITE
EDGE RIDES AGAINST
RIP FENCE

TOP LIP
C

SHOULDER

ALIGN BLADE
TO SHOULDER
OF RABBETED
MITER

TOP
B

SET
BLADE TO
35° ANGLE

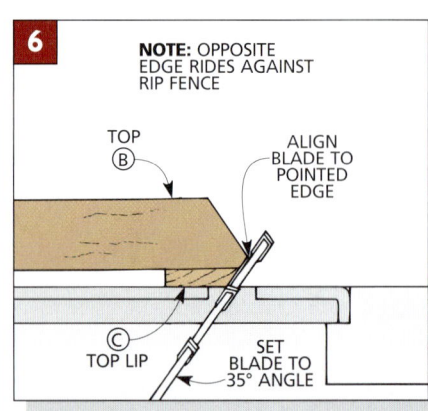

6

NOTE: OPPOSITE
EDGE RIDES AGAINST
RIP FENCE

TOP
B

ALIGN
BLADE TO
POINTED
EDGE

C
TOP LIP

SET
BLADE TO
35° ANGLE

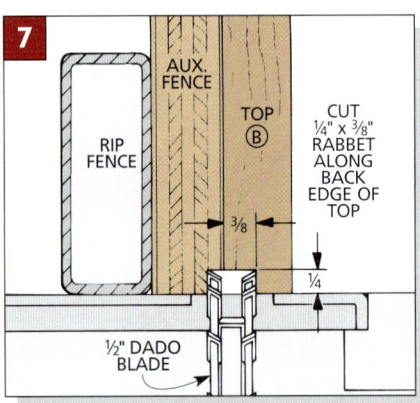

7

AUX.
FENCE

RIP
FENCE

TOP
B

CUT
1/4" x 3/8"
RABBET
ALONG
BACK
EDGE OF
TOP

3/8

1/4

1/2" DADO
BLADE

TOP

After flush trimming the angle on both of the side panels and cutting the rabbet for the back, go ahead and set the panels aside for now so you can continue working on the case top (B).

ATTACH LIP. Before cutting the case top to finished width, I first glued a top lip piece (C) to the underside of the front edge (refer to *Fig. 1a* on page 110 and *Fig. 4*). This top lip serves two purposes. First, it provides a little more heft to the

case top where it meets the door. And if you decide to build the Slant Front Desk with the pigeonhole insert, the lip acts as a stop when the insert is installed inside the assembled case.

RIP TWO BEVELS. After attaching the top lip, rip a 35° bevel along the front edge of the case top *(Fig. 5)*.

Note: The angle of this bevel must be exactly the same as the angle on the two side panels so the door will fit tight to the case when it's closed.

Cut this bevel with the top face down against the table *(Fig. 5)*.

Next, rip an intersecting bevel along the front edge, this time with the bottom face against the table *(Fig. 6)*.

Note: Because of the lip on the front edge, the workpiece won't lie flat on the table for this second cut. That's probably okay — because only the first bevel angle is critical. But, you could use carpet tape to temporarily stick a thin spacer on the other end of the workpiece. Just be sure to remove it after you've ripped the bevel.

RIP TOP TO WIDTH. Now the case top can be ripped to finished width (once again, with the beveled edge against the fence). Be sure to sneak up on the finished width until the top aligns with the sides at the front and back edges (refer to *Figs. 1a and 1b* on page 110).

To accept a plywood panel for the back of the case, cut a rabbet along the lower back edge of the top piece *(Fig. 7)*.

SHELF & FRAMES

When I finished building the case sides and top, I began work on the shelf and the web frames that hold the sides together.

SHELF. The shelf (D) is built from glued-up hardwood stock just like the case sides and top. Then it's ripped to finished width to match the width of the case sides (less the width of the rabbet for the back panel) *(Fig. 8)*.

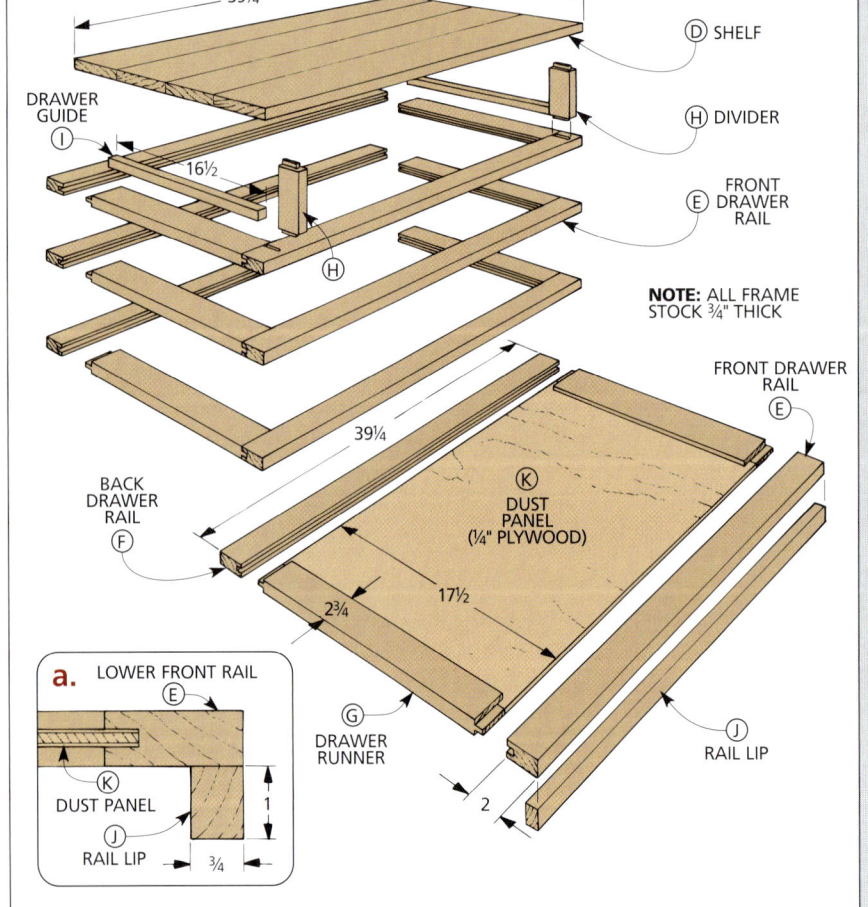

8

39 1/4

20 3/4

D SHELF

DRAWER
GUIDE
I

16 1/2

H DIVIDER

H

E FRONT
DRAWER
RAIL

NOTE: ALL FRAME
STOCK 3/4" THICK

39 1/4

FRONT DRAWER
RAIL
E

BACK
DRAWER
RAIL
F

K
DUST
PANEL
(1/4" PLYWOOD)

17 1/2

2 3/4

G
DRAWER
RUNNER

J
RAIL LIP

2

a. LOWER FRONT RAIL
E

K
DUST PANEL

J
RAIL LIP

3/4

1

You don't always have to cut mortises and tenons to hold a frame together. In fact, a quick and easy way to build a web frame like those used in the desk is with a stub tenon and groove joint. If the frames include a panel (like the dust panel), then the grooves are cut on all four pieces *(Step 1)*. And then stub tenons are cut on the ends of the web frame sides *(Step 2)*. (If all the frames are open, then the grooves are cut on the rails only.)

To get the best fit, center the grooves and tenons by cutting them in multiple passes (flipping between passes).

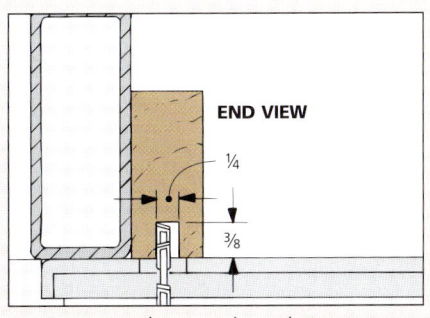

1 A centered groove is cut in two passes. The first pass is roughly centered. Then for the second pass, the workpiece is flipped end for end.

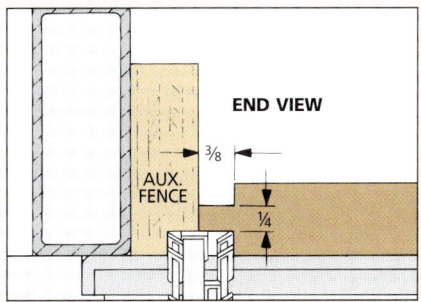

2 Stub tenons are also cut in multiple passes with a dado blade. It's best to sneak up on the height of the blade until the tenon fits snug in the groove.

To determine the finished length of the shelf, measure across the underside of the top, from shoulder to shoulder. To this dimension add the combined depth of the opposing dovetail grooves ($^3/_4$"). Now cut the shelf to this length.

FRAMES. All four web frames are built the same way. Two side drawer runners fit between a front and a back rail with stub tenon and groove joints *(Figs. 8 and 9)*.

Note: Since the back rails and drawer runners will be hidden, I used a less expensive wood (maple). But for the visible front rails, I used cherry.

Start by ripping all the frame pieces to finished width *(Fig. 8)*.

Next, cut the front and back drawer rails (E, F) to finished length to match the length of the shelf (D).

To determine the length of the drawer runners (G), first measure from the front edge of the case side to the shoulder of the rabbet at the rear. Then subtract the width of both drawer rails. To this number add 1" (for the $^1/_2$"-long tenon on each end of the runners), then subtract $^1/_4$" (for an expansion gap where the runners meet the back rail).

GROOVES AND TENONS. The next step is to cut a groove centered on the inside edges of all the frame pieces *(Fig. 9)*.

Note: Cut these grooves to match the thickness of the $^1/_4$" plywood dust panel (K) to be used as a dust (and pest) barrier for the lower panel.

Now cut stub tenons on both ends of all the drawer runners *(Figs. 8 and 9)*.

SLOT MORTISES. A pair of vertical dividers separate the top drawer from the two sliding door supports *(Figs. 8 and 11)*. These dividers have $^1/_4$"-long stub tenons centered on the ends that fit into slot mortises *(Figs. 10 and 11)*.

TOP DIVIDERS. After routing the mortises, rip the two dividers (H) to finished

width to match the front rails *(Fig. 11)*. To determine the length of the dividers, measure between the centers of the top two dovetail grooves and subtract $^1/_4$".

After cutting the dividers to length, cut the stub tenons on the ends *(Fig. 11)*.

DRAWER GUIDES. Next, I cut a pair of drawer guides (I) for the top drawer to ride against *(Figs. 8 and 11)*.

DOVETAIL TONGUES. Then I routed the dovetail tongues that fit the grooves in the case sides (see the Joinery box on page 116). They're routed on the ends of all eight web frame rails (E), the edges of the drawer runners (G) and on both ends of the shelf (D) *(Fig. 12)*.

NOTCHES. Before the front rails and shelf can be glued in place, notches must be cut from the ends of the dovetail tongues *(Fig. 12)*. Also notch the front edge of the tongues on both dividers *(Fig. 11)*. This way they'll fit in the stopped groove you already cut in the sides and the front edge of each workpiece will also end up being flush with the front edge of the case sides.

RAIL LIP. Next, cut a narrow rail lip (J) to fit between the shoulders of the front rail of the bottom dust frame *(Figs. 8 and 8a)*. (This supports molding that gets attached later.)

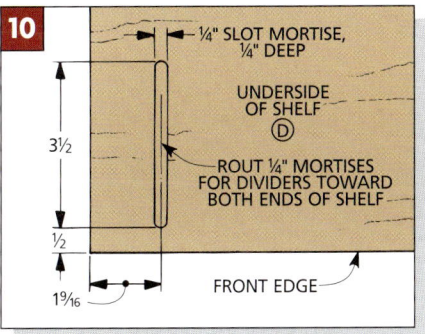

9
DUST PANEL (K)
G DRAWER RUNNER
$^1/_2$
$^1/_2$
E F DRAWER RAIL
CUT STUB TENONS TO FIT GROOVES
CUT GROOVES TO MATCH THICKNESS OF PLYWOOD

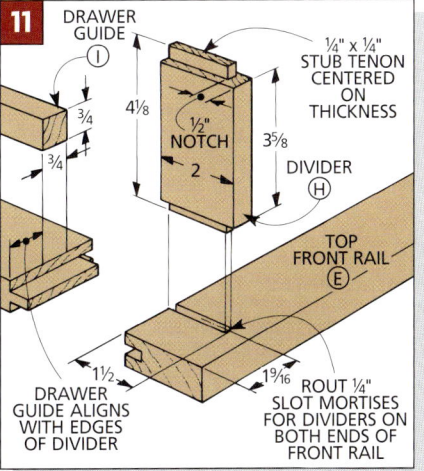

11
DRAWER GUIDE (I)
$^3/_4$
$^3/_4$
$4^1/_8$
$^1/_2$" NOTCH
2
$3^5/_8$
$^1/_4$" x $^1/_4$" STUB TENON CENTERED ON THICKNESS
DIVIDER (H)
TOP FRONT RAIL (E)
$1^1/_2$
$1^9/_{16}$
DRAWER GUIDE ALIGNS WITH EDGES OF DIVIDER
ROUT $^1/_4$" SLOT MORTISES FOR DIVIDERS ON BOTH ENDS OF FRONT RAIL

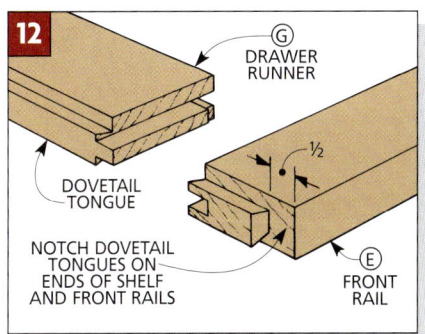

12
G DRAWER RUNNER
$^1/_2$
DOVETAIL TONGUE
NOTCH DOVETAIL TONGUES ON ENDS OF SHELF AND FRONT RAILS
E FRONT RAIL

10
$^1/_4$" SLOT MORTISE, $^1/_4$" DEEP
UNDERSIDE OF SHELF (D)
$3^1/_2$
$^1/_2$
$1^9/_{16}$
ROUT $^1/_4$" MORTISES FOR DIVIDERS TOWARD BOTH ENDS OF SHELF
FRONT EDGE

Sliding dovetails are a two-part joint. Even without glue, the angled sides of the tongue fit the angled walls of the groove exactly. It's an extremely strong way to join two pieces of wood. And they allow the wide side panels on the desk to float independently of the frames during seasonal changes in humidity.

BE PRECISE. Routing both parts of the joint requires precision — a tight fit holds the project together. But the joint shouldn't be *too* tight. (You must be able to assemble the parts.)

SNEAK UP TO TIGHT FIT. The secret to the best fit is sneaking up on the final cut until the tongue just fits the groove. To

help, I built a tall fence to hold the tall pieces on edge while routing (see the Shop Jig article on the facing page).

GROOVES AND TONGUES. First, rout the grooves with a hand-held router, running it against a straightedge (*Step 1*). Then rout the tongues on the router table, sneaking up on a perfect fit (*Step 2*).

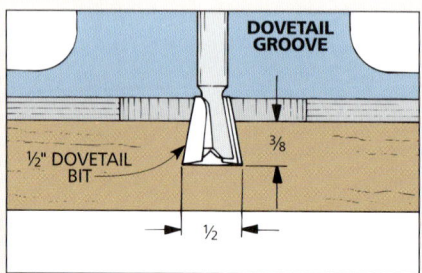

1 Rout dovetail grooves with a hand-held router. Set depth of cut and then run router against a straightedge.

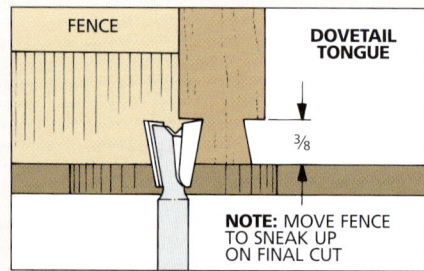

2 Dovetail tongues are routed on the router table. The height of the bit matches depth of the dovetail groove.

CASE ASSEMBLY

Here's where all the parts get joined to create the carcase of the desk.

Note: Because the solid wood sides must be allowed to expand and contract with changes in humidity, the case is assembled with glue only in certain spots (*Fig. 14*). Don't put glue on the tongue of the front rail. (It will scrape off in the dovetail groove.) Instead, apply glue to the front end of the groove. Also, do not apply glue to the tongues on the edges of the drawer runners.

ASSEMBLY. Start assembling the case by sliding the shelf (D) in place in the upper dovetail groove (*Fig. 13*). (Because the dovetail groove is stopped, you'll have to slide the shelf in place from the back edge.) The shelf holds the sides together while the web frames are installed.

There's a sequence that needs to be followed for installing the frames.

With the shelf in place, continue by sliding in the front drawer rail (E) until the front edges are flush. Next, slide in both drawer runners (G) so the tongues at the front fit into the grooved edge of the front rail (*Fig. 14*).

PLYWOOD PANEL. Now cut a dust panel (K) the same length as the drawer runner to fit inside the web frame.

Note: I installed a panel only in the lower web frame. But since the other frames have grooves to accept a panel, you could install a panel in these as well. (Extra panels add weight and cost.)

Finally, slide in the back rail (F). This should fit flush to the shoulder of the rabbet for the back panel.

Note: There should be a $1/4$" gap between the back of each runner and the

front edge of this rail. This lets the case sides move without splitting the frames.

TOP WEB FRAME. The assembly sequence for the top web frame is a little different than for the lower frames. That's because the dividers (H) are glued in the mortises between the shelf and front rail *before* the drawer runners are installed (*Fig. 15*). Here, the extra-long mortises (on the underside of the shelf) permit the tenons to slide in even though the rail and shelf are in place.

Now the remaining sections of the top web frame can be installed just as you did the lower frames. Then install the top (B) between the sides.

COMPLETE ASSEMBLY. Finish building the case by gluing the drawer guides onto the upper frame runner (refer to *Figs. 8 and 11* on pages 114 and 115). Also, glue on the rail lip (*Figs. 8 and 8a*).

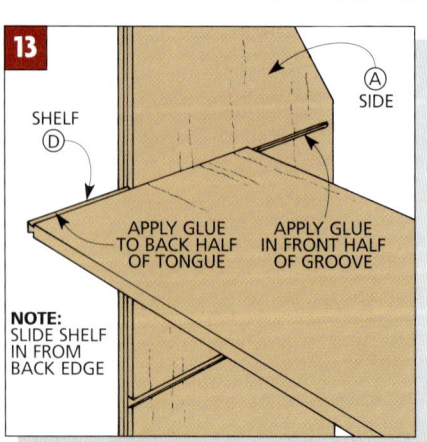

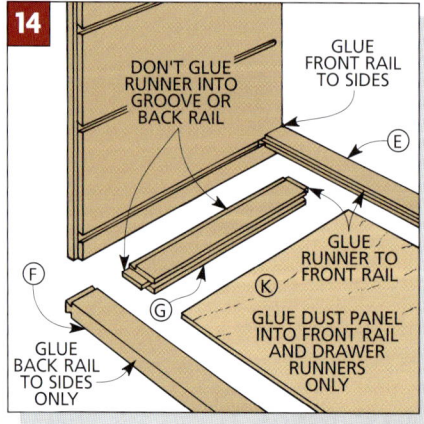

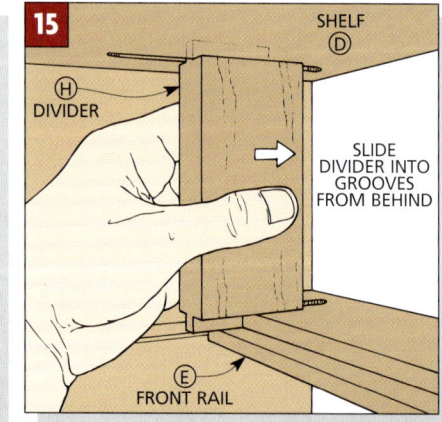

When I started making the joinery for the Slant Front Desk, I wanted to use sliding dovetails on some parts of the desk. But there was a problem.

To rout a dovetailed tongue on the router table, the workpieces must stand on edge (see photo below left). But my router table fence was too short to easily support a large panel while routing. And I ran into the same problem when I tried to rout the ends of the tall, narrow rails.

To solve this, I built a tall fence that clamps to the top of my router table (see photos below). The tall fence offers a lot of support when routing the edges of large panels. And when used with a miter gauge, it works great for routing the ends of long pieces like the drawer rails or even holding pieces at an angle.

BASE AND FENCE. To build the fence, begin by cutting a base from $3/4$"-thick plywood. The length of the base should equal the top of your router table. Then cut a 9"-high fence to the same length as the base (see drawing).

BIT NOTCH. Both the base and fence need a notch for the router bit. To cut the notches, I used a jig saw and then cleaned them up with a drum sander.

MITER GAUGE GROOVE. Next, cut a groove along the fence to guide your miter gauge. (The width of this groove should match the width of the runner on your miter gauge.)

Note: The position of the miter gauge groove is critical — be sure not to cut it too low. As its pushed past the bit, the miter gauge should easily clear even your largest diameter router bit when the bit is set at its *highest* point.

AUXILIARY FENCE. I added an auxiliary fence to hold the miter gauge in a vertical position. This fence is screwed to the miter gauge and hooks behind the tall router table fence. This way, the miter gauge can't fall out of the groove.

Note: When screwing the auxiliary fence to the miter gauge, positon it $1/16$"

above the surface of the router table. This allows the miter gauge to be tilted in either direction to support angled pieces (see photo below right).

BRACES. Now cut two 7" x 7" triangular braces to support the fence and keep it square to the table top (see drawing).

Note: One corner of each triangle must be exactly 90°.

Finally, glue and screw all the pieces for the fence together. Then, after the glue has completely set up, I like to wipe a coat of wax on the face of the fence as well as in the miter gauge groove to get a nice slick surface. Now you should be able to safely rout the dovetail joints.

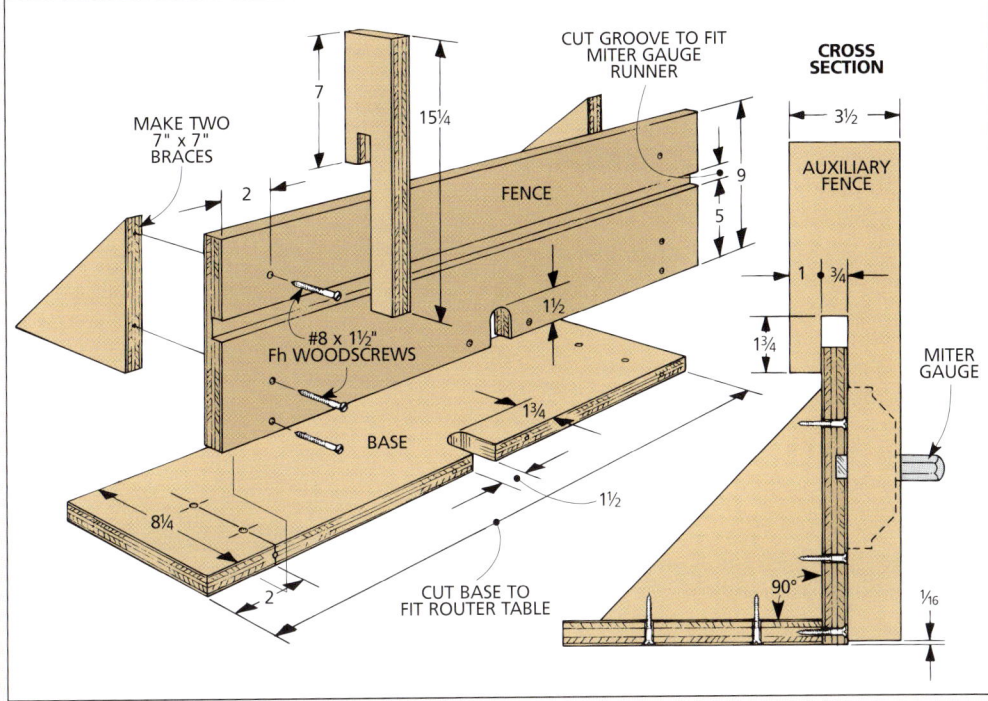

Large Panels. *Routing a dovetail tongue on the ends of a panel can be difficult with a short fence. This tall fence keeps the panel exactly 90° to the router bit.*

Long Pieces. *Long narrow pieces need even more support than panels. Adding a miter gauge with an auxiliary fence keeps the piece from tipping forward or back.*

Angled Pieces. *The miter gauge and an auxiliary fence are able to tilt forward or backward, which really helps when you're routing grooves for splined miters.*

OGEE FEET & MOLDING

The desk is considered a piece of Chippendale-style furniture. It's distinguished by its short, sculptured feet (called ogee bracket feet). In the Technique article starting on page 122, I've detailed the steps needed to build the feet (S), cleats (T), and gussets (U).

MOLDING STRIP. After installing the feet, cut a blank for the molding (L) to finished width and rough length (*Fig. 16*). Then rout a profile along the edge with a $\frac{3}{8}$" roundover bit (*Fig. 16a*).

Now miter the molding to fit around the front and sides of the case. Glue on the front strip, but for the side strips apply glue to the mitered corner only. Anchor the back part of the strips with screws from inside the case through a pair of slotted shank holes (*Fig. 16*).

DOOR & DOOR SUPPORTS

The drop-down door is made up of three pieces — a glued-up panel and two "breadboard" ends (*Fig. 17*).

DOOR ENDS. After the door panel (M) is trimmed to finished size, cut a pair of door ends (N) to length (to match the width of the panel).

JOINERY. Now the breadboard door ends are fastened to the door panel with stub tenon and groove joints (*Figs. 17 and 17b*). (For more on cutting these joints, see page 115.)

To allow the wide panel to expand and contract, the ends are glued only along the middle third of the tongues (*Fig. 17*).

Once you've finished building the fold-down door unit, rout a roundover (with a small shoulder) around all four edges on the outside face (*Figs. 17 and 17a*).

Then, to allow the door to fit inside the door opening, rout a rabbet on the inside face of three edges (*Fig. 17b*). (Don't rabbet the bottom edge.)

DOOR SUPPORTS. Now rip a pair of door supports (O) to width $\frac{1}{16}$" less than the height of the opening to fit between the case and dividers. Then cut the door supports to finished length (*Fig. 18*).

Next, cut a pair of support ends (P) to length to match the width of the supports (*Fig. 18*). Then rip the support ends to finished width, and attach them to the supports with tongue and groove joints.

RELIEF NOTCH. Next I routed a shallow notch along the top edge of each door support (*Fig. 19*). This allows the support to slide with minimal binding.

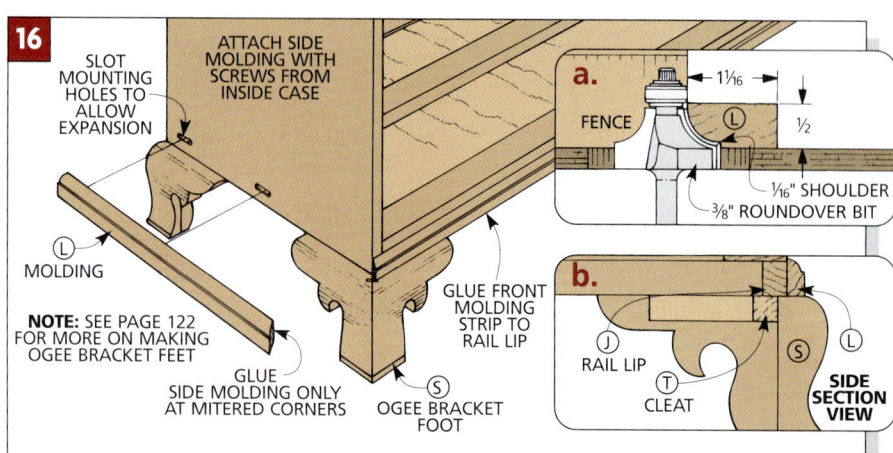

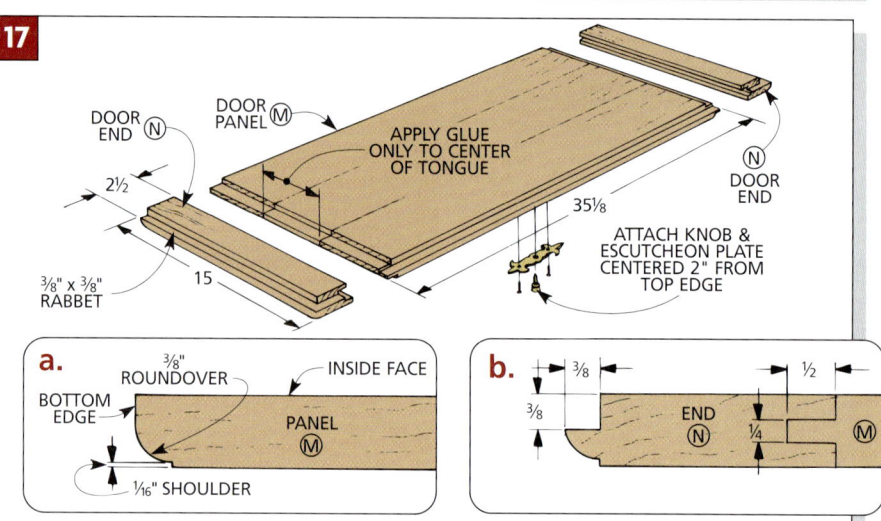

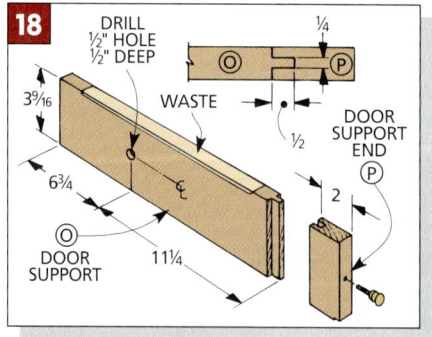

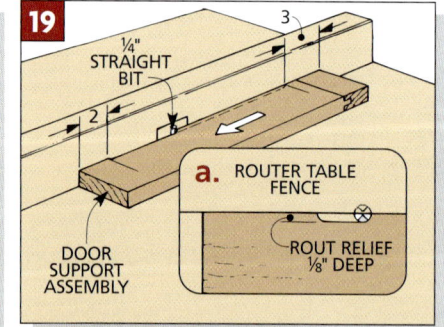

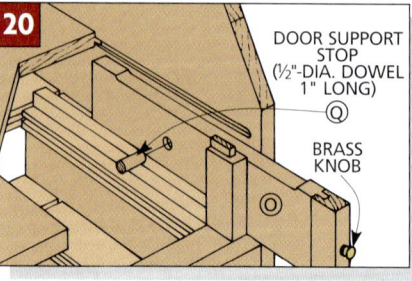

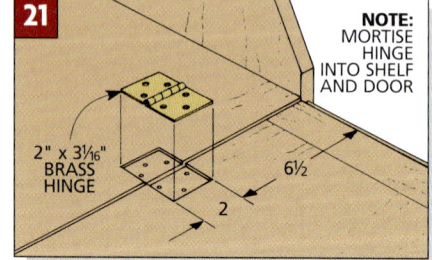

DOWEL PIN AND BRASS KNOB. Now glue a dowel pin into each door support as a stop (*Fig. 20*). Then a small brass knob is attached to the front support end.

INSTALL DOOR. Before starting on the drawers, I installed the door with a pair of brass hinges mounted flush to the surface of both the door and the shelf (*Fig. 21*).

DRAWERS

At this point the project becomes more like an ordinary cabinet with dovetail-joined drawers.

There's one small difference. On most chests of drawers, all the drawers are the same width. On this desk, the drawers are the same width except the top drawer (because of the door supports).

DRAWER PARTS. I began the drawers by cutting the drawer backs (V, Z, CC, FF) $1/8$" smaller in each dimension than the drawer openings *(Fig. 22)*.

Note: I used $1/2$"-thick maple for all the drawer backs and sides.

Next, cut the drawer fronts (W, AA, DD, GG) the same size as each drawer back. (I used $3/4$"-thick cherry.)

After that, cut eight drawer sides (X, BB, EE, HH) to the same height as the fronts and backs.

Note: Cut the sides $1^5/8$" shorter than the depth of the drawer openings. This allows for the stop blocks, plus $1/8$" for the

drawer backs *(Fig. 22a)*. It also allows for a $3/8$" overhang on the front when the drawers are closed *(Fig. 22b)*.

DOVETAIL JOINTS. After cutting all the drawer parts to finished size, rout half-blind dovetails on the ends of each. (To do this, I used a dovetail jig with a router and a $1/2$" dovetail bit.)

Before assembling the drawers, rout a $1/4$"-deep groove around the lower inside face of each drawer part to accept a $1/4$" plywood bottom *(Fig. 22)*.

Note: Measure your plywood and cut the groove to this size — $1/4$" plywood is usually less than $1/4$" thick.

ROUNDOVERS. Now rout a roundover around the face of each drawer front *(Fig. 22b)*. This profile should match the profile around the door *(Fig. 17a)*.

ADD DRAWER BOTTOMS. Now cut the drawer bottoms (Y, II) to size and glue them up. (I used $1/4$" maple plywood with the grain direction running from front to back. Cutting them with the grain running right to left it will take more plywood.)

Half-blind dovetails are customary on a well-built drawer. I routed the joints using a hand-held router and dovetail jig.

GUIDES, STOPS, AND BACK. To keep each drawer centered in its opening, glue $1/16$"-thick hardwood drawer glides (JJ) to the case sides and on the tops of the runners *(Fig. 23)*. Finally, add two $3/4$"-thick drawer stop blocks (KK) to the back rail for each of the drawers *(Fig. 24)*, and screw the back (R) in place.

Note: If you build the pigeonhole unit described in the Designer's Notebook on page 120, you'll want to complete it first before adding the back panel. ■

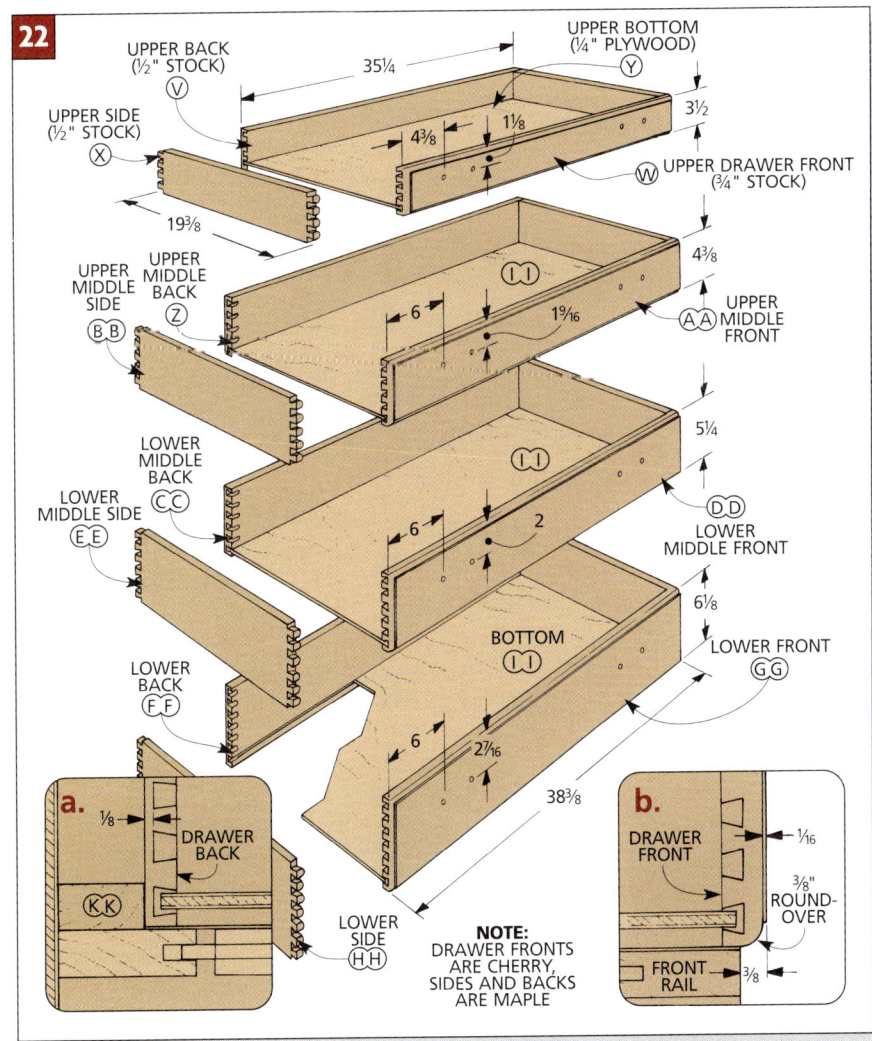

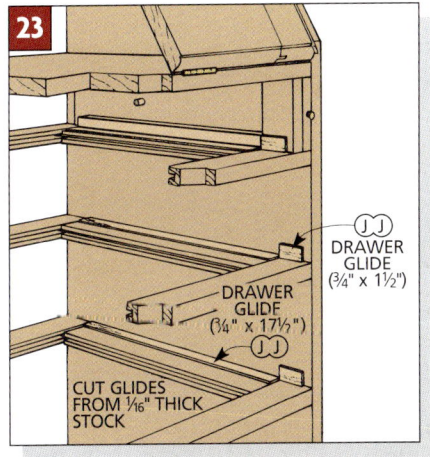

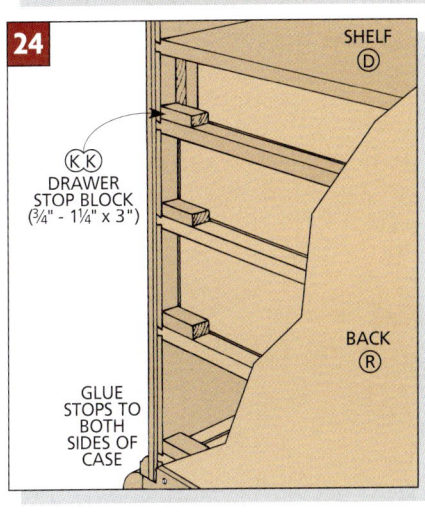

DESIGNER'S NOTEBOOK

Fight clutter on the top of your desk with this Pigeonhole Insert. It's a helpful way to organize all the things you want to keep inside — and you can add a hidden compartment for storing valuables.

CONSTRUCTION NOTES:

■ Start by building the insert to fit the opening behind the drop-down door. First, measure from the back edge of the door lip (C) to the shoulder of the rabbet at the back of the case. Then rip all the case parts to this width ($11^3/_4$").

■ Next, I cut the case top and bottom to length. (They are actually $^1/_{16}$" less than the length of openings so they will slide inside but still be fairly tight.)

■ To determine the length of the sides (MM) and dividers (MM), measure the height of the desk opening and subtract $^1/_2$" (since they fit in dado joints). Then, for ease of installation, I subtracted another $^1/_{16}$" to come up with the final length.

■ The joinery for the Pigeonhole Insert consists of rabbets and dadoes *(Fig. 2)*. I cut opposing pieces on the table saw at the same time. This way, all the joints will be aligned opposite each other.

■ With the dadoes and rabbets complete, dry-assemble the case. Now the middle shelf (NN), and outside shelves (OO) can be cut to length to fit in the case.

■ Next, work on the storage dividers (PP). They're sized to fit between the case top and outside shelves *(Fig. 3)*.

■ And to make it easier to pull files from the compartments, I cut an arc on the front of each divider *(Fig. 3)*.

PIGEONHOLE INSERT

MATERIALS LIST

PIGEONHOLE CASE

LL	Top/Bottom (2)	$^3/_8$ x $11^3/_4$ - $38^7/_{16}$
MM	Sides/Dividers (4)	$^3/_8$ x $11^3/_4$ - $11^{11}/_{16}$
NN	Middle Shelf (1)	$^3/_8$ x $11^3/_4$ - $12^7/_{16}$
OO	Outside Shelves (2)	$^3/_8$ x $11^3/_4$ - $12^5/_8$
PP	Storage Dvdr. (6)	$^3/_8$ x $11^3/_4$ - $9^5/_{16}$

PIGEONHOLE DRAWERS

QQ	Middle Front/Bk. (2)	$^3/_8$ x $4^5/_{16}$ - $12^1/_8$
RR	Middle Sides (2)	$^3/_8$ x $4^5/_{16}$ - $9^3/_4$
SS	Middle Bottom (1)	$^1/_4$ x $11^5/_8$ - $9^1/_2$
TT	Outside Front/Bk. (4)	$^3/_8$ x $1^{15}/_{16}$ - $12^5/_{16}$

UU	Outside Sides (4)	$^3/_8$ x $1^{15}/_{16}$ - $11^1/_2$
VV	Outside Bottom (2)	$^1/_4$ x $11^{13}/_{16}$ - $11^1/_4$

HIDDEN COMPARTMENT

WW	Front (1)	$^3/_8$ x $4^5/_{16}$ - $12^1/_8$
XX	Back (1)	$^3/_8$ x $4^5/_{16}$ - $10^1/_8$
YY	Sides (2)	$^3/_8$ x $4^5/_{16}$ - $1^3/_8$
ZZ	Bottom (1)	$^1/_4$ ply - $1^3/_8$ x $9^3/_4$

HARDWARE SUPPLIES

(20) No. 4 x $^3/_4$" Fh woodscrews
(3) $^1/_2$" x $^1/_2$" brass knobs w/ screws

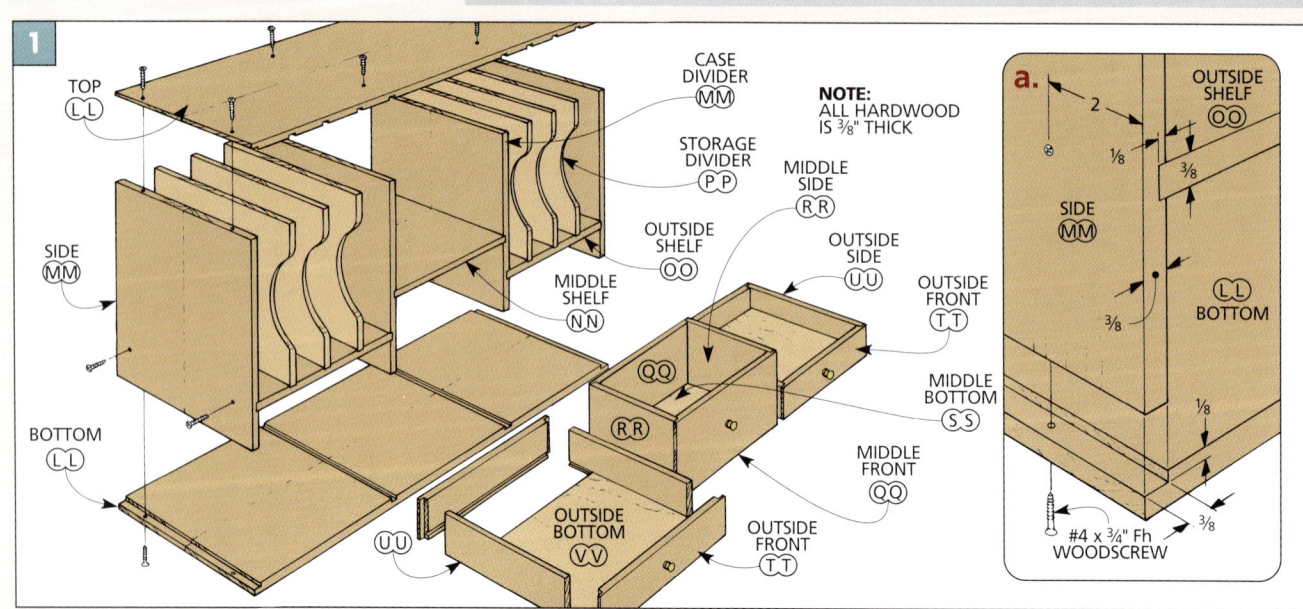

1

TOP
LL

CASE DIVIDER
MM

NOTE:
ALL HARDWOOD
IS $^3/_8$" THICK

STORAGE DIVIDER
PP

MIDDLE SIDE
RR

SIDE
MM

OUTSIDE SHELF
OO

OUTSIDE SIDE
UU

OUTSIDE FRONT
TT

MIDDLE SHELF
NN

QQ

RR

MIDDLE BOTTOM
SS

BOTTOM
LL

UU

OUTSIDE BOTTOM
VV

OUTSIDE FRONT
TT

MIDDLE FRONT
QQ

a.

OUTSIDE SHELF
OO

SIDE
MM

BOTTOM
LL

#4 x $^3/_4$" Fh
WOODSCREW

- Now the case can be assembled with glue and screws *(Fig. 1)*.
- The next thing you'll do is build the drawers. The drawers are made using ³⁄₈"-thick stock. First, cut the fronts and backs (QQ, TT) ¹⁄₁₆" less than the height and width of the openings *(Fig. 4)*. Then rip the sides (RR, UU) to the same height as the front and back pieces *(Fig. 4)*.
- Next, cut a rabbet joint at both ends of each drawer front and back *(Fig. 4a)*. Now the sides can be cut to length so that when the drawers are closed, the fronts fit flush with the case.
- To hold the ¹⁄₄" plywood I used for the drawer bottoms (SS, VV), I cut a groove around the inside of the drawer parts

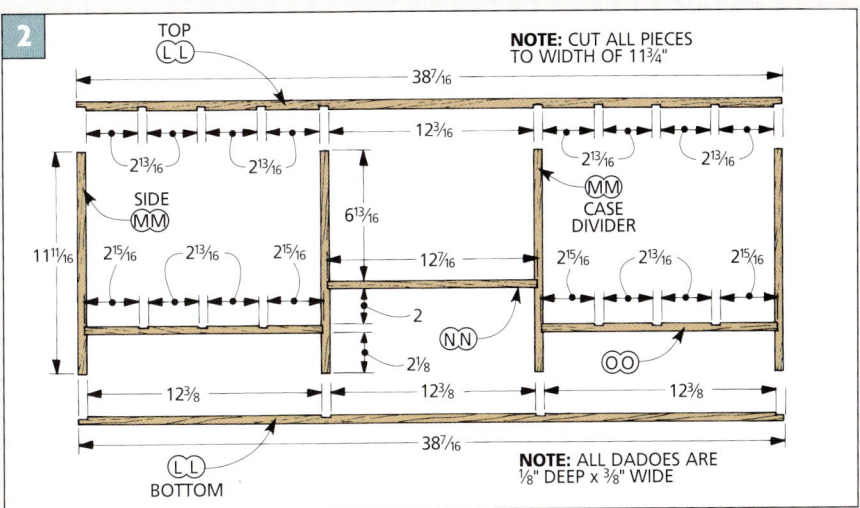

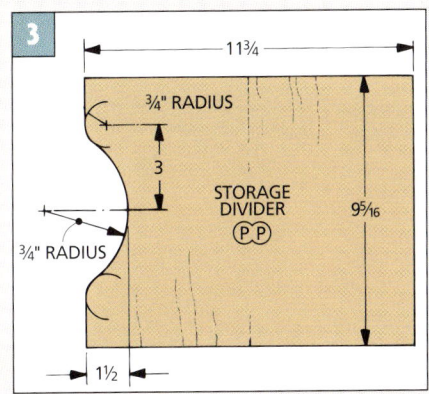

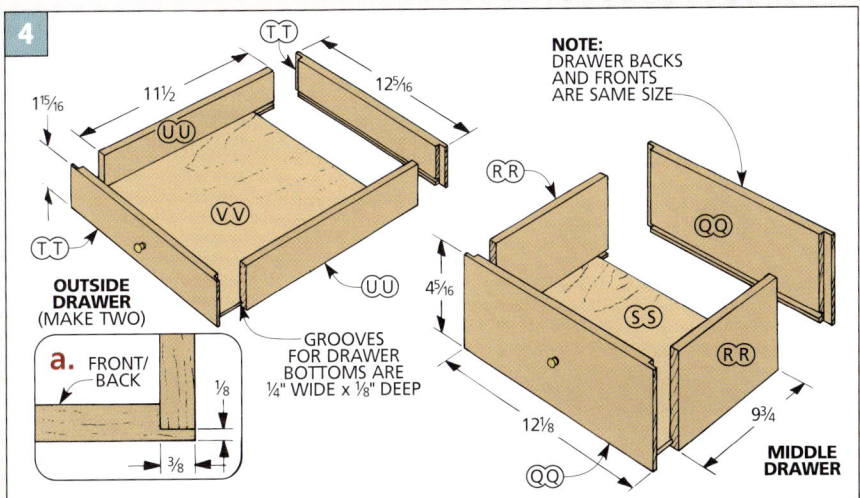

(Fig. 4). Finally, assemble the drawers with glue in the joints and in the grooves.
- Next, I built the hidden compartment behind the middle drawer. I started with the front piece (WW). The height and width of this piece must be exactly the same as the drawer opening.
- When the front is completed it should fit the drawer opening perfectly. Then you can build the back (XX) *(Fig. 5)*.
- Now a dado is cut on the inner side of the front piece and a rabbet on each end of the back piece *(Fig. 5)*. Then the sides (YY) can be cut to width to fit between the rabbets and the dadoes *(Fig. 5)*.

- Finally, cut a groove for the bottom around the inside of all four pieces. Then dry-assemble the drawer to measure for the bottom piece (ZZ) and cut it to size.
- Before assembling the box, sand a roundover on the back corners of the front and back, creating pivot points.

Then, to open the compartment, the "sweet spots" at the sides can be pushed in so the opposite side pivots forward allowing you to pull it out *(Fig. 6)*.

Secret Nook. *To open this compartment you have to know exactly where the "sweet spots" are at the sides. Push either one and the opposite side pivots open.*

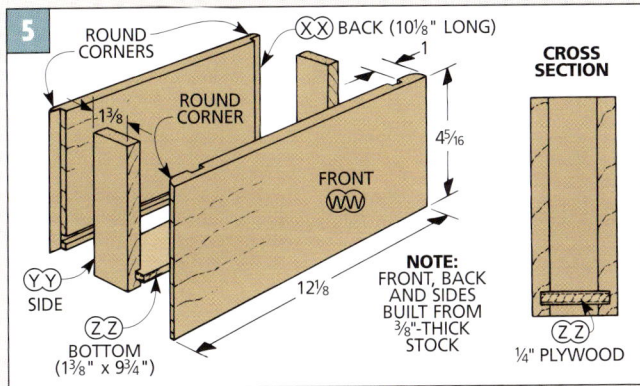

Even though the ogee bracket feet used for both the Slant Front Desk and the Bedside Chest (page 58) look like traditional hand-made ogee feet from two hundred years ago, they're actually much easier to make. Back then, these feet would have been shaped with hand tools — I used a table saw and a band saw.

But as you can see, the final results are the same — great looks and plenty of strength — without all the hard work.

SPLINED MITER JOINT. When you first look at a foot like this, it may be hard to figure how it's made. It's not one big block of wood as you might expect. Instead it's two thick pieces of wood joined with a splined miter joint.

PROFILES. Also, each leg piece has two distinct profiles. There's a large S-shaped ogee (cove) profile cut in the face, and a scalloped (or scrollsawn) cutout to form a support bracket.

POWER TOOLS. In the early days, the S-shaped profile was usually shaped with a big plane that had a huge cutter. But the problem was pushing this tool through the workpiece. It was hard work that required a lot of effort.

Today, most of the hard, physical work can be done in your home workshop with the table saw, a regular saw blade, and a stacked dado set. (I'll have to admit though, I did use a couple of "modern" hand tools for some final shaping.)

And originally the scallop cutout along the edges of the foot was probably cut with a fret (or coping) saw. Here, I used the band saw with a 1/8"-wide blade.

CUTTING A COVE

These ogee feet start out as long, thick blanks. You'll need three blanks — one for the back feet and two for the front.

GLUING UP BLANKS. The blanks are made from two pieces of 3/4"-thick stock glued face-to-face. Once the glue has set up, the blanks are cut to rough size (5 1/4" x 16" for the desk or 4 1/2" x 16" for the chest) (refer to *Step 3*).

CUTTING A COVE. Now work can begin on roughing out the profile. To do this, first set up the table saw to cut a cove on the front of each blank. (The cove is the concave area of the S-shaped profile.)

The blade is actually going to plow through the workpiece at an angle.

SAW SET-UP. To set up the table saw, a fence has to be positioned at an angle to the saw blade. The problem is determining that angle to get a certain width cove. (The cove is 2" wide for the desk, and 1 3/4" wide for the Bedside Chest.)

The best method I've found for setting up the fence to the correct angle is to use a posterboard template that looks like a little window (*Step 2*).

But before you can use the template, you need to raise the blade to match the

depth of the cove. Then mark on strips of masking tape where the teeth of the blade enter and exit the saw table (*Step 1*).

Now take the template and angle it until it touches the entry and exit points (*Step 2*). Then clamp a straightedge (fence) against the template (*Step 2*).

With the fence clamped in place, cut the cove in multiple passes, resetting the blade height after each pass (*Step 3*).

Note: The angle of the fence to the blade changes with an 8"-dia. saw blade. To cut the cove with a benchtop table saw, the angle of the fence will be steeper.

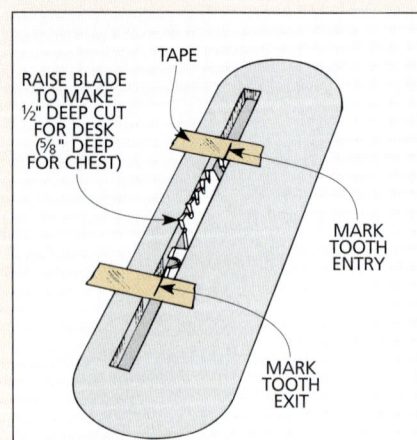

1 To begin, raise the saw blade to the final depth of the cove. Then mark on strips of masking tape where the teeth of the blade enter and exit the saw.

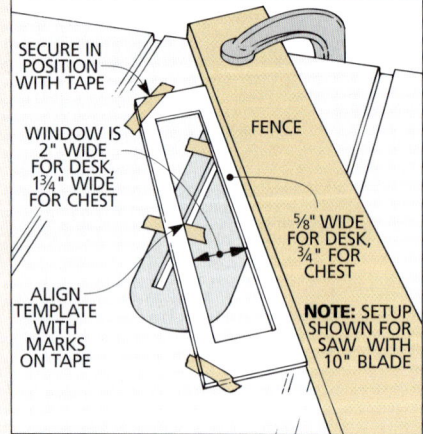

2 Next, make a template with an inside dimension equal to the width of the cove. Angle the template so the inside edges of the template touch the marks.

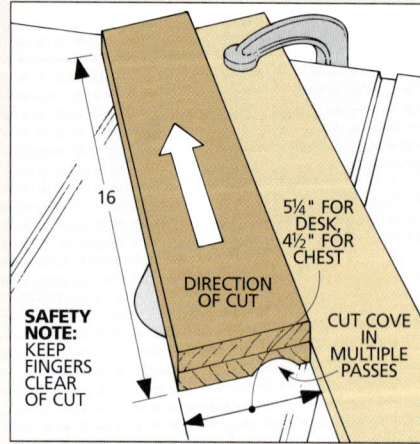

3 Clamp fence in place and raise saw blade to a height of 1/16". Raise blade in 1/16" increments between passes until the full depth of the cove is reached.

ROUGH OUT PROFILE

With the coves cut, the next area of the S-shaped profile to work on is the convex shape near the top outside corner. To complete this part of the profile, two things have to happen. First, the cove must be elongated at one end so there's a smooth transition between it and the face of the blank. And secondly, the top outside corner (above the scalloped area) has to be rounded over.

TRACE PROFILE. Before you start removing any waste, though, it's a good idea to mark what's waste and what's not. To do this, I drew the S-shaped ogee pattern onto a piece of posterboard and cut

it to shape (see page 65 for the chest or page 109 for the desk). Then I traced the template onto the ends of each blank. This will give you a general idea as to what the S-shaped profile will look like once the waste is hogged out.

DADO SET. To elongate the cove, I used a ½"-wide stacked dado set. A rasp or file would work here, but the dado set makes it a lot easier to take out the majority of the waste *(Step 4)*.

FINISHED WIDTH. At this point the blanks could be cut to finished width. But because they started out wider than necessary, the lip below the cove might

be too wide. So before ripping them to finished width, I first ripped the bottom edge of the blanks to leave a small lip *(Step 5)*. Then you can rip the blanks to finished width from the opposite edge.

ROUNDOVER. After the blanks are ripped to width, the roundover located on the top outside corner can be roughed out. Once again I used the table saw to remove most of the waste on this edge *(Step 6)*. (You could also remove the waste with a ¾" roundover bit if you prefer.) Either way, this will leave a rough profile that will have to be smoothed out in the next few steps.

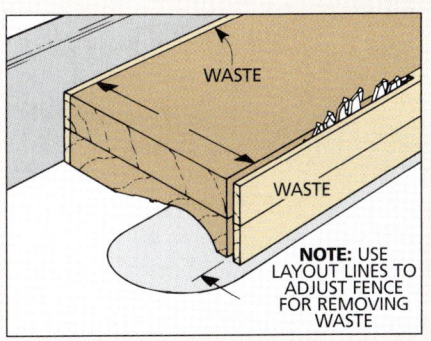

4 Lay out the pattern on the ends. Then elongate the cove with a dado blade. Sneak up on the layout line by adjusting the rip fence, and the blade height and angle between passes.

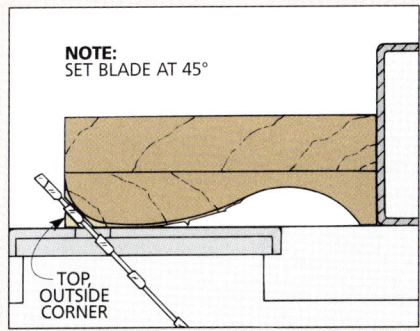

5 Now adjust the fence to rip a strip off the blank, leaving the lip along the bottom edge. Then cut each blank to finished width by ripping the remaining waste from the opposite edge.

6 The final step in roughing out the S-shaped profile is to trim off the top outside corner of each blank. To do this, tilt the saw blade to 45°. Then sneak up on layout line, making several passes.

FACE PROFILE CLEANUP

Up to this point, all the work you've completed at the table saw has been to get the face of the blanks to rough shape. Now it's time to clean up all the unwanted shoulder lines and saw marks left by the saw blade so that you end up with a smooth curve on the face profile.

HANDWORK. This is where the handwork of making bracket feet comes in. And you'll find there's really not much to it. Each blank only has a little material left to remove, and the profiles drawn on the ends will guide you. But don't be too critical. The bracket feet end up far enough apart so that no one will notice if the profiles aren't exactly identical.

OUTSIDE CURVES. The areas that need the most shaping are the outside (convex) curves at the top and bottom of the feet. First, I shaped them with a block plane set to take a thin shaving *(Step 7)*. (But you can also get the job done with a rasp or a Surform-type plane, which looks like a block plane but works like a rasp.)

Start this step by smoothing out the noticeable shoulders. Then simply keep taking thin shavings, following the profile drawn on the end of the blank.

INSIDE CURVES. The inside curves on the feet are even easier to shape. All you

need to do is sand them *(Step 8)*. I wrapped sandpaper around a length of plumbing insulation. I found it provides just enough support and flexibility to sand the curve efficiently. Continue sanding until all the marks are gone.

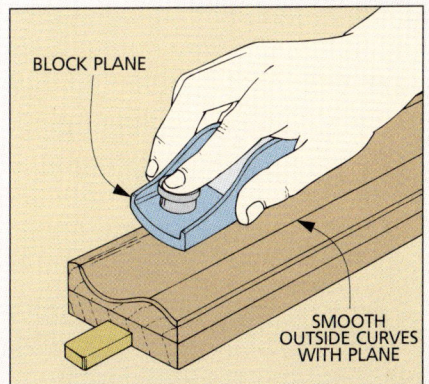

7 On the blank's outside curves, plane any hard lines, removing enough waste to create a gentle curve matching the layout line on the ends.

8 Once the ridges have been planed away, smooth the curve using sandpaper wrapped around a short length of plumbing insulation.

MITER & SPLINE JOINT

Now that the cove profile of each blank is complete, they can be cut into individual pieces. Then one end of each front piece can be mitered.

LABEL BLANKS. But before you get started, label the workpieces (see drawing). For each front foot, you want to glue the ends you cut apart back together. (I joined them with splined miters.)

This way, the grain on the faces of the halves will match up and "wrap around" the foot. Plus, since you've already done

the final shaping, this ensures the profiles of the pieces match as closely as possible. (You may still need to do some light sanding after they're glued together.)

MITER FRONT PIECES. With the parts labeled and cut apart, the next step is to miter one end of each front foot piece *(Step 9)*. The nice thing here is you don't have to worry about an exact length. That will be taken care of when you create the scallop profile later. But I still added a stop block to the auxiliary miter gauge

fence so the piece wouldn't shift as it was being pushed across the blade.

After mitering the pieces, lower the blade and reposition the stop block to cut a kerf for a spline *(Step 10)*. The splines are added mostly to keep the pieces aligned when gluing them together.

BACK FEET. Because these projects are usually placed against a wall, only the front feet are mitered. The back feet are simply supported with a gusset in back (refer to *Steps 17 and 19* on facing page).

FIRST:
CUT BLANKS
IN HALF AND
LABEL PIECES

Front left

Front left

MITER THESE
ENDS FOR
BEST PROFILE
AND GRAIN
MATCH

SECOND:
MITER TWO BLANKS
FOR FRONT FEET
(STEPS 9 & 10)

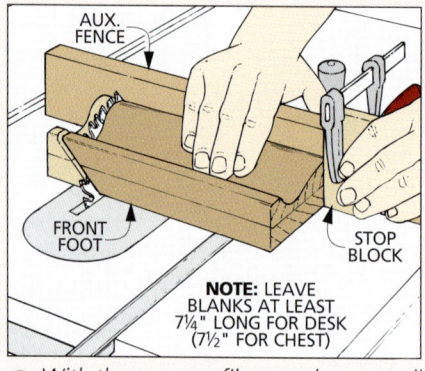

AUX.
FENCE

FRONT
FOOT

STOP
BLOCK

NOTE: LEAVE
BLANKS AT LEAST
7¼" LONG FOR DESK
(7½" FOR CHEST)

9 With the cove profile complete, cut all the blanks in half. Then miter the four pieces that will be used for front feet.

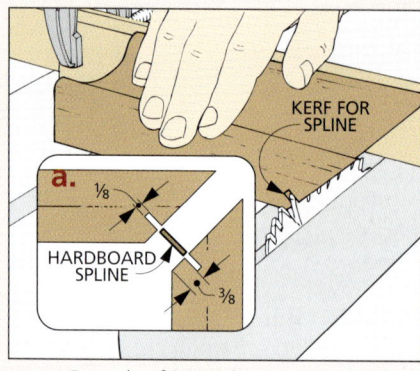

KERF FOR
SPLINE

a. ⅛

HARDBOARD
SPLINE

⅜

10 Cut a kerf in each mitered end for a spline to help align the pieces. Then cut a hardboard spline to fit in the kerf.

SCALLOPED PROFILES

Now the blanks are ready to have the scalloped profile cut out. This fancy cutout makes each foot look like it has a large, overhanging bracket. The work for each is done at the band saw and drill press, but there's a difference in how the patterns are positioned (see pages 65 and 108).

On the Slant Front Desk, the feet are all the same length, so I aligned the template with an edge on each workpiece. But the front and back feet for the Bedside

Chest are different lengths, so the procedure is slightly different. You can find out more about this on the facing page.

DESK SCALLOPED PATTERN. Because the ogee profile is shaped on the front face of each blank, it's easier to lay out the scalloped pattern on the back of each blank. When tracing out the pattern, make sure you're using the correct reference line on the template for the front and back feet *(Step 12 and 13)*.

SHAPE PROFILE. The scalloped profiles are easier to create than the face (cove) profiles. First, I roughed out the profile at the band saw. Then I sanded as much as possible with a drum sander before finishing them with a little hand sanding.

Note: Since I used a ½"-dia. drum sander chucked in the drill press to sand out the scalloped profiles, I found it easier to do this before the feet were glued up into an L-shaped bracket.

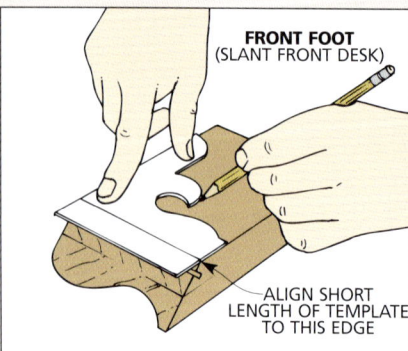

FRONT FOOT
(SLANT FRONT DESK)

ALIGN SHORT
LENGTH OF TEMPLATE
TO THIS EDGE

11 Transfer the scalloped pattern onto the back side of the mitered pieces. Then cut out the shape. Remove the saw blade marks with a ½"-dia. drum sander.

BACK FOOT
(SLANT FRONT DESK)

ALIGN FULL LENGTH
OF TEMPLATE FLUSH
WITH EDGE

12 Next, transfer the scalloped pattern onto the back feet. Make sure template is aligned with edge of workpiece. Then cut to shape and sand out saw marks.

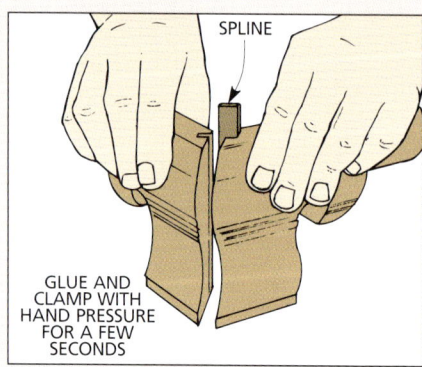

SPLINE

GLUE AND
CLAMP WITH
HAND PRESSURE
FOR A FEW
SECONDS

13 After the scallop is sanded, glue up the L-shaped blanks for the front feet. After the glue is dry, trim the spline flush with the top and bottom of the foot.

BEDSIDE CHEST

The main difference in the feet used for the Bedside Chest is in the shape of the scallop (refer to *Fig. 13* on page 65). But the front and back legs are different as well. With the chest, the front feet are slightly longer (7") (see left drawing) than the back feet ($6\frac{1}{2}$") (right drawing). For both feet, I laid out a line across the back side of each blank and then aligned the template with this line.

With the back feet, the thing to keep in mind is that they're not identical. With contoured faces, they're mirrored images of each other, so make sure you end up with both a right and a left back foot.

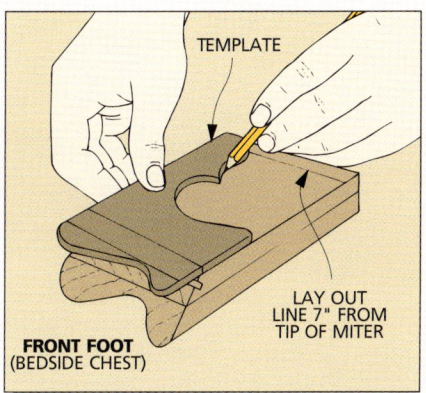

Transfer the scalloped part of the pattern onto the back side of the mitered pieces. Position the template so the feet end up 7" long, laying it out from the tip of the miter.

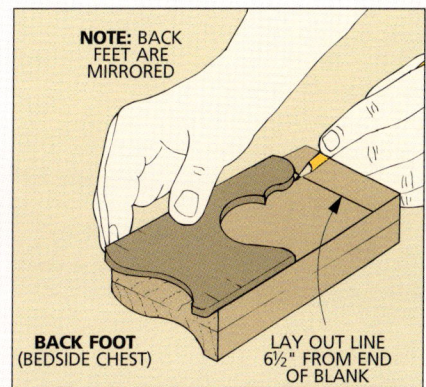

Lay out the scalloped part of the pattern on each back foot blank so it will end up $6\frac{1}{2}$" long. Flip the template for the second foot so the back pieces are mirrored.

MOUNTING THE FEET

Now mount the feet to the bottom of the desk. To do this, first lay the desk down on its back. (To mount the feet on the chest, refer to page 65.)

FRONT FEET. To provide a frame for mounting the front feet, I used two support cleats (T) for each foot *(Step 14)*. First screw the cleats to the desk, then screw the cleats to the feet *(Step 15)*. (Don't worry about the exposed splines in the miter joints. A strip of molding added later will cover them.)

BACK FEET. Since the back feet are only viewed from one side, I used a gusset to help hold each foot in place and offer additional support *(Steps 16 and 17)*.

First, lay out and cut the gusset (U) to size *(Step 16)*. Then cut a rabbet on the back inside edge of each back foot, drill pilot holes and screw the gusset in place.

The cleats used to attach the back feet are offset from one another *(Step 18)*.

Once the back cleats are screwed on, mount the back feet in place *(Step 19)*.

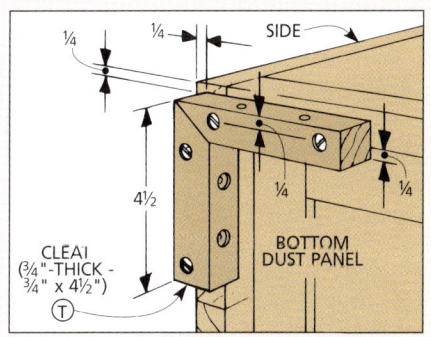

14 *The front feet are held in place by two mitered cleats. After the cleats are cut to size, drill shank holes for mounting to both the desk and the feet.*

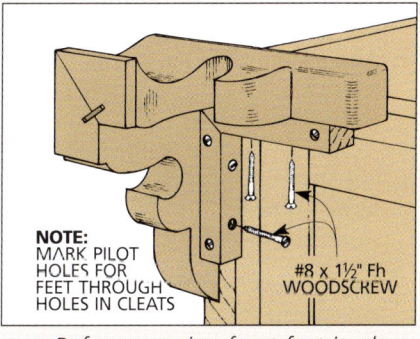

15 *Before screwing front feet in place, first position them on the cleats. Then mark and drill pilot holes in feet so the screws don't split the wood.*

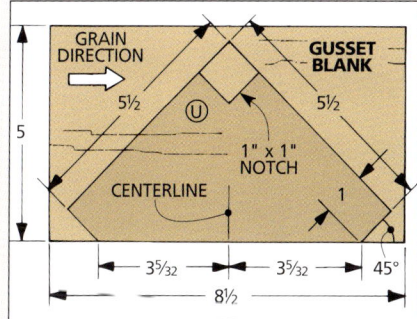

16 *The back feet are supported by two $\frac{3}{4}$"-thick gussets. When laying out the gussets on the workpiece, make sure the grain is oriented for maximum strength.*

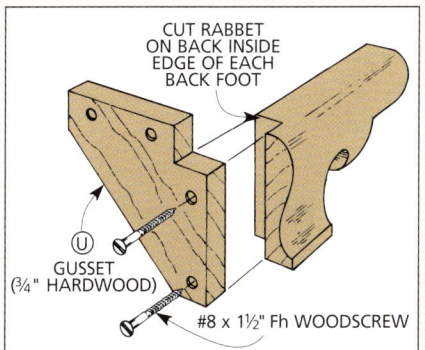

17 *First cut a rabbet on the back inside edge of each back foot. Before screwing gussets (T) to feet, mark and drill pilot holes, then screw the gusset in place.*

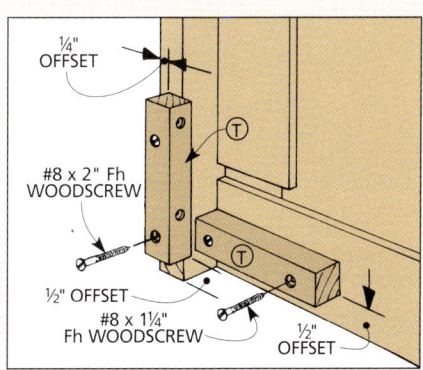

18 *Now cut $\frac{3}{4}$"-square cleats for the back feet and gusset assemblies. Back cleats are positioned so gussets create $\frac{1}{4}$"-wide ledge for the back panel.*

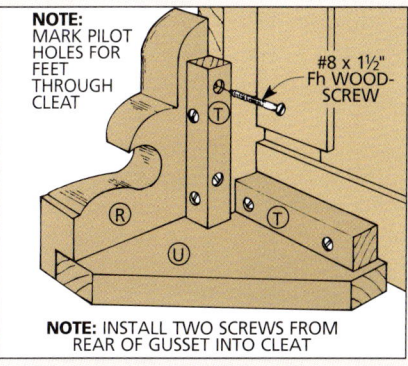

19 *Before screwing back feet in place, position each foot against the cleats. Then mark and drill holes in the feet so the screws don't split the wood.*

One of the first things we take into consideration when designing projects at *Woodsmith* is whether the hardware is affordable and easy to find. Most of the hardware and supplies for the projects in this book can be found at local hardware stores or home centers. Sometimes, though, you may have to order the hardware through the mail. If that's the case, we've tried to find reputable national mail order sources with toll-free phone numbers (see box at right).

In addition, *Woodsmith Project Supplies* offers hardware for a couple of the projects in this book (see below).

WOODSMITH PROJECT SUPPLIES

At the time this book was printed, the following hardware kits were available from *Woodsmith Project Supplies*. The kits include the hardware listed, but you must supply any lumber, plywood, or finish. For current prices and availability, call toll free:

1-800-444-7527

Mantel Clock
(pages 32-37)

This kit contains a brass knob, a silk-screened aluminum clock face, mini-quartz movement, hands, and brass turnbuttons. It also has all the woodscrews you'll need and the felt pads.

.... No. 7512170

Steamer Trunk
(pages 40-49)

This kit contains all the hardware supplies you'll need to build the Steamer Trunk, including brass-plated case corners, L-shaped corner clamps, hinges, large draw catches, a trunk lock with key, lid stays, and handle loops with pegs. Russet leather handles are also provided with this kit.

...................................No. 7512150

KEY: TL16

MAIL ORDER SOURCES

Some of the most important "tools" you can have are mail order catalogs. The ones listed below are filled with special hardware, tools, finishes, lumber, and supplies that can't be found at a hardware store or home center. You should be able to find many of the supplies for the projects in this book in one or more of these catalogs.

It's amazing what you can learn about woodworking by looking through these catalogs If they're not currently in your shop, you may want to have them sent to you. You can order your catalog by phone or online.

Note: The information below was current when this book was printed. August Home Publishing does not guarantee these products will be available nor endorse any specific mail order company, catalog, or product.

THE WOODSMITH STORE

**2200 Grand Avenue
Des Moines, IA 50312
800–835–5084**
Our own retail store filled with tools, hardware, books, and finishing supplies. Though we don't have a catalog, we do send out items mail order. Call for information.

LEE VALLEY TOOLS

**P.O. Box 1780
Ogdensburg, NY 13669-6780
800–871–8158
www.leevalley.com**
Several catalogs actually, with hardware and finishing supplies. A good source of hinges, knobs, pulls, catches, and lid stays.

ROCKLER WOODWORKING & HARDWARE

**4365 Willow Drive
Medina, MN 55340
800–279–4441
www.rockler.com**
A very good hardware catalog, including hinges, knobs, catches, pulls, and lid stays. You'll also find a good supply of clockworks and accessories, plus veneer, and veneering supplies.

VAN DYKE'S RESTORERS

**P.O. Box 278
Woonsocket, SD 57385
800–558–1234
www.vandykes.com**
An amazing collection of reproduction hardware, veneer and veneering supplies, finishing supplies, bracket feet, and lots more.

WOODCRAFT

**560 Airport Industrial Park
Parkersburg, WV 26102-1686
800–225–1153
www.woodcraft.com**
Just about everything for the woodworker including magnetic catches, knobs, hinges, turnbuttons, pulls, and escutcheons. They also stock mechanical clockworks and through-dovetail jigs.

CONSTANTINE'S

**1040 E. Oakland Park Blvd.
Ft. Lauderdale, FL 33334
954–561–1716
www.constantines.com**
One of the original woodworking mail order catalogs. Good selection of knobs, hinges, pulls, veneer and veneering supplies, finishing supplies, and other hardware.

KLOCKIT

**P.O. Box 636
Lake Geneva, WI 53147-0636
800–556–2548
www.klockit.com**
A great mail order source for clockworks and accessories, including clock faces with bezels.

S. LAROSE

**P.O. Box 21208
Greensboro, NC 27420
888–752–7673
www.slarose.com**
Another complete source for clockworks, both mechanical and quartz. They also stock a full line of clock faces, hands, and accessories.

INDEX

AUGUST HOME
PUBLISHING COMPANY

President & Publisher: Donald B. Peschke
Executive Editor: Douglas L. Hicks
Project Manager/Senior Editor: Craig L. Ruegsegger
Creative Director: Ted Kralicek
Art Director: Doug Flint
Senior Graphic Designers: Robin Friend, Chris Glowacki
Assistant Editor: Joel Hess
Editorial Intern: Cindy Thurmond
Graphic Designers: Jonathan Eike, Vu Nguyen

Designer's Notebook Illustrator: Chris Glowacki
Photographer: Crayola England
Electronic Production: Douglas M. Lidster
Production: Troy Clark, Minniette Johnson
Project Designers: Chris Fitch, Ryan Mimick, Ken Munkel, Kent Welsh
Project Builders: Steve Curtis, Steve Johnson
Magazine Editors: Tim Robertson, Terry Strohman
Contributing Editors: Vincent S. Ancona, Jon Garbison, Phil Huber,
Brian McCallum, Bryan Nelson, Ted Raife
Magazine Art Directors: Cary Christensen, Todd Lambirth
Contributing Illustrators: Harlan Clark, Mark Higdon, David Kreyling,
Erich Lage, Roger Reiland, Kurt Schultz, Cinda Shambaugh, Dirk Ver Steeg

Corporate V.P., Finance: Mary Scheve
Controller: Robin Hutchinson
Production Director: George Chmielarz
Project Supplies: Bob Baker
New Media Manager: Gordon Gaippe

For subscription information about
Woodsmith and *ShopNotes* magazines, please write:
August Home Publishing Co.
2200 Grand Ave.
Des Moines, IA 50312
800-333-5075
www.augusthome.com/customwoodworking

Woodsmith® and *ShopNotes*® are registered trademarks of August Home
Publishing Co.

©2003 August Home Publishing Co.
All rights reserved. No part of this book may be reproduced in any form or by
any electronic or mechanical means, including information storage and retrieval
devices or systems, without prior written permission from the publisher, except
that brief passages may be quoted for reviews.
First Printing. Printed in U.S.A.

Oxmoor House®

Oxmoor House, Inc.
Book Division of Southern Progress Corporation
P.O. Box 2463, Birmingham, Alabama 35201

ISBN: 0-8487-2690-1
Printed in the United States of America

To order additional publications, call 1-800-765-6400.
For more books to enrich your life, visit **oxmoorhouse.com**